ELECTROMAGNETIC FIELDS

ELECTROMAGNETIC FIELDS

S. SIVANAGARAJU

Associate Professor
Department of Electrical Engineering
J.N.T.U. College of Engineering (Autonomous)
Kakinada, Andhra Pradesh

S. SRINIVASA RAO

Assistant Professor
Department of Electrical Engineering
G. Pulla Reddy Engineering College (Autonomous)
Kurnool, Andhra Pradesh

PUBLISHING FOR ONE WORLD

NEW AGE INTERNATIONAL (P) LIMITED, PUBLISHERS

New Delhi • Bangalore • Chennai • Cochin • Guwahati • Hyderabad
Jalandhar • Kolkata • Lucknow • Mumbai • Ranchi

Visit us at **www.newagepublishers.com**

Published by New Age International (P) Ltd., Publishers
First Edition: 2008
Reprint: 2009

Branches:

- 36, Malikarjuna Temple Street, Opp. ICWA, Basavanagudi, **Bangalore**. ✆ (080) 26677815
- 26, Damodaran Street, T. Nagar, **Chennai**. ✆ (044) 24353401
- Hemsen Complex, Mohd. Shah Road, Paltan Bazar, Near Starline Hotel, **Guwahati**. ✆ (0361) 2543669
- No. 105, 1st Floor, Madhiray Kaveri Tower, 3-2-19, Azam Jahi Road, Nimboliadda, **Hyderabad**. ✆ (040) 24652456
- RDB Chambers (Formerly Lotus Cinema) 106A, Ist Floor, S.N. Banerjee Road, **Kolkata**. ✆ (033) 22275247
- 16-A, Jopling Road, **Lucknow**. ✆ (0522) 2209578
- 142C, Victor House, Ground Floor, N.M. Joshi Marg, Lower Parel, **Mumbai**. ✆ (022) 24927869
- 22, Golden House, Daryaganj, **New Delhi**. ✆ (011) 23262370, 23262368

ISBN : 978-81-224-2378-5

Rs. 195.00

C-09-04-3620

Printed in India at Ram Printograph, Delhi.
Typeset at Goswami Associates, Delhi.

PUBLISHING FOR ONE WORLD
NEW AGE INTERNATIONAL (P) LIMITED, PUBLISHERS
4835/24, Ansari Road, Daryaganj, New Delhi-110002
Visit us at **www.newagepublishers.com**

Preface

This book is designed for use as a textbook for a course in electromagnetic fields and can be used by electrical engineering students as well as other engineering and technology students. Emphasis is placed on the basic concepts, laws and problem-solving abilities which are common to most courses.

This book is divided into 8 chapters covering duly-recognized areas of theory and study. The chapters begin with basic concepts and their applications to various levels together with illustrative examples. This is followed by sets of short questions and answers, objective questions and review questions.

Chapter 1 is introductory with emphasis on Vector Analysis. Chapter 2 deals with Coulomb's law and electric field intensity. The concepts of electric flux, Gauss's law and their various applications along with divergence are dealt with in Chapter 3. Chapter 4 deals with the concept of energy and potential and also on dielectrics and capacitors. The concepts of static magnetic fields are discussed in Chapter 5. In Chapter 6 presents the different types of magnetic forces existing in different elements. The concept of magnetic potential and inductance is discussed in detail in Chapter 7. Maxwell's equations and time varying fields are covered extensively in Chapter 8.

This book has been particularly written for students who can make self study also. Basic concepts have been written carefully and explained in detail with numerous problems. Special features of this book include objective questions, review questions etc.

The authors express their whole hearted gratefulness to their colleagues of Electrical and Electronics Engineering Department who directly or indirectly have helped the authors to bring this book in the present form. The authors sincerely express thankfulness to the Publishers who have taken the initiative for publishing the book.

Dr. S. Sivangaraju

C. Srinivasa Rao

Contents

CHAPTER

1

Vector Analysis

1.1 INTRODUCTION

Any problem concerning an electrical network can be viewed from two angles:

(*a*) Circuit point of view.

(*b*) Field point of view.

It can be said that a proper understanding of concept of field theory provides a better understanding of analysis of circuit problems. Vector analysis is a valuable mathematical tool for engineers for solving certain type of problems where the conventional methods become lengthy and cumbersome. The analysis with vectors even though little difficult to understand in the beginning, but effectively gives the solution of field quantities both in magnitude and direction. For this reason, a review of vector analysis is provided at the beginning of course.

There is a lot of difference between circuit theory and field theory. Electromagnetic field theory deals directly with the field vectors viz. electric field (E) and magnetic field (H) while circuit theory deals with voltage (V) and current (I) that are the integrated effects of electric and magnetic fields. In general, electromagnetic field problems involve three space variables as a result of which the solutions tend to become comparatively complex. The additional problem that arises due to dealing with vector quantities in three dimension can be overcome by use of vector analysis. The use of vector analysis in the study of electromagnetic field theory thus saves time and provides economy of thought. In addition the vector form gives a clear understanding of physical laws which is described by mathematics.

1.2 SCALAR AND VECTOR QUANTITIES

Quantities associated with electric and magnetic fields are either scalars (or) vectors possessing characteristic properties. A quantity which possesses only magnitude but no direction is called a scalar quantity. For example, physical quantities like mass, area, volume, temperature are scalar quantities. On the other hand, quantities which possess both magnitude and direction are called vector quantities. For examples, velocity, force, acceleration are vector quantities.

In order to distinguish a vector from a scalar, an arrow on the top is used to denote a vector. So for example, $\vec{A}, \vec{B}, \vec{C}$ represent vector quantities and A, B, C represent scalar quantities.

1.3 FIELD

Although a scalar field has no direction, it does have a specific location. Mathematically a field is a function which describes a physical quantity at all points in space. This field can be classified as either scalar (or) vector.

1.3.1 Scalar Field

It implies the distribution of scalar quantity with a definite position in space. The temperature of hot water in a container is an example of scalar field.

1.3.2 Vector Field

If the value of physical function at every point is vector quantity, the field is a vector field. For example, the wind velocity of atmosphere, gravitational force and electric field intensity are vector fields.

1.4 VECTOR OPERATIONS

1.4.1 Addition of Vectors

Consider two vectors $\vec{P}$ and $\vec{Q}$ as shown in Fig. 1.1(*a*). The vector $\vec{R}$ which is called as resultant vector can be obtained by moving a point along $\vec{P}$ and then along $\vec{Q}$ or sum of vectors $\vec{P}$ and $\vec{Q}$, we can written as

$$\vec{R} = \vec{P} + \vec{Q} \qquad \text{...(1.1)}$$

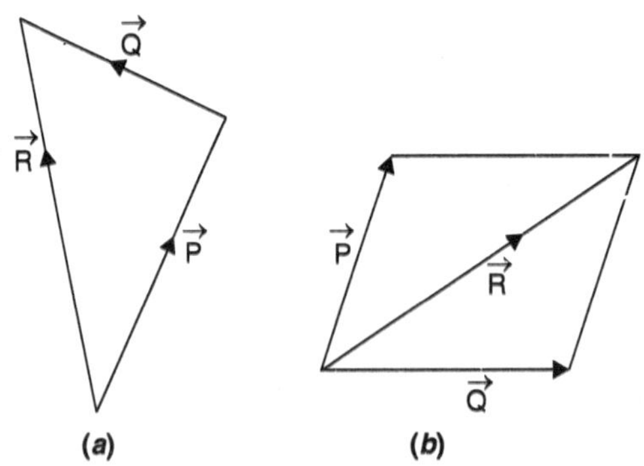

Fig. 1.1

From nature of definition of vector addition, it is apparent that

$$\vec{Q} + \vec{P} = \vec{P} + \vec{Q} \qquad \text{...(1.2)}$$

If the vectors $\vec{P}$ and $\vec{Q}$ are situated as shown in Fig. 1.1 (*b*), then the resultant vector $\vec{R}$ is obtained by completing the parallelogram formed by two vectors.

1.4.2 Subtraction of Vectors

To subtract two vectors the best method which is to be followed is to change the direction of vector and then add. As shown in Fig. 1.2, the direction of vector $\vec{Q}$ is changed which results in magnitude of $(-)\vec{Q}$ and which on addition with $\vec{P}$ give the resultant vector $\vec{R}$ whose value is given by $\vec{R} = \vec{P} - \vec{Q}$.

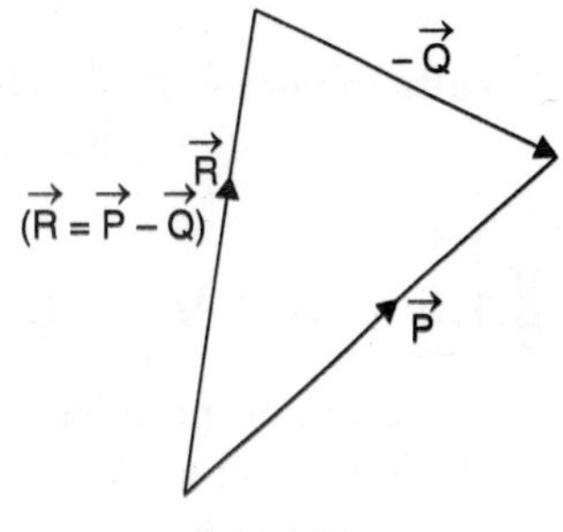

Fig. 1.2

1.4.3 Multiplication of Vectors

The multiplication of vectors has been classified into three categories:

(*a*) When vector is multiplied by a scalar.

(*b*) When a vector is multiplied by another vector resulting in a scalar quantity. This is called as Scalar (or) Dot product.

(*c*) When a vector is multiplied by another vector resulting in vector quantity. This is called as Vector (or) Cross product.

These three classifications of multiplication can be discussed in detail as follows:

1.4.3.1 *Multiplication of Vector by a Scalar*

When a vector quantity is multiplied by a scalar quantity, the magnitude of vector changes, but its direction remains unchanged. Thus it can be shown that if vector $\vec{P}$ is multiplied with a scalar '*a*' it results in vector $\vec{Q}$ with same direction as of $\vec{P}$.

Thus, $$\vec{Q} = a\,\vec{P} \qquad ...(1.3)$$

1.4.3.2 *Scalar (or) Dot Product*

The scalar (or) Dot product of two vectors is the product, of magnitude of vectors multiplied by cosine of the smaller angle between them.

Let $\vec{P}$ and $\vec{Q}$ be two vectors and θ_{pq} denote angle between them.

$\therefore$ $$\vec{P} \cdot \vec{Q} = |P|\,|Q| \cos\theta_{pq} \qquad ...(1.4)$$

Obviously the result is a scalar quantity and also it can be written as

$$\vec{P} \cdot \vec{Q} = \vec{Q} \cdot \vec{P} \qquad ...(1.5)$$

We can come to a decision from equation (1.4) that the vectors have same direction when $\theta = 0$ *i.e.*, $\cos\theta = 1$ and the two vectors are opposite in direction when $\theta = 180°$ *i.e.*, $\cos\theta = -1$ and the two vectors are perpendicular to each other when $\theta = 90°$ *i.e.*, $\cos 90° = 0$.

1.4.3.3 *Vector (or) Cross Product*

It is the product of two vectors, where the magnitudes of vectors are multiplied by sine of the smaller angle between them.

$$\therefore \qquad \vec{P} \times \vec{Q} = P\,Q \sin\theta \qquad \text{...(1.6)}$$

1.4.4 Division of Vectors by Scalar

It is the same as multiplication of vector by reciprocal of scalar. So vector $\vec{P}$ divided by scalar Q results in vector $\vec{P}\,\frac{1}{Q}$. Here also the direction of vector does not change.

1.5 UNIT VECTOR

It is a common practice to express each of component vectors as a product of scalar magnitude and a unit vector. A unit vector is a vector of unit magnitude and having specified direction and is denoted by U.

So the component vector $\vec{P}_x$ can be written as

$$\vec{P}_x = P_x\, U_x \qquad \text{...(1.7)}$$

The unit vector in the direction of vector $\vec{P}$ can be obtained by dividing $\vec{P}$ with its modulus value $|\vec{P}|$.

$$U_p = \frac{\vec{P}}{|\vec{P}|} \qquad \text{...(1.8)}$$

The concept of unit vector is useful in representing a vector in terms of component vectors.

1.5.1 Applications of Unit Vectors

With the help of unit vectors the dot (or) scalar product and cross (or) vector product of two vectors can be worked out by direct multiplication. Let the two vectors $\vec{P}$ and $\vec{Q}$ be given by

$$\vec{P} = \vec{i}\,P_x + \vec{j}\,P_y + \vec{k}\,P_z \qquad \text{...(1.9)}$$

$$\vec{Q} = \vec{i}\,Q_x + \vec{j}\,Q_y + \vec{k}\,Q_z \qquad \text{...(1.10)}$$

Case (*a*): Dot (or) scalar product:

$$\begin{aligned}\vec{P}\cdot\vec{Q} &= (\vec{i}\,P_x + \vec{j}\,P_y + \vec{k}\,P_z)\cdot(\vec{i}\,Q_x + \vec{j}\,Q_y + \vec{k}\,Q_z)\\ &= \vec{i}\cdot\vec{i}\,P_x\,Q_x + \vec{i}\cdot\vec{j}\,P_x\,Q_y + \vec{i}\cdot\vec{k}\,P_x\,Q_z\\ &\quad + \vec{j}\cdot\vec{i}\,P_y\,Q_x + \vec{j}\cdot\vec{j}\,P_y\,Q_y + \vec{j}\cdot\vec{k}\,P_y\,Q_z\\ &\quad + \vec{k}\cdot\vec{i}\,P_z\,Q_x + \vec{k}\cdot\vec{j}\,P_z\,Q_y + \vec{k}\cdot\vec{k}\,P_z\,Q_z\end{aligned}$$

but $\vec{i}\cdot\vec{i} = \vec{j}\cdot\vec{j} = \vec{k}\cdot\vec{k} = 1$, while other product unlike unit vectors will be zero.

$$\therefore \qquad \vec{P}\cdot\vec{Q} = P_x\,Q_x + P_y\,Q_y + P_z\,Q_z \qquad \text{...(1.11)}$$

Case (*b*): Cross (or) vector product:

$$\vec{P} \times \vec{Q} = (\vec{i}\,P_x + \vec{j}\,P_y + \vec{k}\,P_z) \times (\vec{i}\,Q_x + \vec{j}\,Q_y + \vec{k}\,Q_z)$$

$$= \vec{i} \times \vec{i}\,P_x\,Q_x + \vec{i} \times \vec{j}\,P_x\,Q_y + \vec{i} \times \vec{k}\,P_x\,Q_z$$

$$+ \vec{j} \times \vec{i}\,P_y\,Q_x + \vec{j} \times \vec{j}\,P_y\,Q_y + \vec{j} \times \vec{k}\,P_y\,Q_z$$

$$+ \vec{k} \times \vec{i}\,P_z\,Q_x + \vec{k} \times \vec{j}\,P_z\,Q_y + \vec{k} \times \vec{k}\,P_z\,Q_z$$

Since $\vec{i} \times \vec{i} = 0$ and $\vec{i} \times \vec{j} = \vec{k}$ and $\vec{i} \times \vec{k} = -\vec{j}$. This becomes

$$\vec{P} \times \vec{Q} = \begin{vmatrix} \vec{i} & \vec{j} & \vec{k} \\ P_x & P_y & P_z \\ Q_x & Q_y & Q_z \end{vmatrix} \qquad \ldots(1.12)$$

Example 1.1: *Vectors* $A = 3U_x + 5U_y + 6U_z$ *and* $B = 6U_x + 4U_y + 2U_z$ *are situated at point* $P(x, y, z)$*. Find (a)* $\overline{A} + \overline{B}$ *(b)* $\overline{A} \cdot \overline{B}$ *(c) angle between* $\overline{A}$ *and* $\overline{B}$ *(d)* $\overline{A} \times \overline{B}$ *(e) unit normal to plane containing* $\overline{A}$ *and* $\overline{B}$*.*

Solution: Given $\overline{A} = 3U_x + 5U_y + 6U_z$ and $\overline{B} = 6U_x + 4U_y + 2U_z$

(*a*) $\overline{A} + \overline{B} = (3U_x + 5U_y + 6U_z) + (6U_x + 4U_y + 2U_z)$

$= (3 + 6)U_x + (5 + 4)U_y + (6 + 2)U_z$

$= 9U_x + 9U_y + 9U_z$

(*b*) $\overline{A} \cdot \overline{B} = A_x B_x + A_y B_y + A_z B_z$

Given $A_x = 3, A_y = 5, A_z = 6$ and $B_x = 6, B_y = 4$ and $B_z = 2$

$\therefore \quad \overline{A} \cdot \overline{B} = (3 \times 6) + (5 \times 4) + (6 \times 2) = 50$

(*c*) $\overline{A} \cdot \overline{B} = |A|\,|B| \cos\theta_{AB}$

We have $|\overline{A}| = \sqrt{A_x^2 + A_y^2 + A_z^2}$

$= \sqrt{3^2 + 5^2 + 6^2} = 8.366$

$|\overline{B}| = \sqrt{B_x^2 + B_y^2 + B_z^2}$

$= \sqrt{6^2 + 4^2 + 2^2} = 7.48$

Also $\overline{A} \cdot \overline{B} = 50$ (Calculated earlier)

$50 = 8.366 \times 7.48 \times \cos\theta_{AB}$

i.e., $\cos\theta_{AB} = 0.799$

$\theta_{AB} = \cos^{-1} 0.799 = 36.96°$

(*d*) $\overline{A} \times \overline{B} = \begin{vmatrix} U_x & U_y & U_z \\ A_x & A_y & A_z \\ B_x & B_y & B_z \end{vmatrix}$ in determinant form

$= \begin{vmatrix} U_x & Y_y & U_z \\ 3 & 5 & 6 \\ 6 & 4 & 2 \end{vmatrix} = 14U_x + 30U_y + 18U_z$ on simplification.

(*e*) Unit normal vector (U_n)

We have $$U_n = \frac{\bar{A} \times \bar{B}}{|\bar{A} \times \bar{B}|}$$

But $$\bar{A} \times \bar{B} = 14U_x + 30U_y + 18U_z$$

and $$|\bar{A} \times \bar{B}| = \sqrt{14^2 + 30^2 + 18^2} = 37.68$$

$$\therefore \quad U_n = \frac{14U_x + 30U_y + 18U_z}{37.68}$$
$$= 0.37U_x - 0.79U_y + 0.477U_z$$

Example 1.2: *If $\bar{R}_A = 3U_x - 2U_y + 4U_z$, $\bar{R}_B = 4U_x + 5U_y - 7U_z$ and point $C = (6, 2, 3)$. Find (a) $\bar{R}_{AB}$ (b) $|\bar{R}_A|$ (c) $|\bar{R}_B|$ (d) U_A (e) U_B (f) U_{AB} (g) unit vector directed from C towards A.*

Solution: Given $\bar{R}_A = 3U_x - 2U_y + 4U_z$, $\bar{R}_B = 4U_x + 5U_y - 7U_z$ and point $C = (6, 2, 3)$

(*a*) Vector $$\bar{R}_{AB} = \bar{R}_B - \bar{R}_A$$
$$= (4U_x + 5U_y - 7U_z) - (3U_x - 2U_y + 4U_z)$$
$$= U_x + 7U_y - 11U_z$$

(*b*) $|\bar{R}_A| = \sqrt{(3)^2 + (-2)^2 + (4)^2} = 5.38$

(*c*) $|\bar{R}_B| = \sqrt{4^2 + 5^2 + 7^2} = 9.48$

(*d*) Unit vector, $$U_A = \frac{\bar{R}_A}{|\bar{R}_A|}$$
$$= \frac{3U_x - 2U_y + 4U_z}{5.38}$$
$$= 0.55U_x - 0.37U_y + 0.74U_z$$

(*e*) Unit vector, $$U_B = \frac{\bar{R}_B}{|\bar{R}_B|}$$
$$= \frac{4U_x + 5U_y - 7U_z}{9.48}$$
$$= 0.42U_x + 0.52U_y - 0.73U_z$$

(*f*) $$U_{AB} = \frac{\bar{R}_{AB}}{|\bar{R}_{AB}|}$$

$$|\bar{R}_{AB}| = \sqrt{(1)^2 + (7)^2 + (11)^2} = 13.07$$

$$\therefore \quad U_{AB} = \frac{U_x + 7U_y - 11U_z}{13.07}$$
$$= 0.076U_x + 0.53U_y - 0.841U_z$$

(*g*) Unit vector directed from C to A is U_{CA}

Given $C = (6, 2, 3)$ but $\bar{R}_A = 3U_x - 2U_y + 4U_z$

$$U_{CA} = \frac{(3-6)}{5.099}U_x + \frac{(-2-2)}{5.099}U_y + \frac{(4-3)}{5.099}U_z$$

since $\sqrt{(3-6)^2+(-2-2)^2+(4-3)^2} = 5.099$

$$U_{CA} = -0.588U_x - 0.784U_y - 0.196U_z$$

Example 1.3: *Given points A (4, 3, 2), B (2, 1, 4) and C (– 4, 1, 3). Find (a) $\overline{R}_{AB} \cdot \overline{R}_{AC}$ (b) the angle between $\overline{R}_{AB}$ and $\overline{R}_{AC}$ (c) length of projection of $\overline{R}_{AB}$ on $\overline{R}_{AC}$ (d) vector projection of $\overline{R}_{AB}$ on $\overline{R}_{AC}$.*

Solution:

(*a*)
$$\begin{aligned}\overline{R}_{AB} &= \overline{R}_B - \overline{R}_A \\ &= (2U_x + U_y + 4U_z) - (4U_x + 3U_y + 2U_z) \\ &= -2U_x - 2U_y + 2U_z \\ \overline{R}_{AC} &= \overline{R}_C - \overline{R}_A \\ &= (-4U_x + U_y + 3U_z) - (4U_x + 3U_y + 2U_z) \\ &= -8U_x - 2U_y + U_z \\ \overline{R}_{AB} \cdot \overline{R}_{AC} &= (-2)(-8) + (-2)(-2) + (2)(1) = 22\end{aligned}$$

(*b*)
$$\begin{aligned}\cos\theta &= \frac{\overline{R}_{AB} \cdot \overline{R}_{AC}}{|\overline{R}_{AB}||\overline{R}_{AC}|} \\ &= \frac{22}{\sqrt{(2)^2+(2)^2+(2)^2}\sqrt{(8)^2+(2)^2+(1)^2}} \\ &= 0.766 \\ \theta &= \cos^{-1}(0.766) = 40°\end{aligned}$$

(*c*) AP = Scalar projection of $\overline{R}_{AB}$ on $\overline{R}_{AC}$

$$= \overline{R}_{AB} \cdot U_{AC}$$

$$= \overline{R}_{AB} \cdot \frac{\overline{R}_{AC}}{|\overline{R}_{AC}|} = \frac{22}{\sqrt{(8)^2+(2)^2+(1)^2}} = 2.648$$

(*d*) $\overline{R}_{AP}$ = Vector projection of $\overline{R}_{AB}$ on $\overline{R}_{AC} = (\overline{R}_{AB} \cdot U_{AC})\, U_{AC}$

$$= 2.648\left[\frac{-8U_x - 2U_y + U_z}{\sqrt{(-8)^2+(-2)^2+(1)^2}}\right]$$

$$= -2.544U_x - 0.636U_y + 0.318U_z$$

Example 1.4: *Given the field $\overline{G} = \left[\frac{3x}{1+y^2}\right] U_x + (y + 2z + 1)\, U_y + (5x - z^2)\, U_z$. Find (a) unit vector in directions of $\overline{G}$ at P (2, 2, – 3) (b) the angle between $\overline{G}$ and y = 0 plane at Q (2, 0, 4), (c) the value of $\int_{y=0}^{2}\int_{x=1}^{3} \overline{G} \cdot dx\, dy\, U_z$ at the plane z = 1.*

Solution:

(*a*) $\overline{G}$ at point P

$$= \left[\frac{3(2)}{1+2^2}\right] U_x + (2 - 6 + 1)\, U_y + [5(2) - (-3)^2]\, U_z$$

$$= 1.2U_x - 3U_y - U_z$$

$$\therefore \qquad \overline{U} = \frac{|\overline{G}|}{|\overline{G}|} = \frac{1.2U_x - 3U_y - 4U_z}{\sqrt{(1.2)^2 + (-3)^2 + (-4)^2}} = 0.233U_x - 0.583U_y - 0.7782U_z$$

(*b*) $\overline{G}$ at point $Q = 3\,(2)\,U_x + (0 + 4 + 1)\,U_y - (5 \times 2 - 4)\,U_z$

$$= 6U_x + 5U_y - 6U_z$$

The unit vector U_y is normal to $y = 0$ plane.

The angle α between U_y and $\overline{G}$ at point Q is given by

$$\cos\alpha = \frac{\overline{G}\cdot U_y}{|\overline{G}||U_y|} = \frac{(6U_x + 5U_y - 6U_z)\cdot U_y}{\sqrt{(6)^2 + (5)^2 + (-6)^2}} = \frac{5}{9.84} = 0.508$$

$$\therefore \qquad \alpha = \cos^{-1}(0.508) = 59.46°$$

Now $\qquad \alpha + \beta = 90°$

$\therefore \qquad \beta$ = angle between $\overline{G}$ at point Q and $y = 0$ plane

$$= 90° - \alpha = 90° - 59.46 = 30.53°$$

(*c*) $\qquad \overline{G}\cdot dx\,dy\overline{U}_z = (5x - z^2)\,dx\,dy$ as $U_x \cdot U_z = U_y \cdot U_z = 0$

$$\therefore \quad \int_{y=0}^{2}\int_{x=1}^{3} \overline{G}\cdot dx\,dy\,U_z = \int_{y=0}^{2}\int_{x=1}^{3}(5x - z^2)\,dx\,dy$$

$$= 5\left[\frac{x^2}{2}\right]_1^3 6[y]_0^2 - (1)^2[x]_1^3[y]_0^2 = 36.$$

Example 1.5: *Find in rectangular components the unit vector which is (a) in the direction of* $\overline{E}$ *at P (1, 2, – 2) if* $\overline{E} = (x^2 + y^2 + z^2)\left[\frac{x}{\sqrt{y^2+z^2}}U_x + \frac{y}{\sqrt{x^2+z^2}}U_y + \frac{z}{\sqrt{x^2+y^2}}U_z\right]$ *(b) perpendicular to plane passing through M (2, – 6, 6), N (– 1, 2, 1) and Q (1, 2, 3) and having positive x component (c) find the angle between vectors* $\overline{R}_{MN}$ *and* $\overline{R}_{MQ}$.

Solution:

(*a*) $\overline{E}$ at $P = [(1)^2 + (2)^2 + (-2)^2]\left[\frac{U_x}{\sqrt{2^2 + (-1)^2}} + \frac{2U_y}{\sqrt{(1)^2 + (-2)^2}} - \frac{2U_z}{\sqrt{(1)^2 + (2)^2}}\right]$

$$= 3.18U_x + 4.08U_y - 4.08U_z$$

$$\therefore \qquad \overline{U} = \frac{\overline{E}}{|\overline{E}|} = \frac{3.18U_x + 4.08U_y - 4.08U_z}{\sqrt{(3.18)^2 + (4.08)^2 + (-4.08)^2}} = 0.482U_x + 0.619U_y - 0.619U_z$$

(*b*) $\qquad \overline{R}_{MN} = \overline{r}_N - \overline{r}_M = (-U_x + 2U_y + U_z) - (2U_x - 6U_y + 6U_z)$

$$= -3U_x + 8U_y - 5U_z$$

$$\overline{R}_{MQ} = \overline{r}_Q - \overline{r}_M = (U_x + 2U_y + 3U_z) - (2U_x - 6U_y + 6U_z)$$

$$= -U_x + 8U_y - 3U_z$$

$$\overline{R}_{MN} \times \overline{R}_{MQ} = \begin{vmatrix} U_x & U_y & U_z \\ -3 & 8 & -5 \\ -1 & 8 & -3 \end{vmatrix} = 16U_x - 4U_y - 16U_z$$

$$U_N = \pm \frac{\overline{R}_{MN} \times \overline{R}_{MQ}}{|\overline{R}_{MN} \times \overline{R}_{MQ}|} = \pm \frac{16U_x - 4U_y - 16U_z}{\sqrt{(16)^2 + (-4)^2 + (-16)^2}}$$

$$= \pm (0.696U_x - 0.174U_y - 0.696U_z)$$

The unit vector with positive x component $= 0.696U_x - 0.174U_y - 0.696U_z$

(*c*) $$\sin \theta = \frac{\overline{R}_{MN} \times \overline{R}_{MQ}}{|\overline{R}_{MN}||\overline{R}_{MQ}|} = \pm \frac{\sqrt{(16)^2 + (-4)^2 + (-16)^2}}{\sqrt{(-3)^2 + (8)^2 + (-5)^2}\sqrt{(-1)^2 + (8)^2 + (-3)^2}}$$

$$= 0.2698$$

(or) $$\theta = \sin^{-1}(0.2698) = 15.65°.$$

1.6 THREE DIMENSIONAL COORDINATE SYSTEMS

There are three important methods of fixing a point in space by specified lengths, directions and angles and they are:

(*a*) Cartesian coordinate system.
(*b*) Cylindrical coordinate system.
(*c*) Spherical coordinate system.

1.6.1 Cartesian Coordinate System

In the Cartesian coordinate system, three coordinate axes which are mutually at right angles to each other are selected and generally named as *x, y* and *z*-axis. It is the usual practice to select a right handed coordinate system in which a rotation of *x*-axis into *y*-axis through smaller angle would cause a right handed screw to advance in the direction of *z*-axis. As an example, if the three fingers of right hand namely thumb, forefinger and middle finger are held mutually at right angles, then they may be identified as *x, y* and *z*-axis respectively.

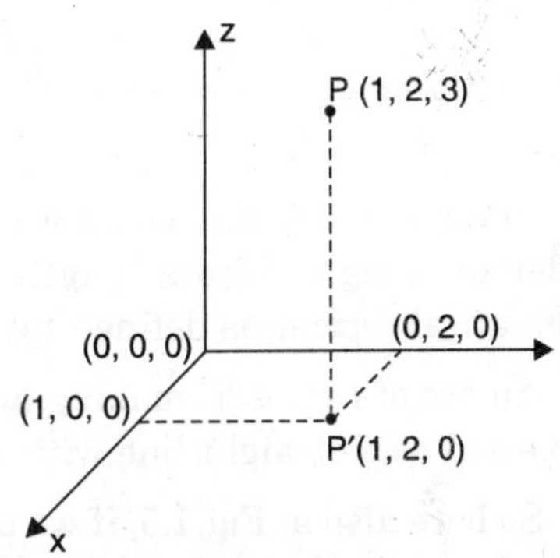

Fig. 1.3 Right-hand system of coordinates

Figure 1.3 shows a right hand system of coordinates (x, y, z). A point (1, 2, 3) is shown in the Fig. 1.3. To locate the point P, the point P' (1, 2, 0) is located first which is the projection of P on $z = 0$ plane ($x - y$ plane). The point P can be located by going three units along z-axis from P'. The above technique is the one known to us in our earlier discussions.

Thus in terms of Cartesian coordinates a unit vector can thus be founded in the following way. Let $\vec{P} = P_x U_x + P_y U_y + P_z U_z$, where P_x, P_y, P_z are the components of $\vec{P}$ along U_x, U_y, U_z respectively. The modulus of vector $|\vec{P}|$ is absolute value of vector.

i.e., $$|\vec{P}| = \sqrt{P_x^2 + P_y^2 + P_z^2} \qquad ...(1.13)$$

1.6.2 Cylindrical Coordinate System

Students prefer to work in the Cartesian system or coordinates since it is easy by virtue of familiarity. But problem with spherical and cylindrical symmetries are not only difficult to work in Cartesian system but the physics of problem will be lost. It is easy to work out a problem with spherical symmetry in spherical coordinate system. Depending upon the symmetry of problem, the suitable coordinate system must be selected to reduce the labour involved in working the problem. In the cylindrical three dimensional coordinate system any point '*P*' in space can be represented by coordinate $P(\rho, \phi, z)$, where ρ is distance of P from z-axis, ϕ is the angle measured from x-axis (any reference line) of line OP' and z is height of P above x-y plane $(P'P = z)$. P' is the projection of P on $z = 0$ plane as shown in the Fig. 1.4.

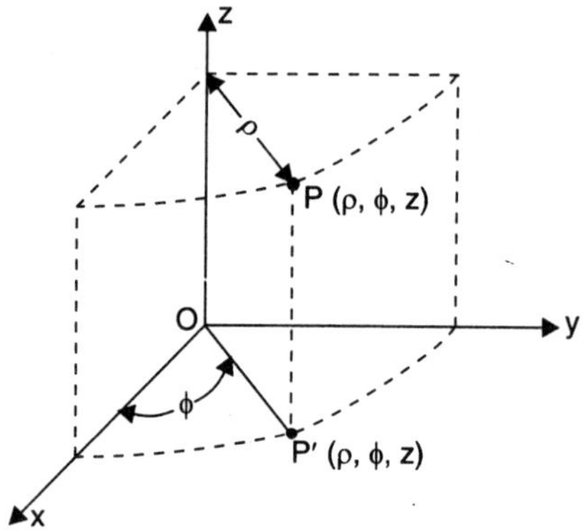

Fig. 1.4 Cylindrical coordinates

In the later chapters, as we start doing problems in electric and magnetic fields, we require transformation of problem from one system of coordinates to another to reduce the labour involved in the solution. Before attempting the transformation of coordinates, it is essential to know the relation between the variables in one system of coordinates to the variables in other system of coordinates. If we now increase (ρ, ϕ, z) by $\rho + d\rho$, $\phi + d\phi$, $z + dz$ we get a small volume in the space in the shape of rectangular parallelopiped as shown in Fig. 1.5.

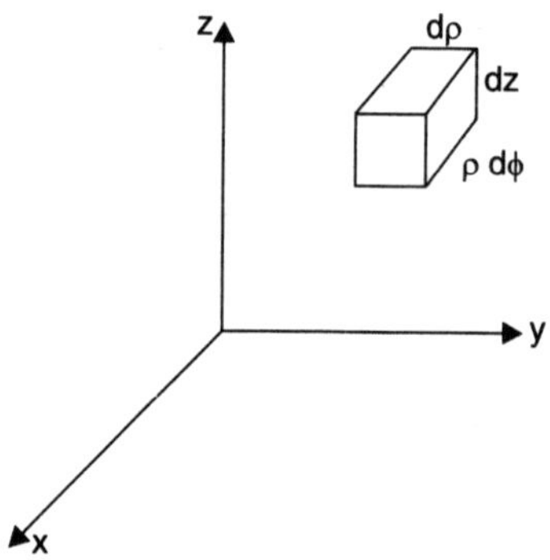

Fig. 1.5

In the Fig. 1.5, $d\rho$ and dz are dimensional lengths but not $d\phi$. So in order to bring to into a length format, we can consider the basic mathematical format as defined by:

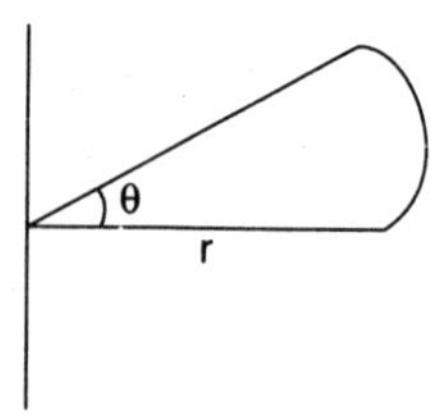

Fig. 1.6

An arc of radius '*r*' making an angle θ with respect to origin will be projected as a straight line with r, θ as shown is Fig. 1.6.

So here also in Fig.1.5, if we consider the curvature of angle to be projected as straight line, we can write it as $\rho\, d\phi$ and as a result the volume becomes equal to $\rho\, d\rho\, d\phi\, dz$.

From Fig. 1.7 we can write the equations which help us to convert cylindrical coordinates into Cartesian coordinates.

$$x = \rho \cos \phi;\ y = \rho \sin \phi;\ z = z \qquad ...(1.14)$$

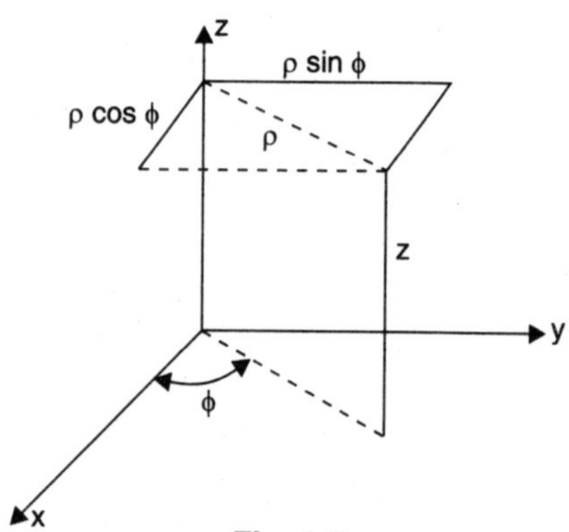

Fig. 1.7

Also from the equation (1.14), it can be written as

$$\rho = \sqrt{x^2 + y^2}\,;\ \phi = \tan^{-1}\frac{y}{x};\ z = z \qquad \text{...(1.15)}$$

The dot product of unit vectors in cylindrical and Cartesian coordinate system can be given as follows:

	U_ρ	U_ϕ	U_z
U_x	$\cos\phi$	$-\sin\phi$	0
U_y	$\sin\phi$	$\cos\phi$	0
U_z	0	0	1

Example 1.6: *Transform each of following vector to cylindrical coordinates at point specified (a) $5U_x$ at P ($\rho = 4$, $\phi = 120°$, $z = -1$) (b) $5U_x$ at Q ($x = 3$, $y = 4$, $z = -1$); (c) $4U_x - 2U_y - 4U_z$ at A ($x = 2$, $y = 3$, $z = 5$).*

Solution:

(*a*) Let $\bar{A} = 5U_x = A_\rho U_\rho + A_\phi U_\phi + A_t U_t$

$$A_\rho = \bar{A}\cdot U_\rho = 5U_x \cdot U_\rho = 5\cos\phi$$

$$A_\phi = \bar{A}\cdot U_\phi = 5U_x \cdot U_\phi = 5\sin\phi$$

$$A_z = \bar{A}\cdot U_z = 5U_x \cdot U_z = 0$$

$$\therefore \quad \bar{A} = 5\cos\phi\, U_\rho - 5\sin\phi\, U_\phi$$

Substituting coordinate of point P in above vector $\bar{A}$ at point $P = -2.5\, U_\rho - 4.33\, U_\phi$

(*b*) For point Q, $\phi = \tan^{-1}\left(\frac{y}{x}\right) = \tan^{-1}\left(\frac{4}{3}\right) = 53.13°$

$$\therefore \quad \bar{A} \text{ at point } Q = 5\cos 53.13° U_p - 5\sin 53.13° U_\phi$$

$$= 3U_\rho - 4U_\phi$$

(*c*) Let $\bar{B} = 4U_x - 2U_y - 4U_z = B_\rho\, U_\rho + B_\phi\, U_\phi + B_z\, U_z$

$$B_\rho = B\cdot U_\rho = (4U_x - 2U_y - 4U_z)\cdot U_\rho$$

$$= 4U_x\cdot U_\rho - 2U_y\cdot U_\rho - 4U_z\cdot U_\rho$$

$$= 4\cos\phi - 2\sin\phi$$

$$B_\phi = B\cdot U_\phi = (4U_x - 2U_y - 4U_z)\cdot U_\phi$$

$$= 4U_x\cdot U_\phi - 2U_y\cdot U_\phi - 4U_z\cdot U_\phi$$

$$= -4\sin\phi - 2\cos\phi$$

$$B_z = \bar{B}\cdot U_z = (4U_x - 2U_y - 4U_z)\cdot U_z = -4$$

$$\bar{B} = (4\cos\phi - 2\sin\phi)\, U_\rho + (-4\sin\phi - 2\cos\phi)\, U_\phi - 4U_z$$

for point A, $\phi = \tan^{-1}\left(\frac{y}{x}\right) = \tan^{-1}\left(\frac{3}{2}\right) = 56.31°$

Substituting ϕ coordinate of point A in above expression

$$\therefore \quad B \text{ at point } A = 0.555U_\rho - 4.438U_\phi - 4U_z$$

Example 1.7: *Give the vector in (a) Cartesian coordinates that extends from P (ρ = 5, ϕ = 15°, z = 2) to Q (ρ = 6, ϕ = 65°, z = 5) (b) Give the vector in cylindrical coordinate at M (x = 4, y = 2, z = 3) that extends to N(1, 5, 5) (c) How far is it from A (100, 60°, – 10) to B (20, 115°, 15)?*

Solution: (*a*) Cartesian coordinate of point P are $x = 5 \cos 15° = 4.829$; $y = 5 \sin 15° = 1.294$; $z = 2$, *i.e.*, P (4.829, 1.294, 2) and that of point Q are $x = 6 \cos 65° = 2.535$; $y = 6 \sin 65° = 5.437$; $z = 5$, *i.e.*, Q (2.535, 5.437, 5).

$$\therefore \quad \overline{R}_{PQ} = \overline{r}_Q - \overline{r}_P$$
$$= (2.535U_x + 5.437U_y + 5U_z) - (4.829U_x + 1.294U_y + 2U_z)$$
$$= -2.294U_x + 4.413U_y + 3U_z$$

(*b*)
$$\overline{R}_{MN} = \overline{r}_N - \overline{r}_M = (1U_x + 5U_y + 5U_z) - (4U_x + 2U_y + 3U_z)$$
$$= -3U_x + 3U_y + 2U_z$$

Let
$$\overline{A} = \overline{R}_{MN} = -3U_x + 3U_y + 2U_z = A_\rho U_\rho + A_\phi U_\phi + A_z U_z$$
$$A_\rho = A \cdot U_\rho = -3U_x \cdot U_\rho + 3U_y \cdot U_\rho + 2U_z \cdot U_\rho$$
$$= -3 \cos \phi + 3 \sin \phi$$
$$A_\phi = A \cdot U_\phi = -3U_x \cdot U_\phi + 3U_y \cdot U_\phi + 2U_z \cdot U_\phi$$
$$= 3 \sin \phi + 3 \cos \phi$$
$$A_z = A \cdot U_z = -3U_x \cdot U_z + 3U_y \cdot U_z + 2U_z \cdot U_z = 2$$
$$\therefore \quad \overline{A} = \overline{R}_{MN} = (-3 \cos \phi + 3 \sin \phi)\, U_\rho + (3 \sin \phi + 3 \cos \phi)\, U_\phi + 2U_z$$

For point M, $\phi = \tan^{-1}\left(\frac{2}{4}\right) = 26.56°$. Substitution this value of ϕ in the above vector we have $\overline{R}_{MN}$ at point $M = -1.341U_\rho + 3.997U_\phi + 2U_z$

(*c*) Cartesian coordinate of point A are $x = 100 \cos 60° = 50$; $y = 100 \sin 60° = 86.60$; $z = -10$

i.e., A (50, 86.60, – 10) and that of point B are $x = 20 \cos 115 = -8.452$;

$y = 20 \sin 115° = 18.126$; $z = 15$, *i.e.*, B (– 8.452, 18.126, 15)

distance of $AB = [(50 + 8.452)^2 + (86.60 - 18.126)^2 + (-10 - 15)^2]^{1/2} = 93.436$

Example.1.8: *Given the points P(ρ = 10, ϕ = 45°, z = 4) and Q (ρ = 5, ϕ = 80°, z = – 3)*

(a) Find the distance $|\overline{R}_{PQ}|$ *(b) Give a unit vector in Cartesian coordinate at P that is directed towards Q (c) Give a unit vector in cylindrical coordinates at P that is directed towards Q.*

Solution: (*a*) Cartesian coordinate of point P are $x = 10 \cos 45° = 7.071$; $y = 10 \sin 45° = 7.071$; $z = 4$, *i.e.*, P (7.071, 7.071, 4) and that of point Q are $x = 5 \cos 80° = 0.8682$; $y = 5 \sin 80° = 4.924$; $z = -3$, *i.e.*, Q (0.8682, 4.924, – 3).

Hence, distance of $PQ = \sqrt{(7.071 - 0.8682)^2 + (7.071 - 4.924)^2 + (4 + 3)^2} = 9.596$

(*b*)
$$\overline{R}_{PQ} = \overline{r}_Q - \overline{r}_P$$
$$= (0.8682U_x + 4.924U_y - 3U_z) - (7.071U_x + 7.071U_y + 4U_z)$$
$$= -6.2028U_x - 2.147U_y - 7U_z$$

$$\therefore \qquad U_{PQ} = \frac{\overline{R}_{PQ}}{|\overline{R}_{PQ}|} = \frac{-6.2028U_x - 2.147U_y - 7U_z}{9.596}$$

$$= -0.6464U_x - 0.2237U_y - 0.7295U_z$$

(c) Let $\overline{A} = \overline{R}_{PQ} = -6.2028U_x - 2.147U_y - 7U_z = A_\rho U_\rho + A_\phi U_\phi + A_z U_z$

$$A_\rho = \overline{A} \cdot U_\rho = -6.202U_x \cdot U_\rho - 2.147U_y \cdot U_\rho - 7U_z \cdot U_\rho$$
$$= -6.202 \cos\phi - 2.147 \sin\phi$$
$$A_\phi = A \cdot U_\phi = -6.2028U_x \cdot U_\phi - 2.147U_y \cdot U_\phi - 3U_z \cdot U_\phi$$
$$= 6.2028 \sin\phi - 2.147 \cos\phi$$
$$A_z = A \cdot U_z = -6.2028U_x \cdot U_z - 2.147U_y \cdot U_z - 3U_z \cdot U_z$$
$$= -3$$

$$\therefore \quad \overline{A} = \overline{R}_{PQ} = (-6.2028 \cos\phi - 2.147 \sin\phi)\, U_\rho + (6.2028 \sin\phi - 2.147 \cos\phi)\, U_\phi - 3\, U_z$$

Substituting the coordinates of point P in above $\overline{A}$ we get $\overline{R}_{PQ}$

at point $P = -5.9042U_\rho + 2.867U_\phi - 3U_z$.

Example 1.9: *(a) Express the field $\overline{F} = 3\,xyz\, U_x - 6\,(x + y + z)\, U_z$ in cylindrical coordinate (variables and components) (b) Find $|\ \overline{F}\ |$ at $P(\rho = 2, \phi = 60°, z = 3)$.*

Solution:

(a)
$$\overline{F} = 3xyz\, U_x - 6\,(x + y + z)\, U_z = F_\rho U_\rho + F_\phi U_\phi + F_z U_z$$
$$F_\rho = \overline{F} \cdot \overline{U}_\rho = 3\,xyz\, U_x \cdot U_\rho - 6\,(x + y + z)\, U_z \cdot U_\rho$$
$$= 3\,xyz \cos\phi - 6\,(x + y + z)\,(0)$$
$$= 3\,(\rho\cos\phi)\,(\rho\sin\phi)\,(z)\cos\phi$$
$$= 3\,\rho^2 z \cos^2\phi \sin\phi$$
$$F_\phi = F \cdot U_\phi = 3\,xyz\, U_x \cdot U_\phi - 6\,(x + y + z)\, U_z \cdot U_\phi$$
$$= 3\,xyz\,(-\sin\phi) - 6\,(x + y + z)\,(0)$$
$$= -3\,(\rho\cos\phi)\,(r\sin\phi)\,(z)\sin\phi$$
$$= -3\rho^2 z \cos\phi \sin^2\phi$$
$$F_z = F \cdot U_z = 3\,xyz\, U_x \cdot U_z - 6\,(x + y + z)\, U_z \cdot U_z$$
$$= 3\,xyz\,(0) - 6\,(x + y + z)\,(1)$$
$$= -6\,(x + y + z)$$
$$= -6\,(\rho\cos\phi + \rho\sin\phi + z)$$

$$\therefore \qquad \overline{F} = 3\rho^2 z \cos^2\phi \sin\phi\, U_\rho - 3\rho^2 z \cos\phi \sin^2\phi\, U_\phi - 6\,(\rho\cos\phi + \rho\sin\phi + z)\, U_z$$

(b) Substituting the coordinate of point P in the above $\overline{F}$, we have

$$\overline{F} = 5.196U_\rho - 9U_\rho - 28.660U_z$$

$$|\ \overline{F}\ | \text{ at } P = \sqrt{(5.196)^2 + (-9)^2 + (-28.660)^2} = 30.486.$$

Example 1.10: *Transform each of the following vector to cylindrical coordinates at the point specified (a) $6U_x$ at P ($\rho = 5$, $\phi = 110°$, $Z = -2$) (b) $6U_x$ at Q ($x = 4$, $y = 6$, $z = -1$) (c) $5U_x - 3U_y - 4U_z$ at A ($x = 3$, $y = 4$, $z = 5$).*

Solution:

(*a*) Let $\bar{A} = 6U_x = A_\rho U_\rho + A_\phi U_\phi + A_z U_z$

$$A_\rho = A \cdot U_\rho = 6\, U_x \cdot U_\rho = 6 \cos \phi$$
$$A_\phi = A \cdot U_\phi = 6\, U_x \cdot U_\phi = -6 \sin \phi$$
$$A_z = A \cdot U_z = 6\, U_x \cdot U_z = 0$$

$\therefore$ $$\bar{A} = 6 \cos \phi\, U_\rho - 6 \sin \phi\, U_\phi$$

Substituting ϕ coordinate of point P in the above vector we have $\bar{A}$ at point P as

$$\bar{A} = -2.052U_\rho - 5.638U_\phi$$

(b) For point Q, $$\phi = \tan^{-1}\left(\frac{y}{x}\right) = \tan^{-1}\left(\frac{6}{4}\right) = 56.30°$$

$\bar{A}$ at point $Q = 6 \cos 56.30°\, U_\rho - 6 \sin 56.30°\, U_\phi = 3.32U_\rho - 4.99U_\phi$

(c) Let $\bar{B} = 5U_x - 3U_y - 4U_z = B_\rho U_\rho + B_\phi U_\phi + B_z U_z$

$$B_\rho = B \cdot U_\rho = (5U_x - 3U_y - 4U_z) \cdot U_\rho$$
$$= 5U_x \cdot U_\rho - 3U_y \cdot U_\rho - 4U_z \cdot U_\rho$$
$$= 5 \cos \phi - 3 \sin \phi$$
$$B_\phi = B \cdot U_\phi = (5U_x - 3U_y - 4U_z) \cdot U_\phi$$
$$= -5 \sin \phi - 3 \cos \phi$$
$$B_z = B \cdot U_z = (5U_x - 3U_y - 4U_z) \cdot U_z = -4U_z$$

$\therefore$ $$\bar{B} = (5 \cos \phi - 3 \sin \phi)\, U_\rho + (-5 \sin \phi - 3 \cos \phi)\, U_\phi - 4U_z$$

for point $\bar{A}$, $\phi = \tan^{-1}\left(\frac{y}{x}\right) = \tan^{-1}\left(\frac{4}{3}\right) = 53.12°$

Substituting ϕ coordinate of point A is the above $\bar{B}$ we have $\bar{B}$ at point A

$$\bar{B} = -0.60U_\rho - 5.799U_\phi - 4U_z$$

1.6.3 Spherical Coordinate System

We have no two-dimensional coordinate system to help us understand the three dimensional spherical coordinate system, as we have for the circular cylindrical coordinate system. In certain respects we can draw on our knowledge of the latitude and longitude system of locating a place on surface of earth, but usually we consider only points on surface and not those below (or) above ground.

Let us start by building a spherical coordinate system on the three Cartesian axes. First define the distance from origin to any point as 'r' as shown in the Fig. 1.8. The second coordinate is an angle θ between z-axis and line drawn from origin to the point in question. The third coordinate ϕ is also an angle and is exactly the same as angle ϕ of cylindrical coordinate. It is the angle between the x-axis and the projection in $z = 0$ plane of

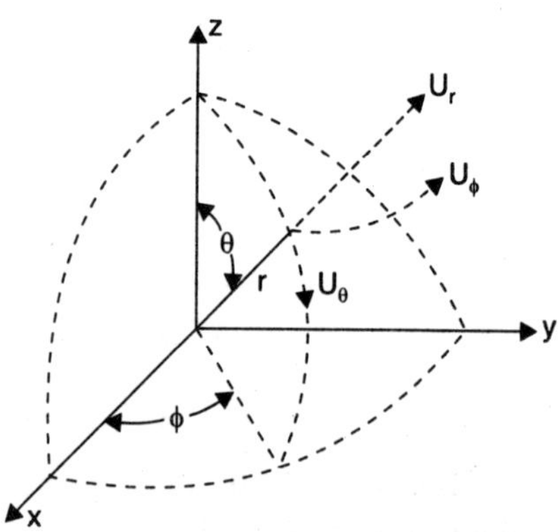

Fig. 1.8 Spherical coordinates

line drawn from origin to the point. The unit vectors corresponding to the three coordinate axes are also represented in the Fig.1.8. They are mutually perpendicular and a differential volume element made of following differential elements $dr, d\theta, d\phi$ can be created. The differential distance 'r' can be written as 'dr'; the differential angle 'θ' can be written as '$r\ d\theta$' and the differential angle 'ϕ' can be written as '$r \sin \theta\ d\phi$' and as a result the volume of differential element can be written as $r \sin \phi\ dr\ d\theta\ d\phi$.

In order to convert the spherical coordinates into Cartesian coordinates the following equations can be used:

$$x = r \sin \theta \cos \phi;\ y = r \sin \theta \sin \phi;\ z = r \cos \theta \qquad ...(1.16)$$

The transformation in the reverse direction is achieved with help of following equations which are derived from eqn. (1.16):

$$r = \sqrt{x^2 + y^2 + z^2}\ ;\ \theta = \cos^{-1} \frac{z}{\sqrt{x^2 + y^2 + z^2}}\ ;\ \phi = \tan^{-1} \frac{y}{x} \qquad ...(1.17)$$

The dot product of unit vectors of spherical and Cartesian coordinate systems can be given as follows:

	U_r	U_θ	U_ϕ
U_x	$\sin \theta \cos \phi$	$\cos \theta \cos \phi$	$-\sin \phi$
U_y	$\sin \theta \sin \phi$	$\cos \theta \sin \phi$	$\cos \phi$
U_z	$\cos \theta$	$-\sin \phi$	0

Example 1.11: *Transform each of the following vectors to spherical coordinates at specified point (a) $3U_x$ at B ($r = 4$, $\theta = 25°$, $\phi = 120°$); (b) $4U_x$ at A($x = 2$, $y = 3$, $z = -1$); (c) $4U_x - 2U_y - 4U_z$ at P($x = -2$, $y = -3$, $x = 4$).*

Solution: (*a*) Let $\bar{A} = 3U_x = A_r U_r + A_\theta U_\theta + A_\phi U_\phi$

$$A_r = \bar{A} \cdot U_r = 3U_x \cdot U_r = 3 \sin \theta \cos \phi$$

$$A_\theta = \bar{A} \cdot U_\theta = 3U_x \cdot U_\theta = 3 \cos \theta \cos \phi$$

$$A_\phi = \bar{A} \cdot U_\phi = 3U_x \cdot U_\phi = -3 \sin \phi$$

$$\therefore \quad \bar{A} = 3 \sin \theta \cos \phi\ U_r + 3 \cos \theta \cos \phi\ U_\theta - 3 \sin \phi\ U_\phi$$

Substituting θ and ϕ coordinates at point B in above vector we have $\bar{A}$, at point B

$$= -0.633U_r - 1.359U_\theta - 2.59U_\phi$$

(*b*) for point A, $r = \sqrt{(2)^2 + (3)^2 + (-1)^2} = 3.742$

$$\theta = \cos^{-1}\left(\frac{-1}{3.742}\right) = 105.5°;\ \phi = \tan^{-1}\left(\frac{3}{2}\right) = 56.3°$$

Substituting θ and ϕ coordinate of point A in above $\bar{A}$, we have $\bar{A}$ at point

$$A = 2.137U_r - 0.2925U_\theta - 3.32U_\phi$$

(*c*) Let $\bar{B} = 4U_x - 2U_y - 4U_z = B_r U_r + B_\theta U_\theta + B_\phi U_\phi$

$$B_r = B \cdot U_r = (4U_x - 2U_y - 4U_z) \cdot U_r$$

$$= 4U_x \cdot U_r - 2U_y \cdot U_r - 4U_z \cdot U_r$$
$$= 4\cos\theta\cos\phi - 2\cos\theta\sin\phi + 4\sin\theta$$
$$B_\phi = B \cdot U_\phi = (4U_x - 2U_y - 4U_z) \cdot U_\phi$$
$$= 4U_x \cdot U_\phi - 2U_y \cdot U_\phi - 4U_z \cdot U_\phi$$
$$= -4\sin\phi - 2\cos\phi$$

$$\therefore \quad \overline{B} = (4\sin\theta\cos\phi - 2\sin\theta\sin\phi - 4\cos\theta)\, U_r + (4\cos\theta\cos\phi - 2\cos\theta\sin\phi + 4\sin\phi)\, U_\theta + (-4\sin\phi + 2\cos\phi)\, U_\phi$$

for point P,
$$r = \sqrt{(-2)^2 + (-3)^2 + (4)^2} = 5.385$$
$$\theta = \cos^{-1}\left(\frac{4}{5}\right) = 42.03^\circ$$

and
$$\phi = 180^\circ + \tan^{-1}\left(\frac{-3}{-2}\right) = 236.3^\circ$$

Substituting θ and ϕ coordinates of point P is above $\overline{B}$ expression we have $\overline{B}$ at point $P = -3.343U_r + 2.266U_\theta + 4.438U_\phi$

Example 1.12: *An electric field intensity is given as* $\overline{E} = \left(\frac{80(\cos\theta)}{r^3} U_r + \frac{40\sin\theta}{r^3} U_\theta\right)$

At the point whose spherical coordinates are $r = 2$, $\theta = 60^\circ$, $\phi = 20^\circ$. *Find (a)* $|\overline{E}|$ *(b) a unit vector in (Cartesian coordinates) in direction of* $\overline{E}$.

Solution:

(*a*)
$$\overline{E} = \frac{80\cos 60^\circ}{r^3} U_r + \frac{40\sin 60^\circ}{r^3} U_\theta$$
$$= 5U_r + 4.330U_\theta$$
$$|\overline{E}| = \sqrt{(5)^2 + (4.330)^2} = 6.614$$

(*b*)
$$\overline{E} = E_x U_x + E_y U_y + E_z U_z$$
$$E_x = \overline{E} \cdot U_x = 5U_r \cdot U_x + 4.330U_\theta \cdot U_x$$
$$= 5\sin\theta\cos\phi + 4.330\cos\theta\cos\phi$$
$$E_y = \overline{E} \cdot U_y = 5U_r \cdot U_y + 4.330U_\theta \cdot U_y$$
$$= 5\sin\theta\cos\phi + 4.330\cos\theta\sin\phi$$
$$E_z = \overline{E} \cdot U_z = 5U_r \cdot U_z + 4.330U_\theta \cdot U_z$$
$$= 5\cos\theta - 4.330\sin\theta$$
$$\overline{E} = (5\sin\theta\cos\phi + 4.330\cos\theta\cos\phi)U_x + (5\sin\theta\sin\phi + 4.330\cos\theta\sin\phi)U_y + (5\cos\theta - 4.330\sin\theta)U_z$$

Substituting $\theta = 60^\circ$ and $\phi = 20^\circ$ in above vector equation

$$\overline{E} \text{ at point } (r = 2, \theta = 60^\circ, \phi = 20^\circ) = 6.094U_x + 2.220U_y - 1.249U_z$$

$$\therefore \quad U_E = \frac{\overline{E}}{|\overline{E}|} = \frac{6.049U_x + 2.220U_y - 1.49U_z}{\sqrt{(6.049)^2 + (2.220)^2 + (-1.49)^2}}$$

$$= 0.923U_x + 0.336U_y - 0.189U_z$$

1.7 GRADIENT OF SCALAR FIELD

Let V denote a scalar field. The vector whose x, y and z-coordinates are $\frac{\partial v}{\partial x}, \frac{\partial v}{\partial y}$ and $\frac{\partial v}{\partial z}$ is termed as gradient of V at point (x, y, z). It is denoted by grad V (or) ∇V. So by definition

$$\text{grad } V \text{ (or) } \nabla V = \frac{\partial v}{\partial x}U_x + \frac{\partial v}{\partial y}U_y + \frac{\partial v}{\partial z}U_z \quad \text{...(1.18)}$$

From the above expression after comparison, we can write as

$$\nabla = \frac{\partial}{\partial x}U_x + \frac{\partial v}{\partial y}U_y + \frac{\partial}{\partial z}U_z \quad \text{...(1.19)}$$

Here ∇ is not a vector but only a differential operator. Hence, gradient operation on a scalar turns it into a vector.

1.8 DIVERGENCE OF VECTOR FIELD

Let V denote a vector field and represented as

$\overline{V} = V_x U_x + V_y U_y + V_z U_z$. The sum of partial derivative viz., $\frac{\partial V_x}{\partial x} + \frac{\partial V_y}{\partial y} + \frac{\partial V_z}{\partial z}$ is termed as divergence of vector V. So evidently **divergence of vector is a scalar**, since above sum is a scalar.

$$\therefore \quad \text{div}\overline{V} \text{ (or) } \nabla \cdot V = \frac{\partial V_x}{\partial x} + \frac{\partial V_y}{\partial y} + \frac{\partial V_z}{\partial z} \quad \text{...(1.20)}$$

If divergence of a vector is zero, the vector is termed as divergence free vector.

1.8.1 Significance of Divergence

If vector V represents the velocity of fluid in motion at any point, it can be shown that $\nabla \cdot V$ represents the rate at which fluid flows out of unit volume enclosing that point.

1.9 LAPLACIAN OF SCALAR FIELD

Let V represent a scalar field

Now
$$\nabla V = \left(\frac{\partial}{\partial x}U_x + \frac{\partial}{\partial y}U_y + \frac{\partial}{\partial z}U_z\right)V \quad \text{...(1.21)}$$

Now determine the divergence of equation (1.21)

$$\nabla \cdot (\nabla V) = \nabla \cdot \left[\frac{\partial}{\partial x} U_x + \frac{\partial}{\partial y} U_y + \frac{\partial}{\partial z} U_z\right] V \quad ...(1.22)$$

$$= \frac{\partial^2 V}{\partial x^2} + \frac{\partial^2 V}{\partial y^2} + \frac{\partial^2 V}{\partial z^2} \quad ...(1.23)$$

So the expression on R.H.S. is termed as laplacian of scalar field V.

1.10 CURL OF A VECTOR FIELD

Consider a vector $\overline{V} = V_x U_x + V_y U_y + V_z U_z$

We define **curl as** $\nabla \times V$, so implementing the above formulae we can get

$$\nabla \times V = \begin{vmatrix} U_x & U_y & U_z \\ \frac{\partial}{\partial x} & \frac{\partial}{\partial y} & \frac{\partial}{\partial z} \\ V_x & V_y & V_z \end{vmatrix}$$

$$\nabla \times V = \left(\frac{\partial V_z}{\partial y} - \frac{\partial V_y}{\partial z}\right) U_x + \left(\frac{\partial V_x}{\partial z} - \frac{\partial V_z}{\partial x}\right) U_y + \left(\frac{\partial V_y}{\partial x} - \frac{\partial V_x}{\partial y}\right) U_z \quad ...(1.24)$$

So we can write as $\left(\frac{\partial V_z}{\partial y} - \frac{\partial V_y}{\partial z}\right), \left(\frac{\partial V_x}{\partial z} - \frac{\partial V_z}{\partial x}\right), \left(\frac{\partial V_y}{\partial x} - \frac{\partial V_x}{\partial y}\right)$ is termed as curl of vector V.

Example 1.13: *Given vector field $\overline{G} = 3x^2 y\, U_x - 4(z - x)\, U_y + 6xyz\, U_z$. Find (a) $\overline{G}$ at P(1, 2, 3) (b) a unit vector in direction of $\overline{G}$ at P (c) the (scalar) equation of surface on which $|\overline{G}| = 90$; (d) the y-coordinate of Q (2, y, 3) if $|\overline{G}_Q| = 90$ and $y > 0$, (e) the distance between P and Q.*

Solution:

(*a*) $\overline{G}$ at $P = 3(1)^2(2)U_x - 4(3-1)U_y + 6(2)(3)U_z$

$= 6U_x - 8U_y + 36U_z$

(*b*) $$\overline{U} = \frac{\overline{G}}{|\overline{G}|} = \frac{6U_x - 8U_y + 36U_z}{\sqrt{(6)^2 + (-8)^2 + (36)^2}} = 0.16U_x - 0.214U_y - 0.96U_z$$

(*c*) $$|\overline{G}| = \sqrt{(3x^2y)^2 + [-4(z-x)]^2 + (6xyz)^2}$$

$$= \sqrt{9x^4y^2 + 16(z-x)^2 + 36x^2y^2z^2} = 90$$

Squaring on both sides we have scalar equation of surface as

$$9x^4 y^2 + 16(z-x)^2 + 36x^2 y^2 z^2 = 8100$$

(*d*) Substitute the coordinates of point Q in above equation of surface

$$\therefore \quad 9(2)^4 y^2 + 16[3-2]^2 + 36(2)^2 y^2 (3)^2 = 8100$$

$$144 y^2 = 6788$$

(or) $y^2 = 47.13$ (or) $y = \pm 6.86$

Since $y > 0, y = 6.86$ is selected.

(*e*) Distance between $(PQ) = \sqrt{(1-2)^2 + (2-6.86)^2 + (3-3)^2} = 4.96.$

Example 1.14: *(a) Find the volume defined by $4 < \rho < 6$, $30° < \phi < 60°$, $2 < z < 5$ (b) what is length of longest straight line that lies entirely within the volume? (c) find the total area of surface.*

Solution:

(*a*)
$$\text{Volume} = \int dv = \int_4^6 \int_{30°}^{60°} \int_2^5 \rho \, d\rho \, d\phi \, dz$$

$$= \left[\frac{\rho^2}{2}\right]_4^6 [\phi]_{30°}^{60°} [z]_2^5 = 15.708$$

(*b*) Suppose points *P* and *Q* are diametrically opposite corners of the volume. Then the lower limits of ρ, ϕ and *z* gives the coordinates of point *P* where as the higher limits of ρ, ϕ and *z* given coordinates of point Q.

∴ $P(4, 30°, 2)$ and $Q(6, 60°, 5)$

Distance between $PQ = \sqrt{(4\cos 30° - 6\cos 60°)^2 + (4\sin 30° - 6\sin 60°)^2 + (2-5)^2}$

$= 4.408$

(*c*) Six surface of volume are located by coordinates

$\rho = 4$ and 6, $\phi = 30°$ and $60°$; $z = 2$ and 5

$$S = S_1 + S_2 + S_3 + S_4 + S_5 + S_6$$

$$= \int_{30°}^{60°} \int_2^5 \rho \, d\phi \, dz \Big|_{\rho=2} + \int_{30°}^{60°} \int_2^5 \rho \, d\phi \, dz \Big|_{\rho=6} + \int_4^6 \int_2^5 d\rho \, dz \Big|_{\phi=30°} + \int_4^6 \int_2^5 d\rho \, dz \Big|_{\phi=60°}$$

$$+ \int_4^6 \int_{30°}^{60°} \rho \, d\rho \, d\phi \, dz \Big|_{z=2} + \int_4^6 \int_{30°}^{60°} \rho \, d\rho \, d\phi \, dz \Big|_{z=2}$$

$$= 4[\phi]_{30°}^{60°} [z]_2^5 + 6[\phi]_{30°}^{60°} [z]_2^5 + [\rho]_4^6 [z]_2^5 + [\rho]_4^6 + [z]_2^5$$

$$+ \left[\frac{\rho^2}{2}\right]_2^6 \left[\phi\right]_{30°}^{60°} + \left[\frac{\rho^2}{4}\right]_4^6 \left[\phi\right]_{30°}^{60°}$$

$= 38.180.$

Example 1.15: *Points A ($r = 90$, $\theta = 90°$, $\phi = 0°$) and B ($r = 90$, $\theta = 90°$, $\phi = 5°$) are located on the surface of 100 m radius sphere (a) what is their separation, using a path on spherical surface? (b) what is their separation, using a straight line path?*

Solution: (*a*) $\bar{d}L = dr\, U_r + rd\,\theta\, U_\theta + r\sin\theta\, d\phi\, U_\phi$

For points *A* and *B*, *r* and θ coordinates are constant. Hence *dr* and *d*θ both are equal to zero. Hence $dL = r\sin\theta\, d\phi\, U_\phi$ (or) $dL = r\sin\theta\, d\phi$

∴
$$L = \int dL = \int_{0°}^{5°} r\sin\theta\, d\phi$$

$= 90 \sin 90° [\phi]_{0°}^{5°}$ where $r = 90$ and $\theta = 90°$

$= 7.86$

(*b*) distance between AB = [(90 sin 90° cos 0° – 90 sin 90° cos 5°)^2

+ (90 sin 90° sin 0° – 90 sin 90° sin 5°)^2 + (90 cos 90° – 90 cos 90°)^2]$^{1/2}$ = 7.847.

Example 1.16: *Given point A (x = 3, y = 4, z = – 2) and B (ρ = 3, φ = – 45°, z = 2), find a unit vector in cylindrical coordinates (a) at point B directed towards point A (b) at point A directed towards point B.*

Solution:

(*a*) Cartesian coordinate of point B are $x = 3\cos(-45°) = 2.12$, $y = 3\sin(-45°) = -2.12$

$z = 2$ *i.e.*, $B\,(2.12, -2.12, 2)$

$$\overline{R}_{BA} = \overline{r}_A - \overline{r}_B = (3U_x + 4U_y - 2U_z) - (2.12U_x - 2.12U_y + 2U_z)$$

Let $\overline{A} = \overline{R}_{BA} = 0.88U_x + 6.12U_y - 4U_z = A_\rho U_\rho + A_\phi U_\phi + A_z U_z$

$$A_\rho = A \cdot U_\rho = 0.88U_x \cdot U_\rho + 6.12U_y \cdot U_\rho - 4U_z \cdot U_\rho$$
$$= 0.88 \cos\phi + 6.12 \sin\phi$$
$$A_\phi = A \cdot U_\phi = 0.88U_x \cdot U_\phi + 6.12U_y \cdot U_\phi - 4U_z \cdot U_\phi$$
$$= 0.88 \sin\phi + 6.12 \cos\phi$$
$$A_z = A \cdot U_z = 0.88U_x \cdot U_z + 6.12U_y \cdot U_z - 4U_z \cdot U_z = -4$$

∴ $\overline{A} = \overline{R}_{BA} = (0.88\cos\phi + 6.12\sin\phi)U_\rho + (0.88\sin\phi + 6.12\cos\phi)U_\phi - 4U_z$

Substituting φ coordinate of point B in the above vector we have $\overline{R}_{BA}$ at point $B = -3.705U_\rho + 4.949U_\phi - 4U_z$

∴ $$\overline{U}_{BA} = \frac{\overline{R}_{BA}}{|\overline{R}_{BA}|} = \frac{-3.705U_\rho + 4.949U_\phi - 4U_z}{\sqrt{(-3.705)^2 + (4.949)^2 + (-4)^2}}$$
$$= -0.503U_\rho + 0.672U_\phi - 0.543U_z$$

(*b*) $$\overline{R}_{AB} = \overline{r}_B - \overline{r}_A = (2.12U_x - 2.12U_y + 2U_z) - (3U_x + 4U_y - 2U_z)$$
$$= -0.88U_x - 6.12U_y + 4U_z$$

Let $\overline{B} = \overline{R}_{AB} = -0.88U_x - 6.12U_y + 4U_z$

$$B_\rho = B \cdot U_\rho = -0.88U_x \cdot U_\rho - 6.12U_y \cdot U_\rho + 4U_z \cdot U_\rho$$
$$= -0.88\cos\phi - 6.12\sin\phi$$
$$B_\phi = B \cdot U_\phi = -0.88U_x \cdot U_\phi - 6.12U_y \cdot U_\phi + 4U_z \cdot U_\phi$$
$$= 0.88\sin\phi - 6.12\cos\phi$$
$$B_z = B \cdot U_z = -0.88U_x \cdot U_z - 6.12U_y \cdot U_z + 4U_z \cdot U_z = 4$$

∴ $\overline{B} = \overline{R}_{AB} = (-0.88\cos\phi - 6.12\sin\phi)U_\rho + (0.88\sin\phi - 6.12\cos\phi)U_\phi + 4U_z$

For point A, $\phi = \tan^{-1}\left(\frac{4}{3}\right) = 53.12°$. Substituting this value of φ in the above vector we have $\overline{R}_{AB}$ at point $A = -5.423U_\rho - 2.969U_\phi + 4U_z$

$$U_{AB} = \frac{\overline{R}_{AB}}{|\overline{R}_{AB}|} = \frac{-5.423U_\rho - 2.969U_\phi + 4U_z}{[(-5.423)^2 + (-2.969)^2 + (4)^2]^{1/2}}$$
$$= -0.736U_\rho - 0.403U_\phi + 0.543U_z.$$

Example 1.17: *A vector field is specified by* $\overline{F} = 4\,(x + y)\sin \pi z\, U_x - (x^2 + y)U_y + \left[\dfrac{20}{x^2 + y^2}\right]U_z$. *Specify the locus of all points at which (a)* $F_x = 0$ *(b)* $F_y = 0$ *(c)* $|F_z| = 1$.

Solution:

(*a*) $$\overline{F} = F_x U_x + F_y U_y + F_z U_z$$

$\therefore$ $$F_x = 4\,(x + y)\sin \pi z = 0$$

$\therefore$ $$x + y = 0 \text{ (or) } \sin \pi z = 0$$

$\therefore$ $$y = -x \text{ (or) } \pi z = 0, \pm \pi, \pm 2\pi \,\ldots\ldots$$

$y = -x$ represents a plane and $z = 0, \pm 1, \pm 2, \pm 3$ represents planes

(*b*) $$F_y = -(x^2 + y) = 0$$

$y = -x^2$. It represents a parabolic cylinder

(*c*) $$F_z = \frac{20}{x^2 + y^2}$$

$$|F_z| = \frac{20}{x^2 + y^2} = 1$$

$$x^2 + y^2 = 20$$

It represents a circular cylinder of radius $\sqrt{20}$ with z-axis as its axis.

Example 1.18: *If* $\overline{A} = 10u_x - 4u_y + 6u_z$ *and* $\overline{B} = 2u_x + u_y$, *find (a) the component of* $\overline{A}$ *along* u_y *(b) the magnitude of 3A – B (c) a unit vector along A + 3B.*

Solution:

(*a*) The component of A along u_y is $A_y = -4$

(*b*) $3A - B = 3(10, -4, 6) - (2, 1, 0)$
$= (30, -12, 18) - (2, 1, 0)$
$= (28, -13, 18)$

Hence $|3A - B| = \sqrt{(28)^2 + (-13)^2 + (18)^2}$
$= 35.74$

(*c*) Let $C = A + 2B = (10, -4, 6) + (4, 2, 0) = (14, -2, 6)$

A unit vector along C is given by

$$u_c = \frac{C}{|C|} = \frac{(14, -2, 6)}{\sqrt{14^2 + (-2)^2 + 6^2}}$$

$$u_c = 0.9113u_x - 0.1302u_y + 0.3906u_z$$

Example 1.19: *Points P and Q are located at (0, 2, 4) and (– 3, 1, 5). Calculate (a) the position vector of P(b) the distance vector from P to Q (c) the distance between P and Q (d) a vector parallel to PQ with magnitude of 10.*

Solution:

(*a*) $\bar{r}_p = 0u_x + 2u_y + 4u_z = 2u_y + 4u_z$

(*b*) $\overline{R}_{pq} = \bar{r}_q - \bar{r}_p = (-3, 1, 5) - (0, 2, 4) = (-3, -1, 1)$

i.e., $\overline{R}_{pq} = -3u_x - u_y + u_z$

(c) Since $\overline{R}_{pq}$ is the distance vector from P to Q the distance between P and Q is the magnitude of this vector *i.e.*, $d = |\bar{r}_{pq}| = \sqrt{9+1+1} = 3.317$.

(d) Let the required vector be A, then it can be written as $A = Au_A$, where $A = 10$ is the magnitude of A. Since A is parallel to PQ, it must have the same unit vector as r_{pq}.

Hence
$$u_A = \pm \frac{\bar{r}_{pq}}{|\bar{r}_{pq}|} = \pm \frac{(-3, -1, 1)}{3.317}$$

and
$$A = \pm \frac{10(-3, -1, -1)}{3.317} = \pm(-9.045u_x - 3.015u_y + 3.015u_z)$$

Example 1.20: *Express vector* $B = \frac{10}{r}u_r + r\cos\theta\, u_\theta + u_\phi$ *in Cartesian coordinates and also find B(– 3, 4, 0).*

Solution:

Using the concept to convert spherical coordinated to cylindrical coordinates it can be written as

$$B_x = \frac{10}{r}\sin\theta\cos\phi + r\cos^2\theta\cos\phi - \sin\phi$$

$$B_y = \frac{10}{r}\sin\theta\sin\phi + r\cos^2\theta\sin\phi - \cos\phi$$

$$B_z = \frac{10}{r}\cos\theta + r\cos\theta\sin\theta$$

But $r = \sqrt{x^2+y^2+z^2}$, $\theta = \tan^{-1}\frac{\sqrt{x^2+y^2}}{z}$ and $\phi = \tan^{-1}\frac{y}{x}$

Hence, $\sin\theta = \frac{\rho}{r} = \frac{\sqrt{x^2+y^2}}{\sqrt{x^2+y^2+z^2}}$; $\cos\theta = \frac{z}{r} = \frac{z}{\sqrt{x^2+y^2+z^2}}$

$$\sin\phi = \frac{y}{\rho} = \frac{y}{\sqrt{x^2+y^2}}; \quad \cos\phi = \frac{x}{\rho} = \frac{x}{\sqrt{x^2+y^2}}$$

Substituting all the above equations in the basic equation it yields,

$$B_x = \frac{10\sqrt{x^2+y^2}}{(x^2+y^2+z^2)} \times \frac{x}{\sqrt{x^2+y^2}} + \frac{\sqrt{x^2+y^2+z^2}}{(x^2+y^2+z^2)} \times \frac{z^2x}{\sqrt{x^2+y^2}} - \frac{y}{\sqrt{x^2+y^2}}$$

$$= \frac{10x}{(x^2+y^2+z^2)} + \frac{xz^2}{\sqrt{(x^2+y^2)(x^2+y^2+z^2)}} - \frac{y}{\sqrt{x^2+y^2}}$$

$$B_y = \frac{10\sqrt{x^2+y^2}}{(x^2+y^2+z^2)} \times \frac{y}{\sqrt{x^2+y^2}} + \frac{\sqrt{x^2+y^2+z^2}}{(x^2+y^2+z^2)} \times \frac{z^2y}{\sqrt{x^2+y^2}} + \frac{x}{\sqrt{x^2+y^2}}$$

$$= \frac{10y}{(x^2+y^2+z^2)} + \frac{yz^2}{\sqrt{(x^2+y^2)(x^2+y^2+z^2)}} + \frac{x}{\sqrt{x^2+y^2}}$$

$$B_z = \frac{10z}{(x^2+y^2+z^2)} - \frac{z\sqrt{x^2+y^2}}{\sqrt{(x^2+y^2+z^2)}}$$

At (– 3, 4, 0) the value of B can be shown as

$$B_x = -\frac{30}{25} + 0 - \frac{4}{5} = -2$$

$$B_y = \frac{40}{25} + 0 - \frac{3}{5} = 1$$

$$B_z = 0 - 0 = 0$$

Thus it can be written as $B = -2u_x + u_y$.

Example 1.21: *Given a vector field* $\overline{D} = r \sin\phi\, u_r - \frac{1}{r} \sin\theta \cos\phi\, u_\theta + r^2 u_\phi$. *Determine (a)* $\overline{D}$ *at P(10, 150°, 330°) (b) the component of* $\overline{D}$ *tangential to the spherical surface r = 10 at P (c) a unit vector at P perpendicular to* $\overline{D}$ *and tangential to the cone θ = 150°*

Solution:

(*a*) At P, $r = 10$, $\theta = 150°$, $\phi = 330°$.

Hence $\overline{D} = 10 \sin 330° u_r - \frac{1}{10} \sin 150° \cos 330°\, u_\theta + 100 u_\phi = (-5, 0.043, 100)$

(*b*) Any vector $\overline{D}$ can be resolved into two orthogonal components: $D = D_t + D_n$ where D_t is tangential to a given surface and D_n is normal to it. In this case, since u_r is normal to the surface $r = 10$. *i.e.*, $D_n = r \sin\phi\, u_r = -5u_r$

Hence $D_t = D - D_n = 0.043u_\theta + 100u_\phi$

(*c*) A vector at P perpendicular to D and tangential to the cone $\theta = 150°$ is the same as the vector perpendicular to both D and u_θ.

$$\text{Hence} \quad \overline{D} \times u_\theta = \begin{vmatrix} u_r & u_\theta & u_\phi \\ -5 & 0.043 & 100 \\ 0 & 1 & 0 \end{vmatrix}$$

$$= -100u_r - 5u_\phi$$

A unit vector along this is given by $u = \dfrac{-100u_r - 5u_\phi}{\sqrt{100^2 + 5^2}} = -0.9988u_r - 0.0499u_\phi$.

Example 1.22: *Show that the cosines of the angle ψ between the vectors A and B is given by the sum of the products of their direction cosines.*

Solution:

Let $\cos\alpha_1$, $\cos\beta_1$, $\cos\gamma_1$ be the direction cosines of A and $\cos\alpha_2$, $\cos\beta_2$, $\cos\gamma_2$ be the direction cosines of B, then $A = A_x u_x + A_y u_y + A_z u_z$

$$= \sqrt{A_x^2 + A_y^2 + A_z^2}\left[\frac{A_x}{\sqrt{A_x^2 + A_y^2 + A_z^2}} u_x + \frac{A_y}{\sqrt{A_x^2 + A_y^2 + A_z^2}} u_y + \frac{A_z}{\sqrt{A_x^2 + A_y^2 + A_z^2}} u_z\right]$$

or $\overline{A} = |A| [\cos \alpha_1 u_x + \cos \beta_1 u_y + \cos \gamma_1 u_z]$

where $A = \sqrt{A_x^2 + A_y^2 + A_z^2}$

$$A_x = A \cos \alpha_1; \quad A_y = A \cos \beta_1; \quad A_z = A \cos \gamma_1;$$

Similarly $\overline{B} = |B| [\cos \alpha_1 u_x + \cos \beta_1 u_y + \cos \gamma_1 u_z]$

$\therefore$ $\overline{A} \cdot \overline{B} = |AB| \cos \psi$

$$= |AB| [\cos \alpha_1 \cos \alpha_2 + \cos \beta_1 \cos \beta_2 + \cos \gamma_1 \cos \gamma_2]$$

where ψ is the angle between vectors A and B

$\therefore$ $\cos \psi = \cos \alpha_1 \cos \alpha_2 + \cos \beta_1 \cos \beta_2 + \cos \gamma_1 \cos \gamma_2$

$$\psi = \cos^{-1}[\cos \alpha_1 \cos \alpha_2 + \cos \beta_1 \cos \beta_2 + \cos \gamma_1 \cos \gamma_2]$$

Example 1.23: *Transform the vector $4u_x - 2u_y - 4u_z$ into spherical coordinates at a point $P(x = -2, y = -3, z = 4)$.*

Solution:

Given the vector $4u_x - 2u_y - 4u_z$ and it is to be transformed into spherical coordinates at point $P(x = -2, y = -3, z = 4)$. The parameters in spherical system can be obtained as follows:

$$r = \sqrt{x^2 + y^2 + z^2} = 5.385; \quad \theta = \cos^{-1} \frac{z}{r} = 42.03°; \quad \phi = \tan^{-1}\frac{y}{x} = 56.31°$$

In this case y and x are both negative, so the point is in 3rd quadrant.

i.e., $\phi = -180° + 56.31° = -123.69°$

$$\cos \theta = 0.7428; \quad \sin \theta = 0.6695; \quad \cos \phi = -0.5547; \quad \sin \phi = -0.8321$$

The components of the vector in r, θ and ϕ direction need to be found out in order to convert the vector into spherical coordinates

$$A_r = A \cdot u_r = (4u_x - 2u_y - 4u_z) \cdot u_r$$
$$= 4 \sin \theta \cos \phi - 2 \sin \theta \sin \phi - 4 \cos \theta$$
$$= -3.342$$
$$A_\theta = A \cdot u_\theta = (4u_x - 2u_y - 4u_z) \cdot u_\theta$$
$$= 4 \cos \theta \cos \phi - 2 \cos \theta \cos \phi - 4 \sin \theta$$
$$= 2.266$$
$$A_\phi = A \cdot u_\phi = (4u_x - 2u_y - 4u_z) \cdot u_\phi$$
$$= -4 \sin \phi - 2 \cos \phi$$
$$= 4.4378$$

So, the vector in spherical coordinates can be written as

$$A(r, \theta, \phi) = -3.342u_r + 2.266u_\theta + 4.4378u_\phi.$$

SHORT QUESTIONS AND ANSWERS

1. **Define scalar quantity and vector quantity.**

   Scalar quantity is one which has only magnitude but no direction. Vector quantity is one which has both magnitude and direction.

2. What is meant dot product and cross product of two vectors?

Dot product and cross product are two methods of vector multiplication. If $\vec{A}$ and $\vec{B}$ are two vectors their dot product is $\vec{A} \cdot \vec{B} = |\vec{A}|\,|\vec{B}| \cos\theta$ where θ/q is the angle between $\vec{A}$ and $\vec{B}$ and the cross product of $\vec{A}$ and $\vec{B}$ is defined as $\vec{A} \times \vec{B} = |\vec{A}|\,|\vec{B}| \sin\theta\, u_n$, where u_n is the unit vector normal to plane containing $\vec{A}$ and $\vec{B}$.

3. Define unit vectors in rectangular coordinate system.

In rectangular coordinate system the unit vector u_x, u_y and u_z are in x, y, and z-coordinate directions. And always they are mutually perpendicular each other.

4. Define unit vectors in cylindrical coordinate system.

In cylindrical coordinate system u_ρ is the unit vector along ρ-coordinate axis, which is measured radially outward from z-axis. The u_ϕ is unit angle in coordinate axis directions, which is measured from positive x-axis in positive director of angle measurement. And u_z is the unit vector in z-coordinate axis direction, which is measured as usual like in rectangular coordinate system. Thus u_ρ, u_ϕ and u_z are three unit vectors in cylindrical coordinate system, which are always mutually perpendicular to each other.

5. Define unit vectors in spherical coordinate system.

In spherical coordinate system u_r, u_ϕ and u_θ are the three unit vectors which are always perpendicular each other. Here, u_r is the unit vector r-axis, which is measured always radially outward from the point of origin. u_ϕ is already defined in cylindrical coordinate system and u_θ is unit angle measured from positive z-axis in anti clock wise direction.

6. What do you mean by "gradient"?

Gradient is the concept of rate of change of a scalar in the given field.

7. What do you mean by "divergence"?

The net out flow of flux per unit volume is called as the divergence. Divergence of vector $\vec{A}$ is defined as $\nabla \cdot \vec{A}$ (Divergence of A) or div $\vec{A}$. From a charged body there is continuous out flow of flux, like sunrays come out continuously from sun. These are the examples of positive divergence, since there is a net out flow. If a vaccum tube is broken there is inflow of air. This is an example of negative divergence since inflow is negative outflow.

8. What is concept of "curl"?

Curl is defined as the net circulation per unit area. Curl of vector $\vec{A}$ is defined as $\nabla \times \vec{A}$ (curl of A) or Curl $\vec{A}$. Consider wind whirl pool which is upward. If a piece of paper is released here, it rotates and finally moves up ward. This is an example of curl.

9. What is concept of laplacian?

Laplacian is of scalar and vector type Laplacian (scalar) is defined as divergence of gradient of scalar and is mathematically represented as $\nabla^2\phi = \nabla \cdot \nabla\phi$, where 'ϕ' is scalar.

And Laplacian vector is defined as follows: $\nabla^2 \vec{A} = \text{grad}\,(\text{div}\,\vec{A}) - \text{curl}\,\vec{A}$.

10. State the condition for the vector to be solenoid.

The vector is said $|\vec{P}|$ to be solenoid $\nabla \cdot \vec{P} = 0$ if and is irrotational if $\nabla \times \vec{P} = 0$.

MULTIPLE CHOICES

1. The electric charges

1. are conserved
2. are quantized
3. exist in pairs
4. have a circular field around it

In the above statements, the following are true

(*a*) 1 only (*b*) 1, 2

(*c*) 1, 4 (*d*) 1, 2, 3

2. The electric charge is transferred from one body to another insulated metal body only when

1. the medium is vaccum
2. the medium is dry air
3. the medium is any dielectric

In the above, the true statements are

(*a*) 1 (*b*) 2, 3

(*c*) 1, 2, 3 (*d*) 1, 3

3. A body can be charged when it is

1. an insulator
2. an insulated metal body
3. held in hand
4. charged in humid environment

The false statements are

(*a*) 1, 2 (*b*) 3

(*c*) 3, 4 (*d*) 1, 4

4. Between a hollow and solid metal sphere, charges reside

(*a*) on the outer surface in both

(*b*) on outer surface in hollow and throughout in solid

(*c*) throughout in both (*d*) none

5. There is a charged metal sphere and thin circular plate. Distribution of charge around the surface is

(*a*) uniform in both

(*b*) uniform in sphere and bulging at the edges in plate

(*c*) non-uniform in both

(*d*) Uniform in circular plate and non-uniform in sphere

6. A lighting conductor on top of a building is made into a pointed spike because

(*a*) rain drops may not collect

(*b*) dust particles may not accumulate

(*c*) charge per unit area becomes very high for lightning to discharge

(*d*) as decoration

7. A charged plate is touched by a metal rod standing on a wooden platform

(*a*) the plate is discharged completely

(*b*) the charge is unaffected in the plate

(*c*) the charge is transferred to the metal rod

(*d*) none

8. A point charge of $+ 3.0 \times 10^{-6}$ coulomb is 12 cm from a second point charge of -1.5×10^{-6} coulomb. The magnitude and direction of the force on each charge is

(*a*) 2 nt directed away from each other

(*b*) 2.81 nt directed towards each other

(*c*) 2.81 nt directed away from each other

(*d*) 1.5 nt towards each other

9. A charged rod attracts bits of uncharged paper. After touching the rod, they jump violently away from it because

(*a*) there is no place in the rod for all bits

(*b*) the charged rod gives a shock to the bits of paper

(*c*) on contact with the rod, the bits of paper acquire the same charge as the rod and are repelled

(*d*) none

10. When charges are applied to a gold leaf electroscope

(*a*) the leaves converge for positive charges

(*b*) the leaves remain stationary

(*c*) the leaves diverge for negative charges

(*d*) the leaves diverge for both positive and negative charges

11. A conductor and an insulator are heated;

(*a*) conductivity increases and insulator unaffected

(*b*) insulating power increases while conducting power remains same

(*c*) insulator decreases in insulating power and conductor decreases in conducting power

(*d*) both are unaffected

12. Match the following:

(*a*) Mercury	1. Insulator
(*b*) Cotton	2. Conductor
(*c*) Sulphur	3. Partial conductor
(*d*) Ivory	4. Partial insulator

13. The unit of ε_0, the permittivity of free space is

(*a*) (coulomb)^2 / newton-metre^2 (*b*) (coulomb)^2 joule-metre

(*c*) farad / metre (*d*) none correct

14. Two positive charges 10 μ coulomb and 15 μ coulomb are separated by a distance of 10 cms with a dielectric of alcohol. Find the force between them. ε_r of alcohol is 20.

(*a*) 180 N (*b*) 135 N

(*c*) 9 N (*d*) 6.75 N

15. F_{12} and F_{13} are two forces of 2 and 3 newtons. F_{13} makes an angle 30° with normal as shown in Fig.1.9. The force acting in x-direction is

(*a*) 5 N (*b*) 3.5 N

(*c*) 1 N (*d*) –1 N

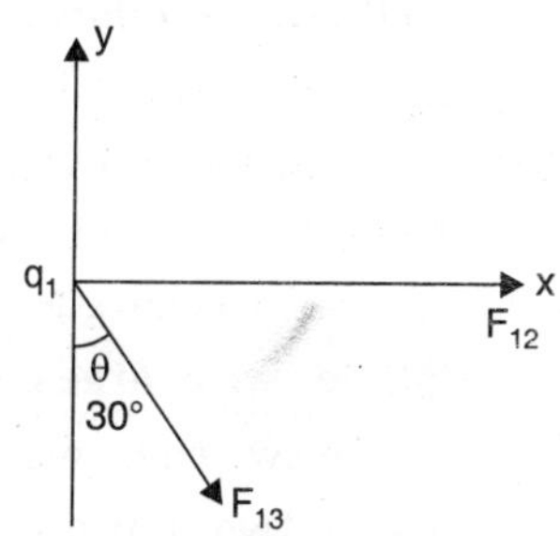

Fig. 1.9

16. In problem 15, the force acting along y-direction is

(*a*) –2.6 N (*b*) 3 N

(*c*) 1 N (*d*) 2.6 N

17. The resultant force F of F_{1x} and F_{1y} in Problem 15 is

(*a*) 5 N (*b*) 3 N

(*c*) 4.36 N (*d*) 2 N

18. A pith ball of 10 g carries a charge of 10 μC. What must be a charge on a ball placed 10 cm directly above which will hold the pith ball in equilibrium?

(*a*) 1 μC (*b*) 0.011 μC

(*c*) –0.011 μC (*d*) 0.1 μC

19. Arrange the following forces in descending order of their strength:

(*a*) electromagnetic (*b*) nuclear

(*c*) gravitational (*d*) radioactive decay

(*a*) 1, 2, 3, 4 (*b*) 1, 3, 2, 4

(*c*) 1, 4, 3, 2 (*d*) 2, 1, 4, 3

20. There are two positive charges separated by a distance

(*a*) the lines of force will occupy the space between them in all manner

(*b*) the electric lines of force will emanate from the two charges and occupy the space without crossing

(*c*) there will be null points in the line joining the two charges

(*d*) the entire space between the charges will be without lines of force

21. The electric field strength of a charge

(*a*) increases with distance

(*b*) decreases with distance

(*c*) decreases with square of distance

(*d*) decreases with cube of distance

22. Electric lines of force do not cross each other because

(*a*) the tangent to the line of force gives the direction of E and if they cross, there will be two tangents for E which is not possible

(*b*) the lines of force mutually repel

(*c*) the lines of force emanate from a point charge

(*d*) they are parallel to each other

23. Two charges of unknown sign and magnitude are a distance d apart. The electric field strength is zero at a point joining them

(*a*) the two charges are of opposite kinds

(*b*) the two charges are of some kinds

(*c*) the two charges, though of same kinds are of equal magnitude

(*d*) the two charges of different kinds have units 1 : 2

24. Identify which of the following quantities is not a vector:
(*a*) force (*b*) momentum
(*c*) toleration (*d*) work

25. Which of the following is not a scalar field:
(*a*) displacement of a mosquito in space
(*b*) high intensity in a drawing room
(*c*) temperature distribution in your classroom
(*d*) atmospheric pressure in a given region

26. The vector projection of $A = 5u_x - 10u_z$ in the direction of u_z is
(*a*) -10 (*b*) $-10u_z$
(*c*) $-10u_x$ (*d*) $-10u_y$

27. $\vec{A} = 10u_x + 5u_y$; $\vec{B} = 2u_z$, $\vec{A} \times \vec{B} =$
(*a*) $20u_x - 20u_y$ (*b*) $-20u_y + 10u_z$
(*c*) $-10u_x + 10u_y$ (*d*) $10u_x - 20u_y$

28. A different volume formed in a cylindrical coordinate system is
(*a*) $dr\, dz$ (*b*) $r\, dr\, d\phi$
(*c*) $r\, d\phi\, dz$ (*d*) all of these

29. The component of $6u_x + 2u_y - 3u_z$ along $3u_x - 4u_y$ is
(*a*) $-12u_x - 9u_y - 3u_z$ (*b*) $30u_x - 40u_z$
(*c*) 10/7 (*d*) 2

KEY

1. (*d*)	**2.** (*c*)	**3.** (*c*)	**4.** (*a*)	**5.** (*b*)	**6.** (*c*)
7. (*c*)	**8.** (*b*)	**9.** (*c*)	**10.** (*d*)	**11.** (*c*)	
12. (*a*) – 2, (*b*) – 3, (*c*) – 1, (*d*) – 4		**13.** (*d*)	**14.** (*d*)	**15.** (*b*)	**16.** (*a*)
17. (*c*)	**18.** (*c*)	**19.** (*d*)	**20.** (*c*)	**21.** (*c*)	**22.** (*a*)
23. (*b*)	**24.** (*d*)	**25.** (*a*)	**26.** (*b*)	**27.** (*b*)	**28.** (*d*)
29. (*d*)					

REVIEW QUESTIONS

1. What is a scalar quantity? Give some examples of scalars.
2. What is a vector quantity? Give some examples of vectors.
3. What do we mean when we say that two vectors are equal?
4. What is the significance of zero vector and is vector addition "closed"?
5. Can the dot product be negative? If yes, what must be the condition?
6. Can you reason why the dot product of two vectors is known as scalar product?
7. How can you determine if two vectors are dependent or independent?

8. Is division of a vector by another vector defined?
9. Give some physical examples of dot product and cross product.
10. Is the projection of a vector on another vector unique?
11. How can you determine the area of parallelogram using vectors and what is meant by right-hand rule?
12. If a vector $\vec{A}$ is given at point $P\left(3, \frac{\pi}{6}, 10\right)$ and vector $\vec{B}$ is given at $Q\left(1, \frac{\pi}{6}, 5\right)$ in cylindrical coordinates, can vector operations be performed without transforming into rectangular coordinates?
13. Two vectors $\vec{A}$ and $\vec{B}$ are given in the spherical coordinate system at $\left(2, \frac{\pi}{2}, \frac{2\pi}{3}\right)$ and $\left(10, \frac{\pi}{2}, \frac{2\pi}{3}\right)$. Can vector operations be performed without making a transformation from spherical to rectangular coordinates?
14. What do we mean by gradient of a scalar function?
15. What does the divergence of a vector signify?
16. What is the significance of a curl of a vector?
17. Which equations will you use to check if a vector is (*a*) continuous (*b*) solenoidal (*c*) rotational (*d*) irrotational and (*e*) conservative? Give some real life examples for each case.
18. How many vector surfaces does a thin sheet of paper possess if we assume that its thickness $\rightarrow 0$?
19. Verify the commutative law for addition of vectors.
20. Show that the necessary and sufficient condition for two non-zero vectors $\vec{A}$ and $\vec{B}$ to be perpendicular is that $\vec{A} \cdot \vec{B} = 0$.
21. Prove that two non-zero vectors are parallel if and only if their cross product is zero.

EXERCISE PROBLEMS

1. (*a*) Express u_ρ in spherical components and variables (*b*) Express u_r in cylindrical components and variables.
2. Given the vector in Cartesian coordinate that extends from $P(r = 4, \theta = 20°, \phi = 10°)$ to $Q(r = 7, \theta = 120°, \phi = 75°)$ (*a*) Give the vector in spherical coordinate at $M(x = 5, y = 1, z = 2)$ that extends to $N(2,4,6)$ (*b*) How far is it from $A(r = 110, \theta = 30°, \phi = 60°)$ to $B(r = 30, \theta = 75°, \phi = 125°)$?
3. A closed surface is defined in spherical coordinates by $3 < r < 5$, $0.1\,\pi < \theta < 0.3\,\pi$, $1.2\,\pi < \phi < 1.6\,\pi$ (*a*) Find the volume enclosed (*b*) Find the distance from $P_1(r = 3, \theta = 0.1\,\pi, \phi = 1.2\,\pi,)$ to $P_2\,(r = 5, \theta = 0.3\,\pi, \phi = 1.6\,\pi)$ (*c*) Find total surface area.

4. Using the coordinate system named, give the vector of point $A(2, -1, -3)$ that extends to $B(1, 3, 4)$ (*a*) Cartesian (*b*) Cylindrical (*c*) Spherical.

5. Find the value of B_z such that the angle between the vector $\bar{A} = 2U_x + U_y + 4U_z$ and $\bar{B} = -2U_x - 1U_y + B_zU_z$ is 45°.

6. Vector $\bar{A} = 3U_x + 4U_y - 5U_z$ and $\bar{B} = -6U_x + 2U_y + 4U_z$ extend out from the origin. Find (*a*) the angle between $\bar{A}$ and $\bar{B}$ (*b*) the distance between the tips of vector (*c*) the unit vector normal to plane containing $\bar{A}$ and $\bar{B}$ (*d*) the area of parallelogram which $\bar{A}$ and $\bar{B}$ are adjacent sides?

7. Give the vector in Cartesian coordinates that extends from P ($\rho = 4$, $\phi = 10°$, $z = 1$) to Q ($\rho = 7$, $\phi = 75°$, $z = 4$) (*a*) Give the vector in cylindrical coordinates at $M(r = 5, y = 1, z = 2)$ that extends to $N(2,4, 6)$? (*b*) How far is it from $A(110, 60°, -20)$ to B (30, 125°, 10)?

8. Find in cylindrical components (*a*) A unit vector P ($\rho = 5$, $\phi = 53.13°$, $z = -2$) in director of $\bar{E} = \rho z \cos\phi\, U_\rho - \rho_z \sin\phi\, U_\phi + \rho a\, U_z$ at P.

9. Given the three points $M(6, 2, -3)$, $N(-2, 3, 0)$ and $P(-4, 6, 5)$ find (*a*) the area of triangle they define (*b*) A unit vector perpendicular to this triangular surface (*c*) a unit vector bisecting the interior angle of triangle at M.

10. Three points P_1, P_2 and P_3 are given by (2, 3, – 2), (5, 8, 3) and (7, 6, 2) respectively. Obtain (*a*) the vector drawn from P_1 to P_2 (*b*) the unit vector along the line from P_1 to P_3.

CHAPTER

2

Coulomb's Law and Electric Field Intensity

2.1 INTRODUCTION

Since a new concept called as Vector Analysis has been brought about in the first chapter, it becomes easy for us to make an attempt to know a few basic principles of electricity. In this chapter the experimental law of Coulomb using vector notation comes first and we shall use this law to solve some simple problems to familiarize with vector analysis. The concepts of electric field intensity, electric field are introduced gradually so that they may be useful to formulate the Maxwell's equation. In this chapter static electric fields in free space or vaccum are discussed. Such fields for example, are normally found in the focusing and deflection system of electrostatic cathode-ray tubes. For all practical purposes, our results are also applicable to air and other gases.

2.2 EXPERIMENTAL LAW OF COULOMB

Knowledge of static electricity was known to man since early times (600 BC). Thales of Miletus was credited first with observing that amber when rubbed attracted light objects to itself. Subsequent experiments found that most substances when rubbed possessed this property to some extent. These experiments reveal that electric charges are fundamental in nature. They are of two types and can be produced by friction when a glass rod is rubbed with a silk cloth. As a result both the glass rod and silk cloth attract pieces of paper because they are electrified. Also there is a force of attraction between glass rod and silk cloth. The charge on glass rod is taken as positive and that on the silk cloth is taken as negative. The electrification of glass rod when rubbed with silk cloth can be explained as due to the migration of electrons from glass rod to the silk cloth. This makes the glass rod positive and silk cloth negative.

Shortly thereafter **Col Charles Coulomb** performed an elaborate series of experiments using a delicate torsion balance, invented by himself, to determine quantitatively the force exerted between two objects each having a static charge of electricity and whose size was small compared to the distance between them and stated the following points:

(*i*) Bodies having like charges repel each other and bodies having unlike charges attract each other.

(*ii*) The magnitude of the force between the charged bodies is proportional to the magnitude of the charges upon the bodies.

(*iii*) The magnitude of the force is inversely proportional to the square of the distance between the charges.

(*iv*) The force depends upon the medium in which the charges are placed.

In brief it can be stated that, Coulomb's law can be expressed as the force between two very small objects separated in a vaccum (or) free space by a distance which is large compared to their size is proportional to charge on each and inversely proportional to square of distance between them which can be shown mathematically as

$$F = K\frac{Q_1 Q_2}{R^2} \quad ...(2.1)$$

where Q_1, Q_2 are positive and negative quantities of charge, R is the separation and K is a proportionality constant $= \frac{1}{4\pi\varepsilon}$

For any medium, $\varepsilon = \varepsilon_0 \varepsilon_r$

where ε = absolute permitivity of medium; ε_r = relative permitivity

ε_0 = permitivity of vaccum

For vaccum (or) free space $\varepsilon_r = 1$;

$\therefore$ $\varepsilon = \varepsilon_0$ and for any other medium $\varepsilon_r > 1$

but $\varepsilon_0 = 8.854 \times 10^{-12}$ Farad/meter

Now $$\frac{1}{4\pi\varepsilon_0} = \frac{1}{4\pi \times 8.854 \times 10^{-12}} = 9 \times 10^9$$

$$\therefore \quad F = 9 \times 10^9 \times \frac{Q_1 Q_2}{R^2} \quad ...(2.2)$$

2.2.1 Vector Form of Coulomb's Law

If Q_1 and Q_2 are both positive charges, the force produced is force of repulsion and this force acts along the line joining Q_1 and Q_2 as shown in the Fig. 2.1. Let charge Q_1 be located at distance R_1 and charge Q_2 at a distance R_2 from origin. In vector form we can write Coulomb's law equation as

$$F = \frac{Q_1 Q_2}{4\pi\varepsilon_0 R_{12}^2} U_{12} \quad ...(2.3)$$

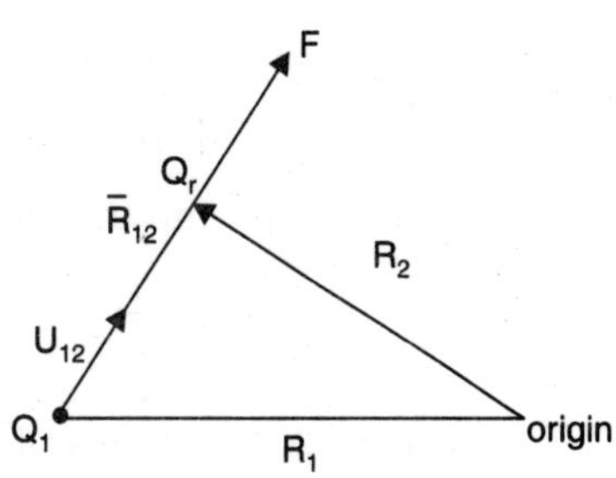

Fig. 2.1

Here U_{12} is a unit vector directed along R_{12}, the direction of force produced.

We have $U_{12} = \frac{\overline{R}_{12}}{|\overline{R}_{12}|}$ and $\overline{R}_{12} = \overline{R}_2 - \overline{R}_1$

Therefore, it can be written as $U_{12} = \frac{\overline{R}_2 - \overline{R}_1}{|\overline{R}_{12}|}$

Example 2.1: *A – 4 mC (milli coulomb) positive charge is located in vaccum at P_1 (6, –4, –8) and 16 μC (micro coulomb) negative charge is at P_2 (2, –8, 4). (a) Find the vector force on negative charge (b) What is the magnitude of force on charge at P_1?*

Solution: Vector force on the negative charge,

$$= \frac{(-4\times10^{-3})(16\times10^{-6})}{4\pi\varepsilon_0\,(13.2)^2}\left[\frac{(2-6)u_x+(-4+8)u_y+(4+8)}{\sqrt{(-4)^2+(-4)^2+(12)^2}}\right]$$

$$= (0.9868u_x + 0.9868u_y - 2.9603u_z)\text{ N}$$

(b) Magnitude of force on charge at P_1 = magnitude of force on charge at

$$P_2 = \sqrt{(0.9868)^2+(0.9868)^2+(2.9603)^2} = 3.2728\text{ N.}$$

Example 2.2: *Determine the force between two charges 4 × 10^{-4} C at A (2, 3, 4) and –2 × 10^{-4} C at B (3, 0, 3) in vaccum.*

Solution: By Coulomb's law, the force F is given by

$$\overline{F} = \frac{Q_1Q_2}{4\pi\varepsilon(R_{12})^2}U_{R_{12}}$$

where $U_{R_{12}}$ = a unit vector in the direction of $\overline{R}_{12}$ and | R_{12} | is the magnitude of vector $\overline{R}_{12}$ (distance between points 1 and 2).

Position vectors $\overline{OP}$ and $\overline{OQ}$ can be written as

$$\overline{OP} = (2-0)u_x + (3-0)u_y + (4-0)u_z$$
$$= 2u_x + 3u_y + 4u_z$$
$$\overline{OQ} = (3-0)u_x + (0-0)u_y + (3-0)u_z$$
$$= 3u_x + 0u_y + 3u_z$$

where point O stands for the origin of the Cartesian coordinate system

Further, $$|\overline{PQ}| = \overline{R_{12}} = \overline{OQ} - \overline{OP}$$
$$= (3-2)u_x + (0-3)u_y + (3-4)u_z$$
$$= u_x - 3u_y - u_z \qquad ...(a)$$

The magnitude of R_{12} gives the distance between the charges,

i.e., $$|\overline{R}_{12}| = \sqrt{1+9+1} = 3.31 \qquad ...(b)$$

And, therefore, the unit vector along vector R_{12} is

$$U_{R12} = \frac{\overline{R}_{12}}{|R_{12}|} = \frac{u_x - 3u_y - u_z}{3.31} \qquad ...(c)$$

Introducing the values of R_{12} and U_{R12} respectively from (b) and (c) in the following equation, it becomes

$$F = \frac{Q_1Q_2}{4\pi\varepsilon R_{12}^2}U_{R12} = \frac{(3\times10^{-4})\times(-2\times10^{-4})}{4\pi\varepsilon_0\times11}U_{R12} = -49.21U_{R12}\text{ N}$$

$$= -49.21\left(\frac{u_x - 3u_y - u_z}{3.31}\right)\text{N} = (-14.86u_x + 44.6u_y + 14.86u_z)\text{ N.}$$

The magnitude of the force is 49.21 N and the direction is specified by the unit vector. The force on Q_2 may also be written as $F_2 = -14.86u_x + 44.6u_y + 14.86u_z$.

Example 2.3: *Four line charges of 30 μC each are located at the four corners of a square, the diagonal of which measures 8 m. Find the force on a 150 μC charge located at 3 m above the centre of the square.*

Solution: Let $|F_1| = |F_2| = |F_3| = |F_4| = |F|$

$$|F| = \frac{Q_1 Q_2}{4\pi\varepsilon_0 R^2} = \frac{30\times10^{-6}\times150\times10^{-6}\,(9\times10^9)}{(5)^2} = 1.62 \text{ N}$$

The resultant force, $\bar{F}_r = \bar{F}_1 + \bar{F}_2 + \bar{F}_3 + \bar{F}_4$

$$= |F|\bar{u}_{1p} + |F|\bar{u}_{2p} + |F|\bar{u}_{3p} + |F|\bar{u}_{4p}$$

$$= |F|\left[\frac{\bar{R}_{1p}}{|R_{1p}|} + \frac{\bar{R}_{2p}}{|R_{2p}|} + \frac{\bar{R}_{3p}}{|R_{3p}|} + \frac{\bar{R}_{4p}}{|R_{4p}|}\right]$$

$$= \frac{|F|}{|R|}[\bar{R}_{1p} + \bar{R}_{2p} + \bar{R}_{3p} + \bar{R}_{4p}]$$

where,
$$\bar{R}_{1p} = \left(0 - \frac{\sqrt{32}}{2}\right)\bar{u}_x + \left(0 + \frac{\sqrt{32}}{2}\right)\bar{u}_y + (3-0)\bar{u}_z$$

$$= -2.83\bar{u}_x + 2.83\bar{u}_y + 3\bar{u}_z$$

Similarly,
$$\bar{R}_{2p} = \frac{\sqrt{32}}{2}\bar{u}_x + \frac{\sqrt{32}}{2}\bar{u}_y + 3\bar{u}_z = 2.83\bar{u}_x + 2.83\bar{u}_y + 3\bar{u}_z$$

$$\bar{R}_{3p} = \frac{-\sqrt{32}}{2}\bar{u}_x - \frac{\sqrt{32}}{2}\bar{u}_y + 3\bar{u}_z$$

$$\bar{R}_{4p} = 2.83\bar{u}_x - 2.83\bar{u}_y + 3\bar{u}_z$$

$$\therefore \quad \bar{R}_{1p} + \bar{R}_{2p} + \bar{R}_{3p} + \bar{R}_{4p} = 12\bar{u}_z$$

$$\therefore \quad \bar{F}_r = \frac{1.62}{5}\times 12\bar{u}_z \approx 3.9\bar{u}_z.$$

2.3 ELECTRIC FIELD

It can be defined as the region around an electric charge or a group of charges, in which no other charge can experience a force.

The force between two charges can be defined in terms of electric field as follows:

(*i*) An electric field is set up by charge Q_1 by itself.

(*ii*) The field of Q_1 acts on Q_2. As a result Q_2 experiences a force, which can be stated as the force on Q_2 is due to the action of field of Q_1 on Q_2.

(*iii*) It can also be imagined that Q_2 sets up the field and this field acts on Q_1 which produces a force on Q_1.

2.4 ELECTRIC FIELD INTENSITY

Let us consider a positive electric point charge Q_1 placed rigidly at the origin of a polar coordinate system. Let another positive point charge Q_2 be brought into the vicinity of Q_1. This charge Q_2 will experience some force which will be directed radially outward and becomes greater as Q_2 approaches Q_1. Therefore it may be said that Q_1 has a field around it where forces may act. The magnitude of the field intensity at any point is measured simply by the force per unit charge at that point. Calling this charge Q_2 a test charge 'Q_t', the force on it is given by Coulomb's law.

i.e.,

$$F_{1t} = \frac{Q_1 Q_t}{4\pi\varepsilon_0 R_{1t}^2} U_{1t} \quad ...(2.4)$$

$$\frac{F_{1t}}{Q_t} = \frac{Q_1}{4\pi\varepsilon_0 R_{1t}^2} U_{1t} \quad ...(2.5)$$

Now $\frac{F_{1t}}{Q_t}$ is the force exerted by charge Q_1 on unit positive charge placed at a distance R_{1t} from it. It is called as electric field intensity denoted by 'E'. According to equation (2.5), the point charge Q_1 is surrounded by an electric field intensity E which is proportional to Q_1 and is inversely proportional to R_{1t}^2. Hence electric field intensity at any point in an electric field may be defined as force which a unit positive charge placed at that point would experience.

2.4.1 Electric Field Intensity Due to Single Charge

It is required to determine field intensity at point $P(x_1, y_1, z_1)$ and let charge Q coulomb's be placed at origin O (0, 0, 0).

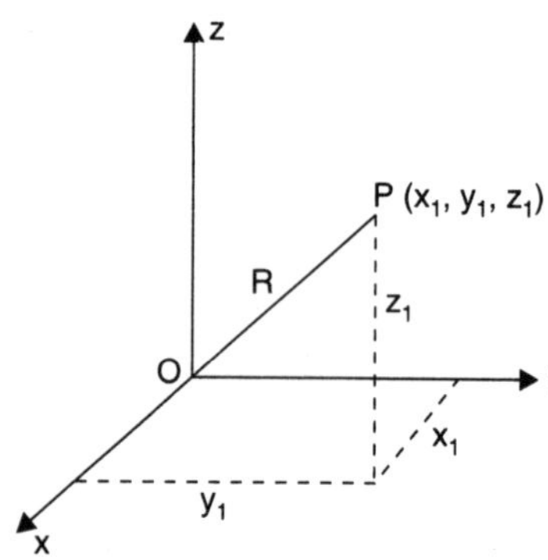

Fig. 2.2

Let R be the directed line segment from origin to P and so we can write the expression of R as $x_1 u_x + y_1 u_y + z_1 u_z$

$$\therefore \quad \bar{R} = x_1 u_x + y_1 u_y + z_1 u_z \quad ...(2.6)$$

So, the magnitude of R can be found as

$$|\bar{R}| = \sqrt{x_1^2 + y_1^2 + z_1^2} \quad ...(2.7)$$

If U_R is unit vector along 'OP', $\bar{U}_R = \frac{\bar{R}}{|\bar{R}|}$

$$\therefore \quad \bar{U}_R = \frac{\bar{R}}{\sqrt{x_1^2 + y_1^2 + z_1^2}} \quad ...(2.8)$$

Now the field intensity at P can be found out by using the equation (2.5)

$$\therefore \quad E = \frac{Q}{4\pi\varepsilon_0 R^2} U_R = \frac{Q}{4\pi\varepsilon_0 (x_1^2 + y_1^2 + z_1^2)} \frac{\bar{R}}{\sqrt{x_1^2 + y_1^2 + z_1^2}}$$

$$= \frac{Q\bar{R}}{4\pi\varepsilon_0 (x_1^2 + y_1^2 + z_1^2)^{3/2}} \text{ N/C} \quad ...(2.9)$$

Suppose if the charge Q is situated at any point other than the origin, then the field intensity at P can be calculated as follows:

Let charge Q be placed at $A(x', y', z')$ and let $P = (x, y, z)$. The expression for R_A and R_p can be written as

$$R_A = x'\,u_x + y'\,u_y + z'\,u_z \quad ...(2.10)$$

$$R_p = x\,u_x + y\,u_y + z\,u_z \quad ...(2.11)$$

$$R_{AP} = (x - x')u_x + (y - y')u_y + (z - z')u_z$$

Magnitude of $\overline{R}_{AP} = |\, R_P - R_A \,|$

$$= \sqrt{(x - x')^2 + (y - y')^2 + (z - z')^2} \quad ...(2.12)$$

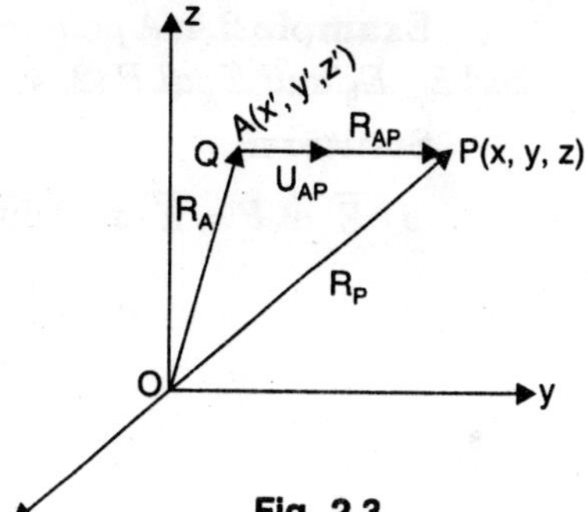

Fig. 2.3

The unit vector U_{AP} can be written as $\dfrac{\overline{R}_{AP}}{|\overline{R}_{AP}|}$

$$U_{AP} = \frac{(x - x')u_x + (y - y')u_y + (z - z')u_z}{\sqrt{(x - x')^2 + (y - y')^2 + (z - z')^2}} \quad ...(2.13)$$

So the field intensity 'E' can be written as

$$E = \frac{Q}{4\pi\varepsilon_0} \frac{R_P - R_A}{|R_P - R_A|^3} \quad ...(2.14)$$

2.4.2 Field Intensity Due to Several Charges (Superposition Principle)

Equation (2.14) gives field intensity at a point, due to a single point change. However there might be number of charges, so it becomes necessary to evaluate field intensity at that point due to each charge. If there are several charges present, each charge will exert a force on the test charge Q_t. The resultant force on Q_t will be the vector sum of the individual forces. So the expression becomes as

$$E = E_1 + E_2 + E_3 + \ldots\ldots + E_n$$

where, $$E_1 = \frac{Q}{4\pi\varepsilon_0 (R_p - R_1)^2} \times \frac{R_P - R_1}{|R_P - R_1|} \quad ...(2.15)$$

Let $$U_1 = \frac{R_P - R_1}{|R_P - R_1|}$$

∴ $$E_1 = \frac{Q}{4\pi\varepsilon_0 (R_p - R_1)^2} U_1$$

Similarly, $$E_2 = \frac{Q_2}{4\pi\varepsilon_0 (R_p - R_2)^2} \times U_2$$

As a result the total electric field intensity can be written as

$$E = \sum_{m=1}^{n} \frac{Q_m}{4\pi\varepsilon_0 (R_P - R_m)^2} \times U_m \quad ...(2.16)$$

The above process of summation of fields involves discrete charge configuration. If charge distribution is continuous rather than discrete, over a certain volume with a charge density ρ then any test charge Q situated in a region under consideration will experience a force due to the charge distribution.

Example 2.4: *A point charge $Q_A = 2\ \mu C$ is at A (0, 0, 2) and $Q_B = -2\ \mu C$ is at B (1, 2, 3). Find E_r, E_θ and E_ϕ at P (2, 4, 6).*

Solution:

(*a*) $\overline{E}$ at P = $\overline{E}$ at P due to charge Q_A + $\overline{E}$ at P due to charge Q_B

$$\overline{E} = \frac{2\times 10^{-6}}{4\pi\varepsilon_0}\frac{(2-0)U_x + (4-0)U_y + (6-2)U_z}{[(2)^2+(4)^2+(4)^2]^{3/2}}$$

$$-\frac{2\times 10^{-6}}{4\pi\varepsilon_0}\frac{(2-1)U_x + (4-2)U_y + (6-3)U_z}{[(1)^2+(2)^2+(3)^2]^{3/2}}$$

$$\overline{E} = (-176.94U_x - 354U_y - 697.6U_z)\text{ V/m}$$

Let $\overline{E} = E_r U_r + E_\theta U_\theta + E_\phi U_\phi$

$$E_r = \overline{E}\cdot U_r = -176.94U_x\cdot U_r - 354U_y\cdot U_r - 697.6U_z\cdot U_r$$
$$= -176.94\sin\theta\cos\phi - 354\sin\theta\sin\phi - 697.6\cos\theta$$

$$E_\theta = \overline{E}\cdot U_\theta = -176.94U_x\cdot U_\theta - 354U_y\cdot U_\theta - 697.6U_z\cdot U_\theta$$
$$= -176.94\cos\theta\cos\phi - 354\cos\theta\sin\phi + 697.6\sin\theta$$

$$E_\phi = \overline{E}\cdot U_\phi = -176.94U_x\cdot U_\phi - 354U_y\cdot U_\phi - 697.6U_z\cdot U_\phi$$
$$= 176.94\sin\phi - 354\cos\phi$$

$\therefore$

$$\overline{E} = (-176.94\sin\theta\cos\phi - 354\sin\theta\sin\phi - 697.6\cos\theta)U_r$$
$$+ (-176.94\cos\theta\cos\phi - 354\cos\theta\sin\phi + 697.6\sin\theta)U_\theta$$
$$+ (176.94\sin\phi - 354\cos\phi)U_\phi$$

For point $P(2, 4, 6)$, $r = \sqrt{2^2+4^2+6^2} = 7.483$

$$\theta = \cos^{-1}\left(\frac{3}{7.483}\right) = 66.36°,\ \phi = \tan^{-1}\left(\frac{6}{3}\right) = 63.43°$$

Substitution the coordinates of P in expression of $\overline{E}$, we have $\overline{E}$ at point P is

$$\overline{E} = [641.90U_r + 480.411U_\theta + 0.058U_\phi]\text{ V/m.}$$

Example 2.5: *Determine electric field E at the origin due to a point charge of 54.9 nC located at (– 4, 5, 3) m in Cartesian coordinates.*

Solution: The vector field intensity at a point charge Q at the origin is given by

$$\overline{E} = \frac{Q}{4\pi\varepsilon_0|R|^2}\cdot\overline{u}_R = \frac{54.9\times 10^{-9}}{4\pi\left(\frac{10^{-6}}{36\pi}\right)(50)}\cdot\left(\frac{4\overline{u}_x - 5\overline{u}_y - 3\overline{u}_z}{\sqrt{50}}\right)$$

$$= (9.882)\left(\frac{4\overline{u}_x - 5\overline{u}_y - 3\overline{u}_z}{\sqrt{50}}\right) = [5.590\overline{u}_x - 6.985\overline{u}_y - 4.191\overline{u}_z]\text{ V/m.}$$

Example 2.6: *A 35 nC point charge is located at P (6, 9, – 6) in free space. Find (a) $\overline{E}$ (r) (b) E at A (– 9, 6, 0) (c) the locus of all points of which $E_x = 1$ V/m.*

Solution:

(*a*)
$$\overline{E}\ (\overline{r}) = \overline{E}\ (x\overline{U}_x + y\overline{U}_y + z\overline{U}_z)$$

$$= 35 \times 10^{-9} \frac{[(x-6)U_x + (y-8)U_y + (z+6)U_z)]}{4\pi\varepsilon_0[(x-6)^2 + (y-8)^2 + (z+6)^2]^{3/2}}$$

$$= \frac{314.571[(x-6)U_x + (y-8)U_y + (z+6)U_z)]}{[(x-6)^2 + (y-8)^2 + (z+6)^2]^{1.5}}$$

(*b*)
$$\overline{E} \text{ at } P\,(-9, 6, 0) = \frac{314.571[(-9-6)U_x + (6-8)U_y + (0+6)U_z)]}{[(-9-6)^2 + (6-8)^2 + (z+6)^2]^{1.5}}$$

$$= -1.093U_x - 0.145U_y + 0.4375U_z \text{ V/m}$$

(*c*)
$$E_x = \frac{314.57[(x-6)}{[(x-6)^2 + (y-8)^2 + (z+6)^2]^{1.5}} = 1$$

(or) $[(x-6)^2 + (y-8)^2 + (z+6)^2]^{1.5} = 179.755\,(x-2)$

This is the equation to locus of all points at which $E_x = 1$ V/m.

Example 2.7: *In free space let $Q_1 = 15$ nC be P_1 (0, – 6, 0) and $Q_2 = 25$ nC be P_2 (0, 0, 6) (a) find $\overline{E}$ at the origin (b) where should a 40 nC point charge be located so that $\overline{E} = 0$ at the origin?*

Solution:

(*a*) $\overline{E}$ at origin = $\overline{E}$ at origin due to change Q_1 + $\overline{E}$ at origin due to Q_2.

$$= \frac{15 \times 10^{-9}}{4\pi\varepsilon_0(6)^2}\left[\frac{(0+6)U_y}{\sqrt{(6)^2}}\right] + \frac{25 \times 19^{-9}}{4\pi\varepsilon_0(6)^2}\left[\frac{(0-6)U_z}{\sqrt{(6)^2}}\right]$$

$$= (3.759U_y - 6.265U_z) \text{ V/m.}$$

(*b*) Let the charge of 40 *nC* be located at (*x, y, z*).

Then $\overline{E}$ at origin = $\overline{E}$ at origin due to change Q_1 + $\overline{E}$ at origin due to change Q_2 + $\overline{E}$ at origin due to change Q_3.

$$= 3.759U_y - 6.265U_z + \frac{40 \times 10^{-9}[(0-x)U_x + (0-y)U_y + (0-z)U_z]}{4\pi\varepsilon_0[x^2 + y^2 + z^2]^{3/2}}$$

$$= \frac{-360.87x}{(x^2+y^2+z^2)^{1.5}}U_x + \left[3.759 - \frac{360.87y}{(x^2+y^2+z^2)^{1.5}}\right]U_y - \left[6.265\frac{360.87z}{(x^2+y^2+z^2)^{1.5}}\right]U_z$$

Since $\overline{E}$ is equal to zero at origin, the coefficient of unit vector must equal to zero. Equating coefficient of U_x equal to zero, we get

$$\frac{-360.87x}{(x^2+y^2+z^2)^{1.5}} = 0 \text{ (or) } x = 0$$

Equating coefficient of U_y equal to zero

$$\left[3.759 - \frac{360.87y}{(x^2+y^2+z^2)^{1.5}}\right] = 0$$

(or) $$\frac{y}{(x^2+y^2+z^2)^{1.5}} = 0.0104 \qquad ...(a)$$

Equating coefficient of U_z equal to zero, we have

$$-6.265 - \frac{360.87z}{(x^2+y^2+z^2)^{1.5}} = 0$$

(or) $$\frac{z}{(x^2+y^2+z^2)^{1.5}} = 0.0173 \qquad ...(b)$$

Since $x = 0$, equations (a) and (b) reduce to

$$\frac{y}{(y^2+z^2)^{1.5}} = 0.0104 \qquad ...(c)$$

and $$\frac{z}{(y^2+z^2)^{1.5}} = -0.0173 \qquad ...(d)$$

Dividing equation (d) by equation (c), we get

$$\frac{z}{y} = -\frac{0.0173}{0.0104} = -1.663$$

(or) $$z = -1.663\,y$$

Substituting this value of z in equation (c), we have

$$\frac{y}{(3.756\,y^2)^{1.5}} = 0.0104 \quad \text{(or)} \quad y^2 = 1.8018$$

(or) $$y = \pm\,1.342$$

Since y value should be positive so $y = 1.342$

Then $$z = -1.663\,(1.342) = -2.232$$

Hence, 40 nC point change should be located at (0, 1.342, – 2.232).

Example 2.8: *A point charge Q_1 located at the origin in free space produces a field. $\overline{E} = 15\,U_r$ V/m for r = 2 m (a) Find Q_1 (b) using this value of Q_1, find $\overline{E}$ at P (3, – 2, 6) in Cartesian components (c) using the same value of Q_1, find $\overline{E}$ at C (3, 3, 3) in cylindrical components.*

Solution:

(a) $\overline{E}$ at (r, θ, ϕ) due to charge Q_1 at origin $= \dfrac{Q_1}{4\pi\varepsilon_0 r^2}\overline{u}_r$ V/m

At $r = 2\ m$, $$\overline{E} = 15\,\overline{u}_r = \frac{Q_1}{4\pi\varepsilon_0 (2)^2}\overline{u}_r$$

$$\therefore \qquad Q_1 = 240\pi\varepsilon_0 \text{ C}$$

(b) $$\overline{E} \text{ at } P = \frac{240\pi\varepsilon_0}{4\pi\varepsilon_0}\left\{\frac{(3-0)\overline{u}_x + (-2-0)\overline{u}_y + (6-0)\overline{u}_z}{[(3)^2+(-2)^2+(6)^2]^{3/2}}\right\}$$

$$= (0.5247\overline{u}_x - 0.3498\overline{u}_y + 1.049\overline{u}_z) \text{ V/m.}$$

(c) $\overline{E}$ at $C = \dfrac{240\pi\varepsilon_0}{4\pi\varepsilon_0}\left\{\dfrac{(3-0)\overline{u}_x + (3-0)\overline{u}_y + (3-0)\overline{u}_z}{[(3)^2 + (-2)^2 + (6)^2]^{3/2}}\right\}$

$= (1.283\overline{u}_x + 1.283\overline{u}_y + 1.283\overline{u}_z)$ V/m

Let $\overline{E} = E_\rho \overline{u}_\rho + E_\phi \overline{u}_\phi + E_z \overline{u}_z$

$$E_\rho = \overline{E} \cdot \overline{u}_\rho = 1.233\overline{u}_x \cdot \overline{u}_\rho + 1.283\overline{u}_y \cdot \overline{u}_\rho + 1.283\overline{u}_z \cdot \overline{u}_\rho$$

$$= 1.283 \cos\phi + 1.283 \sin\phi$$

$$E_\phi = \overline{E} \cdot \overline{u}_\phi = 1.283\overline{u}_x \cdot \overline{u}_\phi + 1.283\overline{u}_y \cdot \overline{u}_\phi + 1.283\overline{u}_z \cdot \overline{u}_\phi$$

$$= -1.283 \sin\phi + 1.283 \cos\phi$$

$$E_z = \overline{E} \cdot \overline{u}_z = 1.283\overline{u}_x \cdot \overline{u}_z + 1.283\overline{u}_y \cdot \overline{u}_z + 1.283\overline{u}_z \cdot \overline{u}_z = 1.283$$

$\therefore \quad \overline{E} = (1.283 \cos\phi + 1.283 \sin\phi)\overline{u}_\rho + (-1.283 \sin\phi + 1.283 \cos\phi)\overline{u}_\phi + 1.283\overline{u}_z$

For point C (3, 3, 3), $\quad \phi = \tan^{-1}\left(\dfrac{3}{3}\right) = 45°$

Substituting this value of ϕ in the above expression for $\overline{E}$, we get

$$\overline{E} \text{ at point } C = (1.8144\overline{u}_\rho + 1.283\overline{u}_z) \text{ V/m.}$$

Example 2.9: *Calculate electric field E at a point P (2, – 3, 3) in free space due to (a) a charge Q_1 = 5 μC at A (0, 0, 0), (b) a charge Q_2 = 3 μC at B (– 2, 3, 5) and (c) both the charges of part (a) and (b) present.*

Solution:

(*a*) The vector from point A to point P is

$$\overline{r}_1 = (2-0)\,\overline{u}_x + (-3-0)\,\overline{u}_y + (3-0)\,\overline{u}_z = 2\overline{u}_x - 3\,\overline{u}_y + 3\,\overline{u}_z$$

$$|\,r_1\,| = \sqrt{4+9+9} = \sqrt{22} = 4.690$$

Unit vector from A to P, $\overline{u}_{r_1} = \dfrac{\overline{r}_1}{|\,r_1\,|} = \dfrac{2\overline{u}_x - 3\overline{u}_y + 3\overline{u}_z}{\sqrt{22}}$

Electric field at P due to charge at A is

$$E_{P_1} = \frac{Q_1}{4\pi\varepsilon_0 |\,r_1\,|^2}\,\overline{u}_{r_1} = \frac{(9\times 10^9)\times 5\times 10^{-6}}{22}\cdot\frac{2\overline{u}_x - 3\overline{u}_y + 3\overline{u}_z}{\sqrt{22}}$$

$$= (872.26\overline{u}_x - 1308.39\,\overline{u}_y + 1308.39\,\overline{u}_z) \text{ V/m.}$$

(*b*) The vector from point B to P is

$$\overline{r}_2 = (2+2)\overline{u}_x + (-3-3)\overline{u}_y + (3-5)\overline{u}_z = 4\overline{u}_x - 6\overline{u}_y - 2\overline{u}_z$$

$\therefore \quad |\,r_2\,| = \sqrt{16+36+4} = 7.5$

Unit vector from B to P, $\overline{u}_{r_2} = \dfrac{\overline{r}_2}{|\,r_2\,|} = \dfrac{4\overline{u}_x - 6\overline{u}_y - 2\overline{u}_z}{7.5}$

Electric field at P due to charge at B is

$$E_{P_2} = \frac{Q_2}{4\pi\varepsilon_0 |r_2|^2}\bar{u}_{r_2} = \frac{(9\times10^9)\times3\times10^{-6}}{(7.5)^2}\cdot\frac{4\bar{u}_x - 6\bar{u}_y - 2\bar{u}_z}{7.5}$$

$$= (256\bar{u}_x - 384\bar{u}_y - 128\bar{u}_z)\text{V/m}.$$

(c) The total field at point P

$$\bar{E}_P = \bar{E}_{P_1} + \bar{E}_{P_2}$$

$$= (872.26\bar{u}_x - 1308.39\bar{u}_y + 1308.39\bar{u}_z) + (256\bar{u}_x - 384\bar{u}_y - 128\bar{u}_z)$$

$$\therefore \quad \bar{E}_P = (1128.26\bar{u}_x - 1692.39\bar{u}_y + 1180.39\bar{u}_z)\text{V/m}.$$

Example 2.10: *Two point charges $Q_1 = 5$ nC and $Q_2 = -4$ nC are kept at (3, 0, 0) and (5, 0, 0). Determine electric field at (5, – 2, 1).*

Solution: Electric field due to 'n' charges is

$$\bar{E} = \frac{1}{4\pi\varepsilon_0}\sum_{i=1}^{n}\frac{Q_i}{|r_i|^2}\bar{u}_{r_1}$$

For $n = 2$;

$$\vec{E} = \frac{\vec{F}}{Q} = \frac{1}{4\pi\varepsilon_0}\left[\frac{Q_1}{|r_1|^2}\bar{u}_{r_1} + \frac{Q_2}{|r_2|^2}\bar{u}_{r_2}\right]$$

$$\vec{r}_1 = (5-3)\bar{u}_x + (-2-0)\bar{u}_y + (1-0)\bar{u}_z$$

$$= 2\bar{u}_x - 2\bar{u}_y + \bar{u}_z$$

$$|r_1| = \sqrt{4+4+1} = \sqrt{9}$$

$$\bar{u}_{r_1} = \frac{\vec{r}_1}{|r_1|} = \frac{2\bar{u}_x - 2\bar{u}_y + \bar{u}_z}{\sqrt{9}}$$

Similarly,

$$\vec{r}_2 = (5-5)\bar{u}_x + (-2-0)\bar{u}_y + (1-0)\bar{u}_z = -2\bar{u}_y + \bar{u}_z$$

$$\bar{u}_{r_2} = \frac{\vec{r}_2}{|r_2|} = \frac{-2\bar{u}_y - \bar{u}_z}{\sqrt{5}}$$

$$\therefore \quad \vec{E} = (9\times10^9)\times\frac{5\times10^{-9}}{(\sqrt{9})^2}\times\frac{2\bar{u}_x - 2\bar{u}_y + \bar{u}_z}{\sqrt{9}} - (9\times10^9)\frac{4\times10^{-9}}{(\sqrt{5})^2}\times\frac{-2\bar{u}_y + \bar{u}_z}{\sqrt{5}}$$

$$= (3.33\bar{u}_x - 3.33\bar{u}_y + 1.66\bar{u}_z) - (-6.44\,\bar{u}_y + 3.224\,\bar{u}_z)$$

$$= (3.33\bar{u}_x + 3.11\bar{u}_y - 1.564\bar{u}_z)\text{V/m}.$$

Example 2.11: *Determine the field strength at a point situated 5 cm away from two equal charge of $Q_1 = +10^9$ C and other charge of $Q_2 = 10^9$ C, if the distance between them is 8 m.*

Solution: According to the data given in the problem the electric field at P due to Q_1

$$\vec{F}_1 = \frac{10^9}{4\pi\varepsilon_0|r_1|^2}\bar{u}_{r_1} \text{ and that due to } Q_2;\ \vec{F}_2 = \frac{10^9}{4\pi\varepsilon_0|r_2|^2}\bar{u}_{r_2}$$

$$\therefore \quad \vec{E}_T = \vec{E}_1 + \vec{E}_2 = \frac{10^9}{4\pi\varepsilon_0|r|^2}[\bar{u}_{r_1} + \bar{u}_{r_2}]$$

$$= \frac{10^9}{\frac{10^9}{36\pi}(5\times10^2)^2}\,[(\cos\theta_1 u_x + \sin\theta_1 u_y) + (\cos\theta_2 u_x - \sin\theta_2 u_y)$$

where $\theta_1 = \cos^{-1}\frac{4}{5}$ and $\sin\theta_1 = \frac{3}{5}$

But as per the problem $\theta_2 = \theta_1$

$$\therefore \quad \vec{E}_T = \frac{9}{25}\times10^{18}\times10^4\,[2\cos\theta_1\, u_x]$$

$$= \frac{900}{25}\times10^{20}\times2\cos\theta_1\, u_x$$

$$= 72\times10^{20}\,\frac{4}{5}\,u_x = 14.4\times10^{20}\times4\,u_x$$

$$\vec{E}_T = 57.6\times10^{20}\,u_x.$$

2.5 ELECTRIC FIELD DUE TO CONTINUOUS CHARGE DISTRIBUTION

In electrostatics the concept of point charges and different types of continuous charge distributions along a line, surface (or) in a volume is dealt with. Normally the line charge density, surface charge density and volume charge density can be denoted by ρ_L (in C/m), ρ_s (in C/m^2) and ρ_v (C/m^3) respectively.

2.5.1 Volume Charge

Each charge element contributes to electric field at an external point, when charge in distributed throughout a specified volume. In order to obtain the total electric field a summation or integration is required.

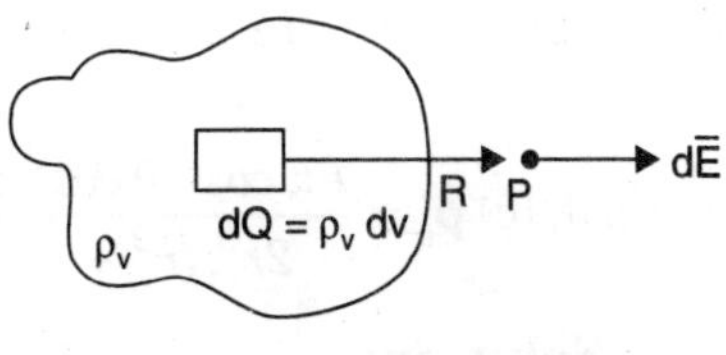

Fig. 2.4 Field intensity due to volume charge

The volume charge density can be defined as

$$\rho_v = \lim_{\Delta v\to0}\frac{\Delta Q}{\Delta v}\ \text{C/m}^3 \qquad ...(2.17)$$

where ΔQ is noted as small amount of charge in small volume Δv

The total charge within a defined volume can be found by taking volume integral throughout the volume.

$$Q = \int_V dQ = \int_V \rho_v\, dv \qquad ...(2.18)$$

where Q is total charge and $\int_V$ represents volume integral.

But it can be written as with reference to volume v in Fig. 2.4 each differential charge dQ produces a differential electric field.

$$d\bar{E} = \frac{dQ}{4\pi\varepsilon_0 R^2}\cdot U_r \qquad ...(2.19)$$

Assuming that the only charge in region is contained within the volume, the total electric field at P is obtained by integration over the volume.

$$\overline{E} = \int_V \frac{\rho_v \, dv}{4\pi\varepsilon_0 R^2} \qquad ...(2.20)$$

Example 2.12: *For x, y and z positive, let $\rho_v = 40\, xyz\ C/m^3$. Find the total charge within the region defined by (a) $0 \le x, y, z \le 2$ (b) $x = 0, y = 0, 0 \le 2x + 3y \le 10, 0 \le z \le 2$.*

Solution:

(a)
$$Q = \int_0^2 \int_0^2 \int_0^2 40\, xyz\, dx\, dy\, dz = 40 \left[\frac{x^2}{2}\right]_0^2 \left[\frac{y^2}{2}\right]_0^2 \left[\frac{z^2}{2}\right]_0^2 = 320 \text{ C.}$$

(b)
$$Q = \int_0^5 \int_0^{\frac{(10-2x)}{3}} \int_0^2 40\, xyz\, dx\, dy\, dz = \int_0^5 40\, x\, dx \left[\frac{y^2}{2}\right]_0^{\frac{10-2x}{3}} \left[\frac{z^2}{2}\right]_0^2$$

$$= \int_0^5 10\, x\, dx \left(\frac{10-2x}{3}\right)^2 (4) = \frac{40}{9}\int_0^5 x\,(100 - 40x + 4x^2)\, dx$$

$$= \frac{40}{9}\int_0^5 (100x - 40x^2 + 4x^3)\, dx = \frac{40}{9}\left[\frac{100x^2}{2} - \frac{40x^3}{3} + \frac{4x^4}{4}\right]_0^5$$

$$= \frac{40}{9}\left[50\,(25) - \frac{40}{3} \times 125 + 625\right] = 925.926 \text{ C.}$$

Example 2.13: *Find the total charge inside each of the volumes indicated (a) $\rho_v = 20\, t^2 e^{-0.2x} \sin \pi y$, $-1 \le x \le 2$, $0 \le y \le 1$, $3 \le z \le 3.6$, (b) $\rho_v = 6xyz^2$, $0 \le \rho \le 2$, $0 \le \phi \le \frac{\pi}{2}$; $0 \le z \le 3$, (c) $\rho_v = \dfrac{6\pi \cos^2\theta \cos^2\pi}{2r^2\,(r^2+1)}$; universe.*

Solution:

(a)
$$Q = \int dQ = \int \rho_v\, dv = \int_{-1}^2 \int_0^1 \int_3^{3.6} 20\, t^2 e^{-0.2x} \sin(\pi y)\, dx\, dy\, dz$$

$$= 20\left[\frac{z^3}{3}\right]_3^{3.6}\left[\frac{e^{-0.2x}}{-0.2}\right]_{-1}^2\left[\frac{-\cos(\pi y)}{\pi}\right]_0^1 = 229.86 \text{ C.}$$

(b)
$$\rho_v = 6\, xyz^2 = 6\,(\rho \cos\phi)(\rho \sin\phi)\, z^2 = 3\rho^2 \sin^2\phi\, z^2$$

$\therefore$
$$Q = \int dQ = \int_0^2 \int_0^{\pi/2} \int_0^3 3\rho^2 \sin^2\phi\, z^2\, \rho\, d\rho\, d\phi\, dz$$

$$= 3\left[\frac{\rho^4}{4}\right]_0^2 \left[\frac{-\cos^2\phi}{2}\right]_0^{\pi/2} \left[\frac{z^3}{3}\right]_0^3 = 108 \text{ C.}$$

(c) For universe, the limits are $0 < \rho < \infty$, $0 \le \theta \le \pi$ and $0 \le \phi \le 2\pi$

$$Q = \int dQ = \int_0^{\infty}\int_0^{\pi}\int_0^{2\pi} \frac{6\pi \cos^2\theta \cos^2\phi}{2r^2(r^2+1)} r^2 \sin\theta \, dr \, d\theta \, d\phi$$

$$= 3\pi\left[\tan^{-1} r\right]_0^{\infty}\left[\frac{-\cos^3\theta}{3}\right]_0^{\pi}\left[\frac{\phi}{2}+\frac{\sin^2\phi}{4}\right]_0^{2\pi} = 31.006 \text{ C.}$$

Example 2.14: *Volume charge density is given by $\rho_v = 20e^{-2000\rho}\, e^{-150z}$ C/m³ (a) find the maximum value of ρ_v in the region $0 \le \rho \le 0.01$ m, $0 \le \phi \le 2\pi$, $0 \le z \le 0.01$ m (b) find the total charge contained is the first octant where x, y, z are positive (c) find 'b' if total charge found is the volume $0 \le \rho \le b$, $0 \le \phi \le \pi/2$, $z \ge 0$ is half the volume found is part (b)?*

Solution:

(a) ρ_v will be maximum if all terms with in ρ_v are maximum. Now 20 $e^{-2000\rho}$ will be maximum at $\rho = 0$ and e^{-150z} will be maximum at $z = 0$.

$\therefore \quad \rho_{v\,max} = 20 \times 1 \times 1 = 20$ C/m³

(b) $$Q = \int_0^{\infty}\int_0^{\pi/2}\int_0^{\infty} 20\, e^{-2000\rho}\, e^{-150z}\, \rho \, d\rho \, d\phi \, dz$$

$$= 20\left[\frac{-\rho e^{-2000\rho}}{2000} - \frac{e^{-2000\rho}}{(2000)^2}\right]_0^{\infty}\left[\frac{-e^{-150z}}{150}\right]_0^{\infty}\left[\phi\right]_0^{\pi/2} = 52.35 \, nC$$

(c) $$\frac{52.35\times 10^{-9}}{2} = \int_0^{b}\int_0^{\pi/2}\int_0^{\infty} 20\, e^{-2000\rho}\, e^{-150z}\, \rho \, d\rho \, d\phi \, dz$$

$$= 20\left[\frac{-\rho e^{-2000\rho}}{20^3} - \frac{e^{-2000\rho}}{20^6}\right]_0^{b}\left[\frac{-e^{-150z}}{150}\right]_0^{\infty}\left[\phi\right]_0^{\pi/2}$$

$$\frac{52.35\times 10^{-9}}{2} = 20\left[\frac{-be^{-2000b}}{(20)^3} - \frac{e^{-2000b}}{20^6} + \frac{1}{20^6}\right]\left(\frac{1}{150}\right)\left(\frac{\pi}{2}\right)$$

$\therefore \quad e^{-2000b}(2000b + 1) = 0.996$

$$2000b + 1 = 0.996e^{2000b}$$

By trail and error procedure the value of b can be found out.

Example 2.15: *Find the charge in the volume defined by $0 \le x \le 1$ m, $0 < y \le 1$ m and $0 \le z \le 1$ m if $\rho = 35x^2y$ μC/m³. What charge occurs for the limits $-1 \le y \le 0$ m?*

Solution:

We know that $dQ = \rho \, dv$

$$\therefore \quad Q = \int_0^1\int_0^1\int_0^1 30\, x^2\, y \, dx \, dy \, dz$$

$$= 30\int_0^1\int_0^1 y \, dy \, dz \int_0^1 x^2 \, dx = 10\int_0^1 dz \int_0^1 y \, dy$$

$$= 5\int_0^1 dz = 5 \, \mu\text{C}$$

Similarly for the change in limits on y

$$Q = \int_0^1 \int_{-1}^0 \int_0^1 30\, x^2\, y\, dx\, dy\, dz = -5\ \mu\text{C}.$$

2.5.2 Surface Charge

The surface charge distribution is described in terms of surface charge density ρ_s C/m^2 if charge is distributed over a surface as shown in Fig. 2.5.

The surface charge density, ρ_s can be defined as

$$\rho_s = \lim_{\Delta s \to 0} \frac{\Delta Q}{\Delta s}\ \text{C/m}^2 \qquad ...(2.21)$$

Where ΔQ is noted as small amount of charge on small surface Δs.

The total charge on a defined surface can be obtained by taking the surface integral over the complete surface.

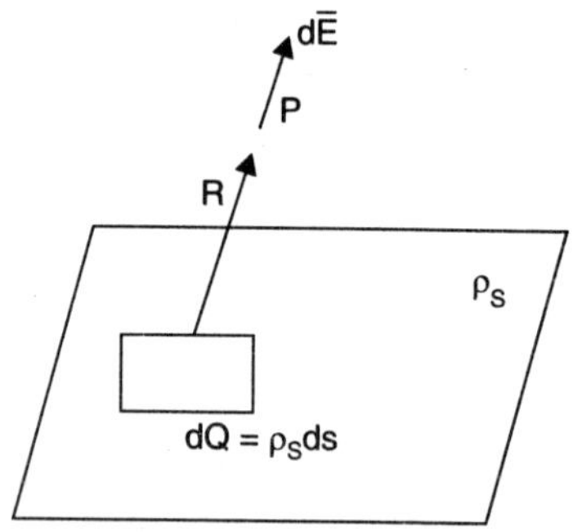

Fig. 2.5 Field intensity due to surface charge

$$Q = \int_S dQ = \int_S \rho_s\, ds \qquad ...(2.22)$$

Where Q is total charge and $\int_S$ represents surface integral .With reference to surface charges in the Fig. 2.5, each differential charge dQ on the sheet results in differential electric field which can be written as

$$d\overline{E} = \frac{dQ}{4\pi\varepsilon_0 R^2} \overline{U}_r \qquad ...(2.23)$$

If the surface charge density is ρ_s (C/m^2) and no other charge is present in region, then total electric field at P can be shown as

$$\overline{E} = \int_S \frac{\rho_s\, ds}{4\pi\varepsilon_0 R^2} \overline{U}_r \qquad ...(2.24)$$

Example 2.16: *Find out the electric field E at a distance x from the centre of a uniform charged sphere of radius 5 cm and with charge density 1 C/m^2 for (i) x = 3 cm (ii) x = 5 cm (iii) x = 8 cm.*

Solution:

(*a*) For $x = 0.03$ m *i.e.*, inside the charged sphere, the field inside is zero.

(*b*) For $x = 0.05$ m

$$\overline{E} = \frac{Q\ \mathit{as}}{4\pi\varepsilon x^2} \frac{Q_s\, 4\pi\,(0.05)^2}{4\pi \times \left(\frac{10^{-9}}{36\pi}\right)(0.05)^2}$$

$$= Q_s\, 36\pi \times 10^9 = 36\pi \times 10^9\ \text{V/m}.$$

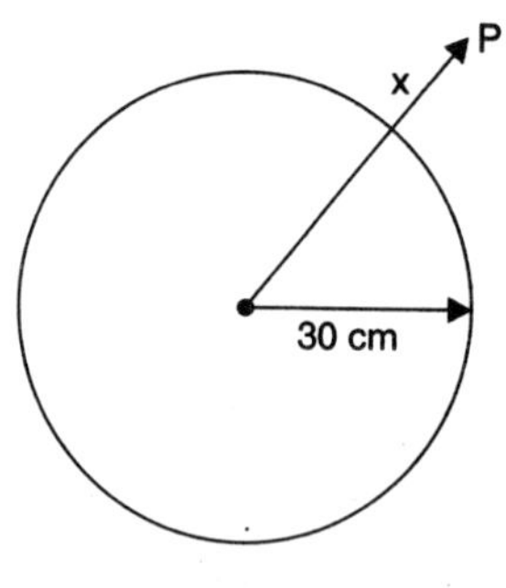

Fig. 2.6

(*c*) For $x = 0.08$ m

$$\vec{E} = \frac{Q_s\, 4\pi\,(0.05)^2}{4\pi \times \left(\dfrac{10^{-9}}{36\pi}\right)(0.08)^2} = 36\pi \times 10^9 \times \frac{4}{9} = 14.06\,\pi \times 10^9 \text{ V/m.}$$

2.5.3 Line Charge

The line charge distribution is described in terms of line charge density ρ_L C/m if charge is distributed over a line as shown in Fig. 2.6. The line charge density ρ_l is defined as

$$\rho_L = \lim_{\Delta l \to 0} \frac{\Delta Q}{\Delta l} \text{ C/m} \qquad \text{...(2.25)}$$

where ΔQ is small amount of charge in elemental length Δl.

With reference to line charge density ρ_l shown in Fig. 2.7, each differential charge dQ along the line produces a differential electric field.

$$d\overline{E} = \frac{dQ}{4\pi\varepsilon_0 R^2}\,\overline{U}_r \qquad \text{...(2.26)}$$

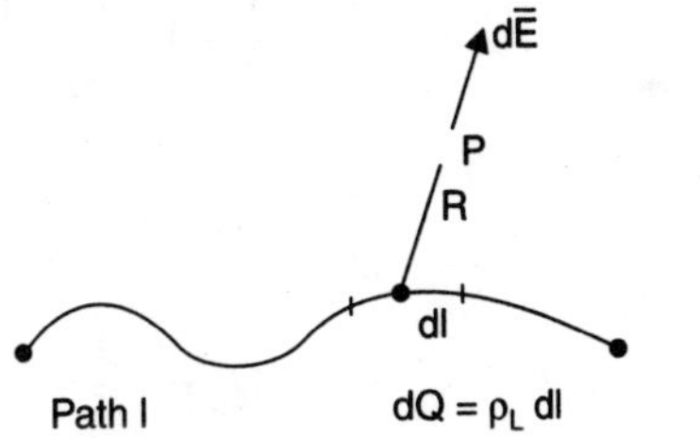

Fig. 2.7 Field intensity due to line charge

If the line charge density is ρ_l (C/m) and no other charge is in the region, then the total electric field at P is

$$\overline{E} = \int_L \frac{\rho_l\, dL}{4\pi\varepsilon_0 R^2}\,\overline{U}_r \qquad \text{...(2.27)}$$

Example 2.17: *It is required to keep three equal point charges placed at the corners of an equilateral triangle in equilibrium by placing a point charge at the centre of the triangle. Find out the value of the point charge.*

Solution: Let Q_0 be the charge required to be placed at the centre of the equilateral triangle of each side l shown in Fig. 2.8.

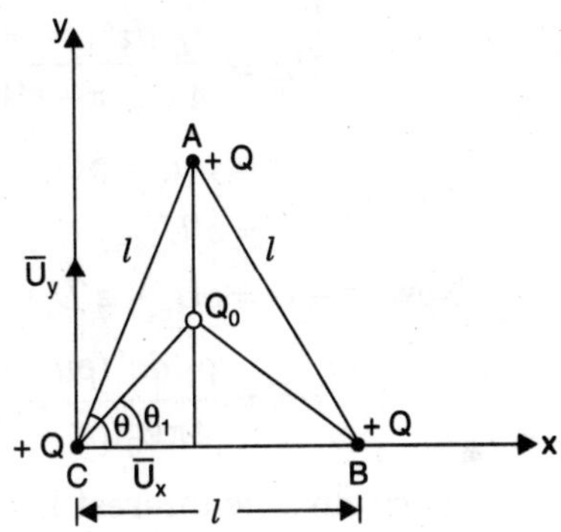

Fig. 2.8

Now, the net force on the charge at center is given by

$$\vec{F}_1 = \vec{F}_{12} + \vec{F}_{13} + \vec{F}_{14}$$

where $$\vec{F}_{12} = \frac{Q^2}{4\pi\varepsilon_0 l^2}\,[\cos 60° \,(U_x) + \sin 60° \,(U_y)]$$

$$\vec{F}_{13} = \frac{Q^2}{4\pi\varepsilon_0 l^2}\,(-U_x)$$

and $$\vec{F}_{14} = \frac{Q\,Q_0}{4\pi\varepsilon_0 \left(\dfrac{2}{3}l^t\right)^2}\,[\cos 30° \,(U_x) + \sin 30° \,(U_y)]$$

where $$l' = l \sin 60° = \frac{\sqrt{3}}{2}\,l$$

$$= \frac{QQ_0}{4\pi\varepsilon_0 \frac{4}{9} \times \frac{3}{4} l^2} \left[\frac{\sqrt{3}}{2} U_x + \frac{1}{2} U_y \right]$$

$$\therefore \quad \vec{F}_{14} = \frac{QQ_0}{4\pi\varepsilon_0 l^2} \left[\frac{3\sqrt{3}}{2} U_x + \frac{3}{2} U_y \right]$$

$$\therefore \quad \vec{F} = \frac{Q^2}{4\pi\varepsilon_0 l^2} \left[-\frac{1}{2} U_x - \frac{\sqrt{3}}{2} U_y - U_x + \frac{Q_0}{Q} \left(\frac{3\sqrt{3}}{2} U_x + \frac{3}{2} U_y \right) \right]$$

$$= 0 \text{ for equilibrium}$$

$$\left(-\frac{1}{2} - 1 + \frac{Q_0}{Q} \frac{3\sqrt{3}}{2} \right) U_x = 0$$

or
$$\frac{Q_0}{Q} \frac{3\sqrt{3}}{2} = 1 + \frac{1}{2} = \frac{3}{2}$$

$$\therefore \quad Q_0 = \frac{Q}{\sqrt{3}}.$$

2.6 ELECTRIC FIELD INTENSITY DUE TO A LINE CHARGE

Let a line charge be assumed extending along z-axis in a cylindrical coordinate system from $-\infty$ to ∞. Now, it is required to find out electric field intensity E at any point 'P' due a line charge ρ_L which spreads from $-\infty$ to ∞.

From Fig. 2.9, $dQ = \rho_L \, dz'$ at $(0, 0, z')$, since the charge is being considered at that point on the z-axis,

$$dE = \frac{\rho_L \, dz' \, (r - r')}{4\pi\varepsilon_0 |r - r'|^3} \quad \text{...(2.28)}$$

$$r = y\, u_y = \rho U_\rho$$

$$r' = z' U_z$$

Now $r - r' = \rho U_\rho - z' U_z$

$$dE = \frac{\rho_l \, dz' \, (\rho u_\rho - z' u_z)}{4\pi\varepsilon_0 (\rho^2 + z'^2)^{3/2}} \quad \text{...(2.29)}$$

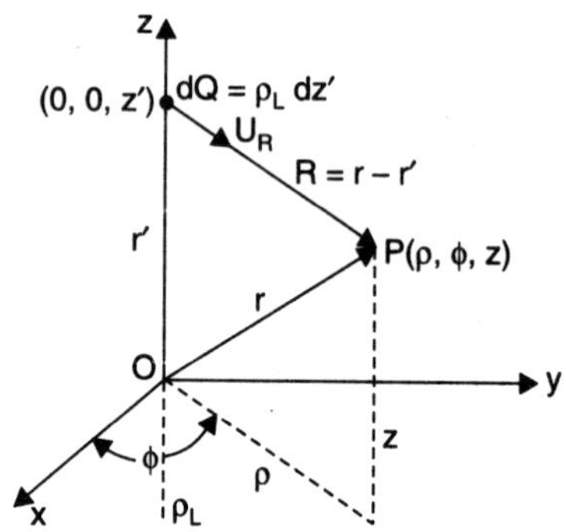

Fig. 2.9

Since our sole interest is in finding out E_ρ due to a line charge it can be simplified and also since here field charges exist due to charge ρ but not ϕ (or) z we can write as

$$dE_\rho = \frac{\rho_L \, \rho \, dz'}{4\pi\varepsilon_0 (\rho^2 + z'^2)^{3/2}} \quad \text{...(2.30)}$$

$$E_\rho = \int_{-\infty}^{\infty} \frac{\rho_L \, \rho \, dz'}{4\pi\varepsilon_0 (\rho^2 + z'^2)^{3/2}} \quad \text{...(2.31)}$$

By solving the above equation we obtain

$$E_\rho = \frac{\rho_L}{2\pi\varepsilon_0\,\rho} \qquad ...(2.32)$$

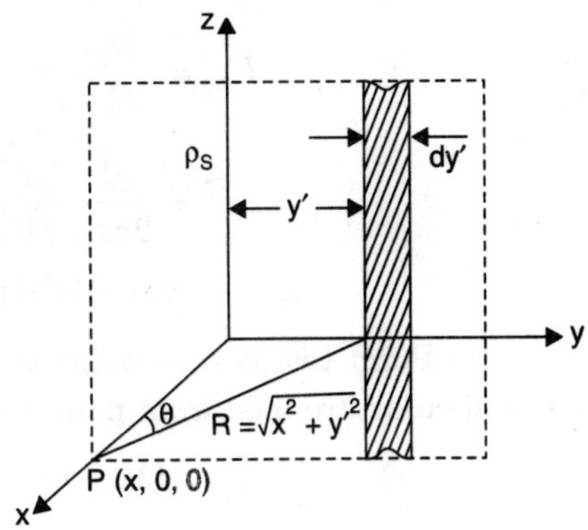

Fig. 2.10

If the point ρ is considered on the *y-z* plane instead of taking it on the *z*-axis then the electric field intensity can be determined as follows:

$$\bar{r} = \rho U_\rho + zU_z$$

$$r' = z'U_z$$

$$\bar{R} = r - r' = \rho U_\rho + (z - z')U_t$$

$$|R| = \sqrt{\rho^2 + (z - z')^2}$$

From equations (2.9) and (2.25), the electric field intensity can be found out as

$$E = \int_{-\infty}^{\infty} \frac{\rho_L\, dz'\,[\rho U_\rho + (z - z')U_z]}{4\pi\varepsilon_0[\rho^2 + (z - z')^2]^{3/2}} \qquad ...(2.33)$$

$$= \frac{\rho_L}{4\rho\varepsilon_0}\left[\int_{-\infty}^{\infty} \frac{\rho dz'\, U_\rho}{[\rho^2 + (z - z')^2]^{3/2}} + \int_{-\infty}^{\infty} \frac{(z - z')U_z\, dz'}{[\rho^2 + (z - z')^2]^{3/2}}\right]$$

Here the unit vector can change direction but can't be changed in magnitude. So direction of U_ρ does not change with z' but it changes with ϕ.

$$E = \frac{\rho_L}{4\rho\varepsilon_0}\left[U_\rho \int_{-\infty}^{\infty} \frac{\rho dz'}{[\rho^2 + (z - z')^2]^{3/2}} + U_z \int_{-\infty}^{\infty} \frac{(z - z')U_z\, dz'}{[\rho^2 + (z - z')^2]^{3/2}}\right]$$

$$= \frac{\rho_L}{4\rho\varepsilon_0}\left[U_\rho \frac{2}{\rho} + U_z\,(0)\right] = \frac{\rho_L}{2\pi\varepsilon_0\rho}\, U_\rho \qquad ...(2.34)$$

Note that the field reduces inversely with the distance to the charged line, as compared with point charge and it leads to a field only one per cent the previous strength, whereas the field which decreases with square of the distance moving ten times as far from a point charge reduces the field to 10 per cent of its former value. The field of an infinity long fluorescent tube thus decays inversely as the first power of radial distance to tube.

Example 2.18: *An infinity long, uniform line charge is located at y =6, z = 10, if ρ_L = 40 n C/m, find $\bar{E}$ at (a) the origin and (b) at P_B (0, 12, 2).*

Solution:

(*a*)
$$\bar{E} = \frac{\rho_l}{2\pi\varepsilon_0 r}\, U_{12}$$

Point two is that point at which $\bar{E}$ is desired = (0, 0, 0)

Point one is the foot of perpendicular dropped from point two on line charge ($x = 0, y = 6, z = 10$)

$$\therefore \qquad \bar{R}_{12} = (0 - 6)U_y + (0 - 10)U_z = -6U_y - 10U_z$$

$$r = |\bar{R}_{12}| = \sqrt{(-6)^2 + (-10)^2} = \sqrt{136}$$

$$U_{12} = \frac{\overline{R}_{12}}{|R_{12}|} = \frac{-6U_y - 10U_t}{\sqrt{136}}$$

$$\overline{E} = \frac{40 \times 10^{-9}}{2\pi\varepsilon_0 \sqrt{136}} \left[\frac{-6u_y - 10u_t}{\sqrt{136}}\right]$$

$$= (-31.764U_y - 52.941U_z) \text{ V/m}$$

(*b*) Point two is that point at which $\overline{E}$ is desired = (0, 12, 2) . Point one is the foot of perpendicular dropped from point two on line charge = (x = 0, y = 6, z = 10).

$$\therefore \quad \overline{R}_{12} = (12 - 6)U_y + (2 - 10)U_z = 6\overline{U}_y - 8\overline{U}_z$$

$$r = |\ \overline{R}_{12}\ | = \sqrt{(6)^2 + (-8)^2} = \sqrt{100}$$

$$U_{12} = \frac{\overline{R}_{12}}{|R_{12}|} \frac{6U_y - 8U_z}{\sqrt{100}}$$

$$\overline{E} = \frac{40 \times 10^{-9}}{2\pi\varepsilon_0 \sqrt{100}} \left[\frac{6U_y - 8U_t}{\sqrt{100}}\right]$$

$$= (43.2\ U_y - 57.6\ U_z) \text{ V/m.}$$

Example 2.19: *A line charge density of 45 n C/m is located is free space on the line y = 1, z = 3 (a) find* $\overline{E}$ *at P (5, – 1, 4) (b) what point charge* Q_A *should be located at Q (– 3, 4, 1) to cause* E_y *to be equal to zero at P?*

Solution: $\overline{E} = \frac{\rho_L}{2\pi\varepsilon_0} U_{12}$

Point two is that point at which $\overline{E}$ is desired = (5, – 1, 4) point one is the foot to perpendicular dropped from point two on the line charge = (5, 1, 3)

$$\overline{R}_{12} = (-1 - 1)U_y + (4 - 3)U_z = -2U_y + U_z$$

$$r = |\ \overline{R}_{12}\ | = \sqrt{(-2)^2 + (1)^2} = \sqrt{5}$$

$$\overline{U}_{12} = \frac{\overline{R}_{12}}{|R_{12}|} = \frac{-2U_y + U_z}{\sqrt{5}}$$

$$\overline{E} = \frac{45 \times 10^{-9}}{2\pi\varepsilon_0 \sqrt{5}} \left[\frac{-2U_y + U_z}{\sqrt{5}}\right] = (-324.78\overline{U}_y + 162.39\overline{U}_z) \text{ V/m}$$

$$\overline{E} \text{ due to } Q_A = \frac{Q_A}{4\pi\varepsilon_0} \left[\frac{(5+3)U_x + (-1-4)U_y + (4-1)U_z}{[(5+3)^2 + (-1-4)^2 + (4-1)^2]^{3/2}}\right]$$

$$\text{Hence, } E_y \text{ (total)} = -324.78 + \frac{Q_A(-5)}{4\pi\varepsilon_0 (98)^{1.5}} = 0$$

$$\therefore \quad Q_A = -6.986\ \mu\text{C}.$$

Example 2.20: *A uniform line charge ρ_{L1} = 25 nC/m is at y = – 2, z = 0 while ρ_{L2} = – 25 nC/ m is symmetrically located at y = 2, z = 0, find $\overline{E}$ as a function of z at y = 0.*

Solution:

In $y = 0$ plane any point is given by $(x, 0, z)$

$$E_1 = \frac{\rho_l}{2\pi\varepsilon_0 d} U_{1L}$$

Point two is that point at which E is desired *i.e.*, $(x, 0, z)$ and Point one is the foot of perpendicular dropped from point two on line charge = $(x, -2, 0)$.

$$\therefore \quad R_{12} = (0 + 2)\, U_y + (z - 0)\, U_z = 2U_y + zU_z$$

$$d = |\overline{R}_{12}| = \sqrt{2^2 + z^2} \text{ and } U_{12} = \frac{2U_y + zU_z}{\sqrt{2^2 + z^2}}$$

$$\overline{E}_1 = \frac{25 \times 10^{-9}}{2\pi\varepsilon_0}\left[\frac{2U_y + zU_z}{4 + z^2}\right]$$

$$E_2 = \frac{-25 \times 10^{-9}}{2\pi\varepsilon_0}\left[\frac{-2U_y + zU_z}{4 + z^2}\right]$$

$$\overline{E} = \overline{E}_1 + \overline{E}_2 = 25 \times 10^{-9}\,\frac{[2U_y + zU_z + 2U_y - zU_z]}{2\pi\varepsilon_0\,(4 + z^2)} = \frac{1804.37}{4 + z^2}\, U_y \text{ V/m.}$$

Example 2.21: *A circular ring of thin conducting wire is set in a horizontal plain in air. The ring has a radius r meter and has charge Q C. A test charge Q_t C is placed partially above the ring at a distance of 'd' meter from the ring.*

(a) Develop the expression for force on.

(b) Show that for d >> r, the ring modeled by a point charge Q.

(c) What is the electric intensity at Q_t?

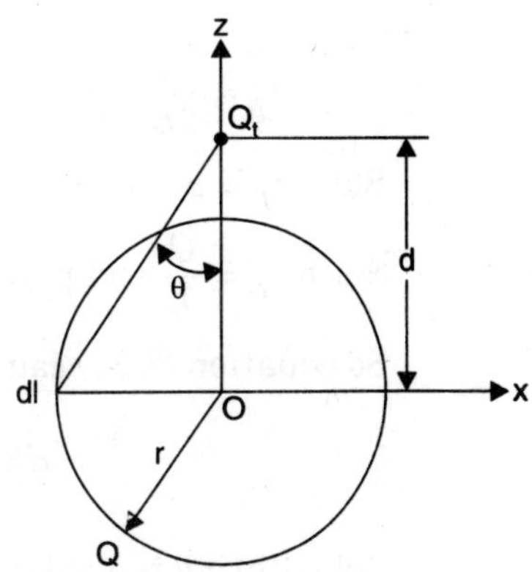

Fig. 2.11(a)

Solution: Due to symmetry x-components of the force will cancel out and the component of force along z-direction for small length dl, is

$$(a) \quad dE_z = \frac{Q_1\, dl\, Q_t}{4\pi\varepsilon\, x^2} \cos\theta\,;$$

where, $$Q_1 = \frac{Q}{2\pi r} \text{ and } \cos\theta = \frac{d}{x}$$

$$E_z = \int dE_z = \int \frac{\left(\frac{Q}{2\pi r}\right) Q_t}{4\pi\varepsilon\, r^2}\left(\frac{d}{x}\right) dl$$

$$= \int \frac{Q\,Q_t}{4\pi\varepsilon\, x^3}\left(\frac{dl}{2\pi r}\right) = \frac{Q\,Q_t\, d}{4\pi\varepsilon\,(d^2 + r^2)^{3/2}}\,\frac{1}{2\pi r}\int dl$$

$$E_z = \frac{Q\,Q_t\, d}{4\pi\varepsilon\,(d^2 + r^2)^{3/2}}\left(\frac{1}{2\pi r}\right) 2\pi r = \frac{Q\,Q_t\, d}{4\pi\varepsilon\,(r^2 + d^2)^{3/2}}$$

$\therefore \qquad \vec{E} = E_z U_z$

(b) $$\vec{F}_z = \frac{QQ_t\, d}{4\pi\varepsilon (r^2 + d^2)^{3/2}} U_z$$

Now for $d >> r$ $$\vec{F}_z = \frac{QQ_t\, d}{4\pi\varepsilon\, d^2} U_z = \frac{QQ_t}{4\pi\varepsilon\, d^2}$$

Hence, the ring may be considered as a point charge Q, if $d >> r$.

(c) Electrical intensity at Q_t

$$E = Q_t \frac{Qd}{4\pi\varepsilon (r^2 + d^2)^{3/2}} U_z.$$

2.7 ELECTRIC FIELD INTENSITY DUE TO A SURFACE CHARGE

Another basic configuration which is to be considered is the infinite sheet of charge having uniform density of ρ_s C/m^2. Let the sheet of charge be placed in y-z plane at a distance y' from z-axis. The differential width is taken as dy'. Let P be a point on y-axis and R is the distance of P from sheet of charge. Now only the x-components remain because sheet of charge is on y-z plane and field point 'P' is along x-axis, and so the y and z-components of field intensity at 'P' cancel out each other

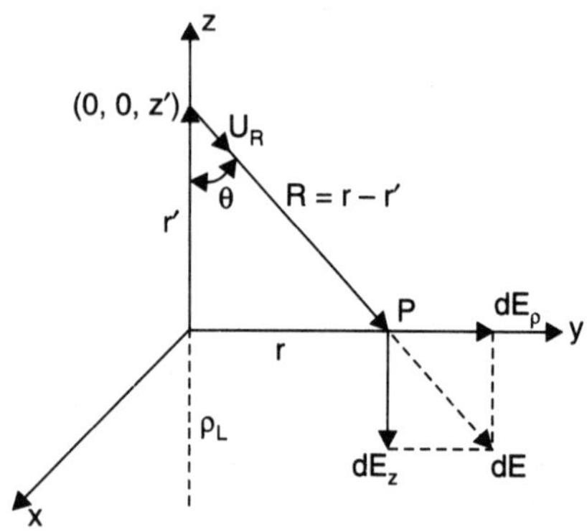

Fig. 2.11(b)

Consider the differential width of charge shown in Fig. 2.11 (b). From equation (2.34) it can be written as

$$E_\rho = \frac{\rho_L}{2\pi\varepsilon_0\, \rho} \qquad ...(2.35)$$

But $\rho_L = \rho_s\, dy'$

Since $\rho_L = \dfrac{Q}{L}$ and $\rho_s = \dfrac{Q}{S} = \dfrac{Q}{L\,dy'}$ and Ldy' = surface area over length L.

So equation (2.35) can be written as

$$dE_x = \frac{\rho_s\, dy'}{2\pi\varepsilon_0 \sqrt{x^2 + y'^2}} \cos\theta \qquad (\because x = \rho \cos\theta)$$

Substituting the value of cos θ from Fig. 2.11(b).

$$\therefore \qquad dE_x = \frac{\rho_s}{2\pi\varepsilon_0} \frac{x\,dy'}{x^2 + y'^2}. \qquad ...(2.36)$$

Adding all the effects of strips, we get

$$E_x = \frac{\rho_s}{2\pi\varepsilon_0} \int_{-\infty}^{\infty} \frac{x\,dy'}{x^2 + y'^2} = \frac{\rho_s}{2\pi\varepsilon_0} \left(\tan^{-1}\frac{y}{x}\right)_{-\infty}^{\infty}$$

$$\therefore \qquad E_x = \frac{\rho_s}{2\varepsilon_0} \qquad \text{...(2.37)}$$

$$E = \frac{\rho_s}{2\varepsilon_0}$$

If point 'P' was chosen on negative x-axis then $E_x = \frac{-\rho_s}{2\varepsilon_0}$ because the field is always directed away from the positive charge. This difficulty in sign is usually overcome by specifying a unit vector U_N which is normal to the sheet and directed away from it.

Then $$E = \frac{\rho_s}{2\varepsilon_0} U_N \qquad \text{...(2.38)}$$

Example 2.22: *Find the force on a point charge 50 μC at (0, 0, 5)m due to a charge 450 π μC that is uniformly distributes over a circular disc of radius 5 m.*

Solution: The electric field intensity of a circular disc of radius 'a' situated in the x–y-plane at $z = 0$, with its centre at the origin is given by

$$\vec{E} = \frac{\rho_s h}{2\varepsilon_0}\left(\frac{-1}{\sqrt{a^2+h^2}} + \frac{1}{h}\right)$$

The charge density,

$$\rho_s = \frac{Q}{\pi a^2} = \frac{450\pi \times 10^{-6}}{\pi \times 5^2} = 18 \times 10^{-6} \text{ C/m}^2$$

Therefore, $$E = \frac{18 \times 10^{-6}}{2 \times 8.854 \times 10^{-12}}\left(\frac{-1}{\sqrt{5^2+5^2}} + \frac{1}{5}\right) = 58956.40 \text{ V/m}$$

But, $$E = \frac{F}{Q} \Rightarrow F = E\,Q$$

$$F = 58956.40 \times 50 \times 10^{-6} = 2.9 \text{ N.}$$

2.8 ELECTRIC FIELD INTENSITY DUE TO A CHARGED RING

Let a wire of negligible cross section be bent in the form of circular ring and also let the wire be uniformly charged with charge density ρ_L C/m. It is required to determine the electric field at point P (0, 0, h) meters from center of ring shown in Fig. 2.12. The ring is assumed to be formed by several point charges. Let $d\overline{E}$ be the electric field at point charge dQ and also let the charge contained by element of length dL be dQ.

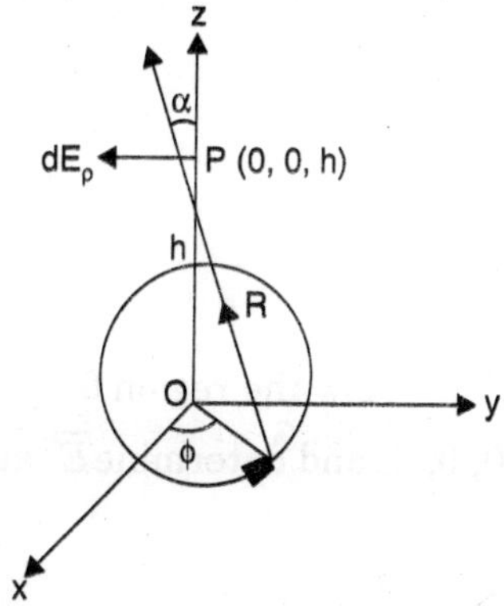

Fig. 2.12

In this case $dl = \rho\, d\phi$

$$\overline{R} = \rho\,(-U_\rho) + h\,(U_z)$$

$$R = |\,\overline{R}\,| = \sqrt{[\rho^2 + h^2]}$$

$$U_R = \frac{\overline{R}}{|\overline{R}|}$$

(or)
$$\frac{U_R}{R^2} = \frac{\overline{R}}{|\overline{R}|^3} = \frac{(-\rho U_\rho + hU_z)}{[\rho^2 + h^2]^{3/2}}$$

We know that the electric field intensity $\overline{E}$ at r is given by

$$\overline{E} = \int \frac{\rho_L\, dl}{4\pi\varepsilon_0 R^2} U_R \quad \text{...(2.39)}$$

Hence,
$$\overline{E} = \frac{\rho_L}{4\pi\varepsilon_0} \int_{\phi=0}^{2\pi} \rho\, d\phi \frac{(-\rho U_\rho + hU_z)}{[\rho^2 + h^2]^{3/2}} \quad \text{...(2.40)}$$

When electric field vectors are resolved, radial components get cancelled and normal components get added. Therefore the direction of resultant electric field $\overline{E}$ at 'P' is normal to plane of ring. Thus contribution of $\overline{E}_\rho$ add up to zero so that $\overline{E}$ has only z-components.

i.e.,
$$\overline{E} = \frac{\rho_L \cdot \rho_h U_z}{4\pi\varepsilon_0 (\rho^2 + h^2)^{3/2}} \int_0^{2\pi} d\phi \quad \text{...(2.41)}$$

$$\overline{E} = \frac{\rho_L\, \rho_h}{2\varepsilon_0 (\rho^2 + h^2)^{3/2}} \overline{U}_z$$

Substitute
$$\frac{h}{\sqrt{\rho^2 + h^2}} = \cos\alpha \quad \text{and} \quad \sqrt{\rho^2 + h^2} = R$$

$\therefore$
$$\overline{E} = \frac{\rho_L\, \rho_h \cos\alpha}{2\varepsilon_0 R^2} \overline{U}_z \quad \text{...(2.42)}$$

(or)
$$\overline{E} = \frac{\rho_L\, \rho_h}{2\varepsilon_0 R^3} \quad \text{...(2.43)}$$

Example 2.23: *Volume charge density is given by $\rho_v = 10^{-5} e^{-100r} \sin\theta\ C/m^3$ for $0 \le r \le 1$ cm and $\rho_v = 0$ for $r > 1$ cm. Estimate $\overline{E}$ at $r = 1$ m, $\theta = 90°$, $\phi = 0$ by thinking in terms of positive charge.*

Solution: Total charge within the region $0 \le r \le 1$ cm is given by

$$Q = \int_0^{0.01} \int_0^{\pi} \int_0^{2\pi} 10^{-5} e^{-1000r} \sin\theta\, r^2 \sin\theta\, dr\, d\theta\, d\phi$$

$$= 10^{-5} \left[\frac{-r^2 e^{-100r}}{100} - \frac{re^{-100r}}{5000} - \frac{e^{-100r}}{500000} \right]_0^{0.01} \left[\frac{\theta}{2} - \frac{\sin^2\theta}{4} \right]_0^{\pi} [\phi]_0^{2\pi}$$

$$= 10^{-5} \times 1.606 \times 10^{-7} \times 0.5 \times \pi \times 2\pi = 1.585 \times 10^{-11}\ \text{C}$$

Now the region $0 \le r \le 1$ cm is a spherical region centered at origin. So let is locate Q at (0, 0, 0) and determine $\overline{E}$ at ($r = 1$ m, $\theta = 90°$, $\phi = 0$), $\overline{E}$ at ($x = 1, y = 0, z = 0$).

$$\overline{E} = \frac{1.585 \times 10^{-11}}{4\pi\varepsilon_0 (\sqrt{1})^2} \left[\frac{(1-0)U_x}{\sqrt{(1)^2}} \right] = 0.142 U_x\ \text{V/m}$$

Example 2.24: *A line charge $\rho_L = 44\ nC/m$ is located along the line $x = 3$, $y = 5$ in free space. (a) Find $\overline{E}$ at P (2, 6,– 8) (b) If the surface $x = 4$ contains a uniform surface density $\rho_s = 36\ nC/m^2$ at what point in the z = 0 plane is $\overline{E}_{total} = 0$?*

Solution: (*a*) We know that $\overline{E} = \dfrac{\rho_L}{4\pi\varepsilon_0 r}\overline{U}_{12}$ and Point two is the point at which E is desired *i.e.*, (2, 6,– 8). Point one is the foot of perpendicular dropped from point two on line charge *i.e.*, (3, 5,– 8).

$$\overline{R}_{12} = (2-3)U_x + (6-5)U_y = -\overline{U}_x + \overline{U}_y$$

$$r = |\overline{R}_{12}| = \sqrt{(-1)^2 + (1)^2}$$

$$U_{12} = \frac{-U_x + U_y}{\sqrt{2}}$$

$$\overline{E} = \frac{44 \times 10^{-9}}{2\pi\varepsilon_0\sqrt{2}}\left[\frac{-U_x + U_y}{\sqrt{2}}\right] = (-396.96U_x + 396.96U_y)\ \text{V/m}$$

(*b*) Any point in the $z = 0$ plane is given by $(x, y, 0)$. Applying the procedure of part (*a*) the electric field intensity at point $(x, y, 0)$ due to line charge is

$$\overline{E} = 44 \times 10^{-9}\frac{(x-3)U_x + (y-5)U_y}{2\pi\varepsilon_0[(x-3)^2 + (y-5)^2]}$$

$\overline{E}$ due to surface charge, $\overline{E} = \dfrac{36 \times 10^{-9}}{2\varepsilon_0}(\pm U_x)$

$$= 2039.66\,(\pm U_x)\ [+U_x, \text{ if } x > 4 \text{ and } -U_x \text{ if } x < 4]$$

$$\therefore \qquad E_{total} = \left\{\frac{793.92\,(x-3)}{[(x-3)^2 + (y-5)^2]} \pm 2039.66\right\}U_x + \frac{793.92\,(y-5)}{[(x-3)^2 + (y-5)^2]}\overline{U}_y$$

If E_{total} is equal to zero, then coefficient of U_x and U_y must be equal to zero. At $y = 5$, the coefficient of U_y is equal to zero. Now, the coefficient of U_x is

$$\frac{793.92\,(x-3)}{(x-3)^2 + (5-5)^2} \pm 2039.66 = 0$$

(or)

$$x - 3 = \pm\frac{793.92}{2039.66} = \pm 0.3892$$

Hence, $x = 3.3892$ and 2.6108 *i.e.*, the points are (3.3892, 5, 0) and (2.6108, 5, 0). In both the points x coordinate is less than four and when $x < 4$, the x component is given by $\dfrac{793.92\,(x-3)}{(x-3)^2 + (y-5)^2} - 2039.66$ and this component is zero only when $x = 3.3892$. When $x = 2.6108$, the x component is not equal to zero. So, the correct point is (3.3892, 5, 0).

Example 2.25: *Find the electric field at point (2, 2, 2) due to two point charges Q_1 (4×10^{-9}C) and Q_2 (3×10^{-9}C) located with coordinates (0, 2, 2) and (2, 0, 2) respectively in free space.*

Solution: According to the data provided in the problem

$$\overline{R}_{1t} = (x_2 - x_1)u_x + (y_2 - y_1)u_y + (z_2 - z_1)u_z$$

$$\overline{R}_{1t} = (2-0)u_x + (2-2)u_y + (2-2)u_z$$

$$\therefore \qquad u_{R1t} = \frac{\overline{R}_{1t}}{|\overline{R}_{1t}|} = \frac{2u_x}{|R_{1t}|} = u_x$$

Similarly,

$$\overline{R}_{2t} = (2-2)u_x + (2-0)u_y + (2-2)u_z$$

$$\therefore \qquad u_{R2t} = u_y$$

The electric field at point P due to charge Q_1 (alone) is

$$\overline{E}_1 = \frac{Q_1}{4\pi\varepsilon_0 R_{1t}} u_{R1t} = \frac{4\times 10^{-9}}{4\pi \times \frac{10^{-9}}{36\pi}\times 2} u_x = 18.04\, u_x$$

And that due to charge Q_2 is

$$\overline{E}_2 = \frac{Q_2}{4\pi\varepsilon_0 R_{2t}} u_{R2t} = \frac{3\times 10^{-9}}{4\pi \times \frac{10^{-9}}{36\pi}\times 2} u_y = 13.5\, u_y$$

Therefore, the resultant field E at P will be given by the vector sum of E_1 and E_2 that is,

$$\overline{E} = \overline{E}_1 + \overline{E}_2 = 18.04\, u_x + 13.5\, u_y$$

$$= \sqrt{(18.04)^2 + (13.5)^2} \angle \tan^{-1}\left(\frac{13.5}{18.04}\right)$$

$$= 22.53 \angle \tan^{-1} 0.748 \text{ V/m}$$

The vector E will be in the plane of the top surface and will lead E_1 *i.e.*, u_x by angle $\tan^{-1}(0.748)$.

Example 2.26: *Three equal positive charges of 2Q C each are located at three corners of a square, side 'a' meter. Determine the magnitude and direction of the resulted force experienced by a test charge if placed at the vacant corner of the square.*

Solution: Figure 2.13, shows the geometry of the problem. Let us assume a point test charge '+ $2Q_t$' placed at the vacant cornet 'O'. Let u_x and u_y be the unit vectors along the positive x and positive y-axes respectively.

The force experienced by Q_t due to charge 'Q' placed at A is

$$F_{Qt,\,QA} = \frac{4QQ_t}{4\pi\varepsilon_0 (OA)^2}(-u_x) = \frac{-QQ_t}{\pi\varepsilon_0 a^2} u_x \qquad ...(a)$$

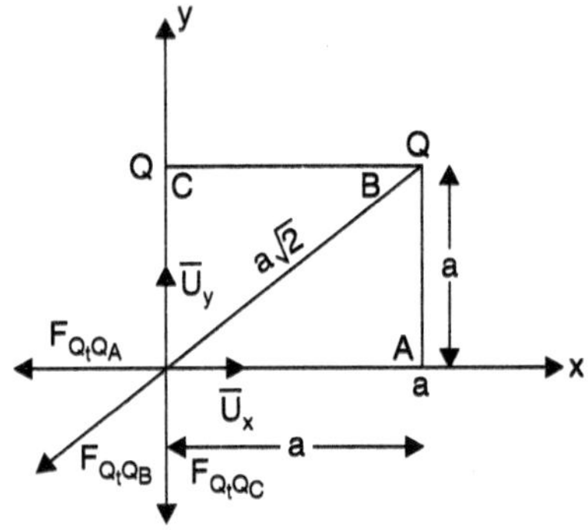

Fig. 2.13

The force due to the charge at B, that is, $F_{Qt,\,QB}$ is also repulsive and will act along the direction shown in Fig. 2.13. The magnitude of this force is

$$F_{Qt,QB} = \frac{4QQ_t}{4\pi\varepsilon_0 (OB)^2} = \frac{QQ_t}{\pi\varepsilon_0 (a\sqrt{2})^2} = \frac{QQ_t}{2\pi\varepsilon_0 a^2} \qquad ...(b)$$

Resolving $F_{Qt,QB}$ along the x and y-axes, we have

$$F_{(Qt,\,QB)_x} = (F_{QtQB} \cos\theta)\,(-u_x) \qquad ...(c)$$

and
$$F_{(Qt,\,QB)_y} = F_{QtQB} \sin\theta\,(-u_y) \qquad ...(d)$$

Therefore, the force $F_{Qt\,QB}$ becomes

$$F_{Qt\,QB} = F_{Qt\,QB}\,[-\cos\theta\, u_x - \sin\theta\, u_y]$$

$$= \frac{QQ_t}{2\pi\varepsilon_0 a^2}\,[-\cos\theta\, u_x - \sin\theta\, u_y]$$

where, $\theta = 45°$ and $\cos\theta = \sin\theta = \dfrac{1}{\sqrt{2}}$

or
$$F_{QtQB} = \frac{QQ_t}{2\pi\varepsilon_0 a^2}\left[\frac{u_x}{\sqrt{2}} - \frac{u_y}{\sqrt{2}}\right] \qquad ...(e)$$

Similarly, the force due to the charge placed at C is

$$F_{Qt\,QB} = \frac{QQ_t}{2\pi\varepsilon_0 a^2}\,(-u_y) \qquad ...(f)$$

Adding equations (*a*), (*e*) and (*f*), the resultant force experienced by the test charge Q_t placed at corner O is

$$F = \frac{-QQ_t}{\pi\varepsilon_0 a^2}\,u_x + \frac{QQ_t}{2\pi\varepsilon_0 a^2}\left[\frac{u_x}{\sqrt{2}} - \frac{u_y}{\sqrt{2}}\right] + \frac{QQ_t}{2\pi\varepsilon_0 a^2}\,(-u_y).$$

Example 2.27: *Two point charges $Q_1 = 5$ nC and $Q_2 = -4$ nC are kept at (3, 0, 0) and (5, 0, 0). Determine electric field at (5, –2, 1).*

Solution: Electric field due to 'n' charges is $\overline{E} = \dfrac{1}{4\pi\varepsilon_0}\displaystyle\sum_{i=1}^{n} \frac{Q_i}{|r_i|^2}\,\overline{u}_{ri}$

For $n = 2$;
$$\overline{E} = \frac{\overline{F}}{Q} = \frac{1}{4\pi\varepsilon_0}\left[\frac{Q_1}{|r_1|^2}\,\overline{u}_{r_1} + \frac{Q_2}{|r_2|^2}\,\overline{u}_{r_2}\right]$$

$$\overline{r}_1 = (5-3)\overline{u}_x + (-2-0)\overline{u}_y + (1-0)\overline{u}_z$$

$$= 2\overline{u}_x - 2\overline{u}_y + \overline{u}_z$$

$$|r_1| = \sqrt{4+4+1} = \sqrt{9}$$

$$\overline{u}_{r_1} = \frac{\overline{r}_1}{|r_1|} = \frac{2\overline{u}_x - 2\overline{u}_y + \overline{u}_z}{\sqrt{9}}$$

Similarly,
$$\overline{r}_2 = (5-5)\,\overline{u}_x\,(-2-0)\,\overline{u}_y + (1-0)\overline{u}_z$$

$$= -2\overline{u}_y + \overline{u}_z$$

$$\bar{u}_{r_2} = \frac{\bar{r}_2}{|r_2|} = \frac{-2\bar{u}_y + \bar{u}_z}{\sqrt{5}}$$

$$\therefore \quad \bar{E} = (9 \times 10^9) \times \frac{5 \times 10^{-9}}{(\sqrt{9})^2} \times \frac{(2\bar{u}_x - 2\bar{u}_y + \bar{u}_z)}{\sqrt{9}} - (9 \times 10^9) \frac{5 \times 10^{-9}}{(\sqrt{5})^2} \times \frac{(-2\bar{u}_y + \bar{u}_z)}{\sqrt{5}}$$

$$= (3.33\bar{u}_x - 3.33\bar{u}_y + 1.66\bar{u}_z) - (-6.44\bar{u}_y + 3.224\bar{u}_z)$$

$$= (3.33\,\bar{u}_x + 3.11\bar{u}_y - 1.564\bar{u}_z) \text{ V/m.}$$

Example 2.28: *Find the force on a point charge of 45 μC at (0, 0, 6) m due to a charge of 450 π μC that is uniformly distributed over the circular disc r ≤ 6 m, z = 0 m.*

Solution: The change density, $\rho_s = \frac{Q}{A} = \frac{450\pi \times 10^{-6}}{\pi (6)^2} = 0.125 \times 10^{-4}$ C/m²

In cylindrical coordinates $R = -ru_r + 6u_z$

Then each differential charge results in a differential force

$$dF = \frac{45 \times 10^{-6} (\rho_s \, r \, dr \, d\phi)}{4\pi \left(\frac{10^{-9}}{36\pi}\right)(r^2 + 36)} \frac{(-ru_r + 6u_z)}{\sqrt{r^2 + 36}}$$

The radial components will cancel and that u_z is constant.

Hence,
$$F = \int_0^{2\pi} \int_0^6 \frac{(45 \times 10^{-6})(0.125 \times 10^{-4})\, 6r \, dr \, d\phi}{4\pi \left(\frac{10^{-9}}{36\pi}\right)(r^2 + 36)^{3/2}} u_z$$

$$= 190.75 \int_0^6 \frac{r \, dr}{(r^2 + 36)^{3/2}} u_z = 190.75 \left[\frac{-1}{\sqrt{r^2 + 36}}\right]_0^6 u_z = 9.321\, U_z \text{ N.}$$

Example 2.29: *Find the force on charge Q_1 = 100 μC due to a charge Q_2 = – 300 μC where Q_1 is at (0, 1, 2) m and Q_2 at (3, 0, 0)m.*

Solution:

Given that the two charges Q_1 = 100 μC and Q_2 = 300 μC

$$\bar{R}_{21} = (0 - 3)u_x + (1 - 0)u_y + (2 - 0)u_z$$

$$= -3u_x + u_y + 2u_z$$

$$|\bar{R}_{21}| = \sqrt{-3^2 + 1^2 + 2^2}$$

$$= \sqrt{14} = 3.74$$

So, unit vector can be written as $u_{21} = \dfrac{-3u_x + u_y + 2u_z}{3.74}$

So, the force on charge Q_1 due to charge Q_2 can be written as

$$F_1 = \frac{Q_1 Q_2}{4\pi\varepsilon_0 |R_{21}|^2} \times u_{21}$$

$$= \frac{(100 \times 10^{-6})(-300 \times 10^{-6})}{4\pi \times 8.854 \times 10^{-12} \times (3.74)^2} \times \left(\frac{-3u_x + u_y + 2u_z}{3.74}\right)$$

$$= -5.154\,(-3u_x + u_y + 2u_z)$$

$$= (15.46u_x - 5.15u_y - 10.31u_z) \text{ N.}$$

Example 2.30: *Two point charges of equal mass m, charge Q are suspended at a common point by two threads of negligible mass and length l. Show that at equilibrium the inclination angle α of each thread to the vertical is given by $Q^2 = 16\pi\,\varepsilon_0\, mg\, l^2 \sin^2\alpha \tan\alpha$. If α is very small, show that* $\alpha = \sqrt[3]{\dfrac{Q^2}{16\pi\varepsilon_0\, mg\, l^2}}$.

Solution:

Consider a system of charges where F_e is the electric force, T is the tension in each thread and mg is the weight of each charge.. At any point it can be written as

$$T \sin\alpha = F_e\,; \quad T\cos\alpha = mg$$

Hence it can be written as $\dfrac{\sin\alpha}{\cos\alpha} = \dfrac{F_e}{mg} = \dfrac{1}{mg} \times \dfrac{Q^2}{4\pi\varepsilon_0\, r^2}$

But it can be written as $r = 2l \sin\alpha$

So $\quad Q^2 \cos\alpha = 16\pi\,\varepsilon_0\, mg\, l^2 \sin^3\alpha$

Or it can be written as $Q^2 = 16\pi\,\varepsilon_0\, mg\, l^2 \sin^2\alpha \tan\alpha$

When α is very small $\tan\alpha \approx \alpha \approx \sin\alpha$ and so the above equation can be written as

$$Q^2 = 16\pi\varepsilon_0\, mg\, l^2 \alpha^3$$

Or
$$\alpha = \sqrt[3]{\frac{Q^2}{16\pi\varepsilon_0\, mg\, l^2}}$$

Example 2.31: *The fine sheet $0 \le x \le 1$, $0 \le y \le 1$ on the $z = 0$ plane has a charge density $\rho_s = xy\,(x^2 + y^2 + 25)^{3/2}$ nC/m². Find*

(a) the total charge on the sheet.

(b) the electric field at (0, 0, 5).

(c) the force experienced by a –1 mC charge located at (0, 0, 5).

Solution:

$$(a)\ Q = \int \rho_s\, ds = \int_0^1 \int_0^1 xy\,(x^2 + y^2 + 25)^{3/2}\, dx\, dy\ nC$$

Since $x\,dx = \frac{1}{2}d\,(x^2)$, we now integrate with respect to x^2.

$$Q = \frac{1}{2}\int_0^1 y\int_0^1 (x^2 + y^2 + 25)^{3/2}\, d(x^2)\, dy$$

$$= \frac{1}{5}\int_0^1 \frac{1}{2}\left[(y^2+26)^{5/2} - (y^2+25)^{5/2}\right] d(y)$$

$$= \frac{1}{35}\left[(27)^{7/2} + (25)^{7/2} - 2(26)^{7/2}\right]$$

$$Q = 33.15\ nC.$$

(*b*) $$E = \int \frac{\rho_s\, ds\, U_R}{4\pi\varepsilon_0\, r^2} = \int \frac{\rho_s\, ds\,(r - r')}{4\pi\,\varepsilon_0\, |r - r'|^3}$$

where $r - r' = (0, 0, 5) - (x, y, 0) = (-x, -y, 5).$

Hence, $$E = \int_0^1\int_0^1 \frac{10^{-9}\, xy\,(x^2+y^2+25)^{3/2}\,(-xu_x - yu_y + 5u_z)\, dx\, dy}{4\pi \times \dfrac{10^{-9}}{36\pi}(x^2+y^2+25)^{3/2}}$$

$$= 9\left[-\int_0^1 x^2\, dx \int_0^1 y\, dy\, u_x - \int_0^1 x\, dx\int_0^1 y^2\, dy\, u_y + 5\int_0^1 x\, dx\int_0^1 y\, dy\, u_z\right]$$

$$= 9\left(\frac{-1}{6}, \frac{-1}{6}, \frac{5}{4}\right)$$

$= (-1.5, -1.5, 11.25)$ V/m.

(*c*) $F = QE = (1.5, 1.5, -11.25)$ m Ns.

Example 2.32: *Calculate electric field E at a point N (3, – 4, 2) in free space caused by (a) a charge $Q_1 = 2\mu C$ at P_1 (0, 0, 0) (b) a charge $Q_2 = 3\ \mu C$ at P_2 (– 1, 2, 3) (c) both the charges of part (a) and (b) present.*

Solution:

(*a*) The vector from point P_1 to point N is

$$\bar{r}_1 = (3-0)U_x + (-4-0)U_y + (2-0)U_z$$
$$= 3U_x - 4U_y + 2U_z$$

Unit vector from P_1 to N (direction of the electric field E_{n1}).

$$U_{r1} = \frac{3U_x - 4U_y + 2U_z}{\sqrt{29}}$$

Field at N due to charge at P_1, $E_{n1} = \dfrac{Q_1}{4\pi\varepsilon_0\,|\bar{r}_1|^2}U_{r1}$

$$= \frac{2 \times 10^{-6}}{4\pi\varepsilon_0 \times 29} \cdot \frac{3U_x - 4U_y + 2U_z}{\sqrt{29}}$$

$$E_{n1} = (345U_x - 460U_y + 230U_z)\text{ V/m.}$$

(*b*) The vector from point P_2 to N is

$$r_2 = (3 + 1)U_x + (-4 - 2)U_y + (2 - 3)U_z = 4U_x - 6U_y - U_z$$

Unit vector from P_2 to N (direction of electric field E_{n2})

$$U_{r2} = \frac{4U_x - 6U_y - U_z}{\sqrt{53}}$$

Field at N due to charge at P_2, is given by $E_{n2} = \dfrac{Q_2}{4\pi\varepsilon_0 |\bar{r}_2|^2} U_{r2}$

$$= \frac{3 \times 10^{-6}}{4\pi\varepsilon_0 \times 53} \times \frac{4U_x - 6U_y - U_z}{\sqrt{53}}$$

$$E_{n2} = (280U_x - 419U_y - 69.9U_z)\text{ V/m.}$$

(*c*) The total field at point N

$$E_N = E_{n1} + E_{n2}$$

$$= 345U_x - 460U_y + 230U_z + 280\,U_x - 419U_y - 69.9U_z$$

$$E_N = (625U_x - 879U_y + 160.1U_z)\text{ V/m.}$$

SHORT QUESTIONS AND ANSWERS

1. Define electric field.

The region surrounding an electric charge or a group of charges, in which no other charge experiences a force, is called electric field.

2. Define the term electric field intensity.

It is defined as the vector force on a unit positive test charge.

3. Write the expression for electric field intensity due to a volume charge.

$$\bar{E} = \int_V \frac{\rho_v\, dV}{3\pi\varepsilon_0 R^2}$$

4. Write the expression for electric field intensity due to a line charge.

$$\bar{E} = \int_L \frac{\rho_l\, dL}{4\pi\varepsilon_0 R^2} \bar{U}_r$$

5. Define Coulomb's law.

If Q_1 and Q_2 are two point charges separated by a distance '*r*', then the force acting between them is proportional to product of the two charges and inversely proportional to square of the distance between them.

6. **Write the expression for electric field intensity due to single charge.**

$$E = \frac{Q}{4\pi\varepsilon_0 R^2} U_R$$

7. **Write the expression for electric field intensity due to several charges.**

$$E = \sum_{m=1}^{n} \frac{Q_m}{4\pi\varepsilon_0 (R_p + R_m)^2} U_m$$

8. **Write the expression for electric field intensity due to charged ring.**

$$\bar{E} = \frac{\rho_L \rho_h}{2\varepsilon_0 R^2}$$

9. **Write the expression for electric field intensity due to surface charge.**

$$E = \frac{\rho_s}{2\varepsilon_0}$$

10. **Write the expression for electric field intensity due to line charge.**

$$E = \frac{\rho_L}{2\pi\varepsilon_0 \rho} \times U_r$$

MULTIPLE CHOICES

1. There is a charge A of 20 units and another B of 5 units. The null point N will be located at
   (a) $2AN = BN$ (b) $AN = 2BN$
   (c) $AN = BN$ (d) $AN = \frac{1}{2} BN$
2. The diagram shows in Fig. 2.14, the electric lines of force :

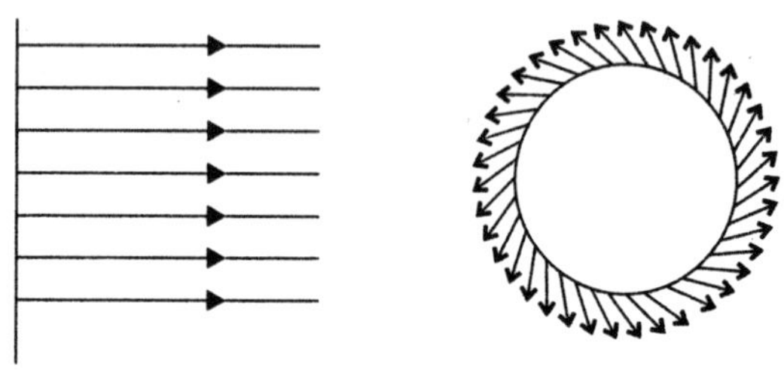

(a) (b)

Fig. 2.14

There is greater crowding of lines in B than A because
(a) A is a line, L is a circle
(b) charges accumulate more at points of curvature
(c) lines of force depart perpendicular to the surface
(d) none of above

3. The electric field strength E of a dipole varies
   (*a*) directly as square of the distance (*b*) inversely as the square of the distance
   (*c*) inversely as distance (*d*) inversely as cube of the distance
4. The electric field strength at a distance 50 cm is 2N/C. The magnitude of charge is
   (*a*) 0.056×10^{-9} C (*b*) 0.5×10^{-12} C
   (*c*) 5×10^{-9} C (*d*) 0.005×10^{-9} C
5. The unit of electric field strength E is
   (*a*) volts/metre (*b*) N/C
   (*c*) joules/ C–metre (*d*) all of the above
6. There are two charges of +10 and +30 e.s.u. There is a point P between them where the electric field strength is zero. Then the point P
   (*a*) is midway between them
   (*b*) lies on the line joining them
   (*c*) the unit point charge is in equilibrium
   (*d*) the force acting on the unit charge is equal and opposite
7. Two point charges of + 2.0×10^{-7} and + 8.5×10^{-8} C are 12 cm a part. The electric field each produces at the site of the other is
   (*a*) 1250×10^2 N/C, 530×10^2 N/C (*b*) 1.25×10^3 N/C, 5.3×10^2 N/C
   (*c*) 1250 N/C, 530 N/C (*d*) 530×10^2 N/C, 1250×10^2 N/C
8. In the above problem, the force acting on each is
   (*a*) 0.5 N attractive (*b*) 1.1×10^{-2} N/C
   (*c*) 1.1×10^{-2} N repulsive (*d*) 1.1×10^{-2} N attractive
9. A tube of force is
   (*a*) the force exerted by a steel tube
   (*b*) the fictitious tube of electric lines of force over a surface
   (*c*) force exerted by a tube of water
   (*d*) all of the above
10. A charged sphere A is placed inside an insulated hollow conducting sphere B, but not at the centre. Choose the correct sketch of the lines of force as shown in Fig. 2.15.

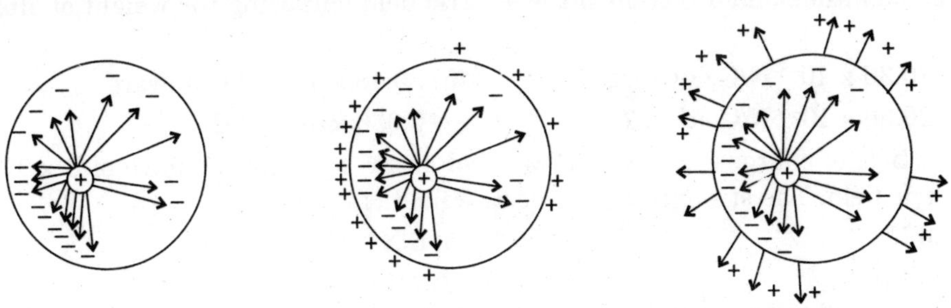

(*a*) (*b*) (*c*)

Fig. 2.15

11. A charge + 2×10^{-6} C is situated in air at the origin of a rectangular coordinate system. E at a point 4 m from the origin on x-axis is

(*a*) 1/1125 V/m
(*b*) 1125 N/C along *x*-axis
(*c*) 1125 V/m
(*d*) 112.5 N/C

12. The magnitude of Coulomb's force of repulsion between two alpha particles (helium ions) in space 10 mm apart is

(*a*) 92.03×10^{-25} N
(*b*) 9.216×10^{-25} N
(*c*) 46.08×10^{-25} N
(*d*) 10×10^{-27} N

13. The electric field strength between two charges $-5q$ and $+2q$ separated by 50 cm is zero at

(*a*) mid-way between them
(*b*) perpendicular bisector
(*c*) beyond the plus charge
(*d*) beyond the negative charge

14. The magnitude and direction of E at a point mid-way between two charges $+2.0 \times 10^{-7}$ C, 15 cm apart are

(*a*) $+3673 \times 10^2$ N/C beyond any charge
(*b*) -3673×10^2 N/C between the charges
(*c*) 7346×10^2 N/C mid-way
(*d*) zero, mid-way in the line joining the charges facing each other

15. $E = 0$ between any two charges on the line joining them but not between. The magnitude and sign of the charges and location of the null point are

(*a*) the same
(*b*) opposite sign
(*c*) opposite sign and beyond the line depending on the magnitude of the charges
(*d*) none of the above

16. An electron is placed mid-way between two equal and opposite charges. The electron will

(*a*) be thrown out of the field
(*b*) spin in the field
(*c*) remain stationary
(*d*) oscillate between the charges

17. The magnitude and direction of the electric field balancing the weight of electron is

(*a*) 5.6×10^{-11} N/C upward
(*b*) 5.6×10^{-11} N/C downward
(*c*) 6×10^{-12} N/C downward
(*d*) 5.6×10^{-11} N/C at $\angle 45°$

18. The magnitude and direction of the electric field balancing the weight of alpha particle is

(*a*) 20.36×10^{-8} N/C downward
(*b*) 20.48×10^{-8} N/C upward
(*c*) 20.36×10^{-8} N/C at $\angle 45°$
(*d*) 2.036×10^{-8} N/C

19. A particle of charge -2×10^{-9} C is acted on by a downward force of 3×10^{-6} N in an electric field. The strength of the electric field is

(*a*) 6 N
(*b*) 6 N – C
(*c*) 6 N/C
(*d*) 1.5×10^3 N/C

20. In the above problem, a proton is placed in the field. The magnitude and direction of the electric force on the proton is

(*a*) 2.4×10^{-16} N upward
(*b*) 2.4×10^{-16} N downward
(*c*) 2.4×10^{-19} N
(*d*) 1.6×10^{-19} N downward

21. In problem 20, the ratio of the gravitational force to electric force is

(*a*) 1.5×10^{10} (*b*) 3×10^{8}

(*c*) 0.67×10^{10} (*d*) 6.8×10^{-11}

22. An electron moves with a speed of 5×10^{6} cm/sec in an electric field of 1×10^{3} N/C. The acceleration, distance and time before coming to rest are

(*a*) 175.6×10^{12} m/sec^2, 7×10^{-6} m, 2.85×10^{-10} sec

(*b*) 177.8×10^{12} cm/sec^2, 7 cm, 2.8×10^{-8} sec

(*c*) 177.8×10^{12} m/sec^2, 7 m, 2.8×10^{-8} sec

(*d*) 180.5×10^{12} m/sec^2, 8 cm, 3.1×10^{-8} sec

KEY

1. (*b*)	**2.** (*b*)	**3.** (*d*)	**4.** (*a*)	**5.** (*d*)	**6.** (*a*)
7. (*a*)	**8.** (*c*)	**9.** (*b*)	**10.** (*c*)	**11.** (*b*)	**12.** (*a*)
13. (*c*)	**14.** (*d*)	**15.** (*c*)	**16.** (*c*)	**17.** (*a*)	**18.** (*b*)
19. (*d*)	**20.** (*a*)	**21.** (*d*)	**22.** (*a*)		

REVIEW QUESTIONS

1. What do we mean when we say that an object is charged?
2. State and Explain Coulomb's law in electrostatics.
3. What do we mean when we say that the net charge is conserved in a closed system?
4. State Coulomb's law in your own words.
5. Explain the experimental law of Coulomb's.
6. Explain with the help of sketch, the principle of super position as applied to electric fields.
7. What is electric field intensity?
8. Explain the concept of electric field and write down expression for the electric field due to a point charge.
9. Derive the expression for the electric field intensity due to a line charge.
10. In the strictest sense, what must be the definition of a point charge? What are the other possible charge distributions?
11. Prove that newtons per coulomb are dimensionally the same as volts per meter.
12. Define surface charge density in terms of volume charge density.
13. Define line charge density in terms of surface charge density.
14. Derive the expression for the field due to a surface charge.
15. What is a line of force?
16. Explain electric fields due to continuous charge distributions.

EXERCISE PROBLEMS

1. Four infinite uniform sheets of charge are located as follows: 20 pC/m^2 at $y = 7$, -8 pC/m^2 at $y = 3$, 6 pC/m^2 at $y = -1$ and -18 pC/m^2 at $y = -4$. Find $\overline{E}$ at the point (*a*) $P_A(2, 6, -4)$, (*b*) $P_B(0, 0, 0)$ (*c*) $P_c(-1, -1, 1.5)$ (*d*) $P_D(10^6, 10^6, 10^6)$.

2. A regular tetrahedron has vertices at P_1 (2, 0, 0), P_2 ($-$ 1, $\sqrt{3}$, 0), P_3 ($-$ 1, $-\sqrt{3}$, 0) and P_4 (0, 0, 2 $\sqrt{2}$). Charges of 1 mC are located at each of four vertices. If the configuration is located in free space (*a*) find the magnitude of force on each charge (*b*) The angle between any two edges of tetrahedron.

3. Eight 25 nC point charges is free spare are located symmetrically on a circle of radius 0.2 m centered at the origin in the $z = 0$ plane (*a*). At what point on the *z*-axis is $|\overline{E}|$ a maximum? What is $|\overline{E}|_{max}$?

4. Eight point charge of 1 nC each are located at the corners of a cube in free space that is 1 m on a side. Find $|\overline{E}|$ at the center of (*a*) the cube (*b*) face (*c*) an edge.

5. Let $P_v = (x + 2y + 3z)$ C/m^3 in the cubical region $0 \leq x, y, z \leq$ 1mm and $\rho_v = 0$ outside the cube (*a*) what is the total charge contained within the cube? (*b*) set up the volume integer that will give $\overline{E}$ (x, 0, 0) for $x >$ 1mm. Do not integrate?

6. A uniform line charge density of 20 nC/m lies on the *z*-axis between $z = 1$ and $z = 3$ m. No other charge is present. Find $\overline{E}$ at (*a*) the origin (*b*) P (4, 0, 0).

7. That portion of *z*-axis for which $-2 < z < 2$ m carries a non uniform charge density of $z^2 + 1$ n C/m is free space. There are no other charges any where. Find *E* at (*a*) P_A (1, 0, 0) (*b*) P_B (2, 0, 0).

8. A sheet to charge $\rho_s = 2$ nC/m^2 is present at the plane $x = 3$ in free space and a line charge $\rho_l = 20$ nC/m, is located at $x = 1, z = 4$ (*a*) find the magnitude of electric field intensity at origin (*b*) find the direction of *E* at *P* (4, 5, 6) (*c*) what is the force per meter length on line charge?

9. In the rectangular region $-2 < x < 2, -3 < y < 3, z = 0$ the surface charge density is given by $\rho_s = (x^2 + y^2 + 1)^{3/2}$. If no other charge is present, Find $\overline{E}$ at *P* (0, 0, 1).

10. The circular region $\rho < a, t = 0$ carries a uniform surface charge density ρ_s. Find $\overline{E}$ at *P* (0, 0, *h*), $h > 0$.

11. A line charge density of 35 nC/m is located is free space on the line $y = 2, z = 3$ (*a*) find $\overline{E}$ at *P* (6, $-$ 1, 2) (*b*) what point charge Q_A should be located at *Q* ($-$ 3, 2, 1) to cause E_y to be equal to zero at *P*?

12. Calculate electric field E at a point *P* (5, $-$ 4, 3) in free space due to (*a*) A charge $Q_1 = 5\ \mu$C at *A* (0, 0, 0), (*b*) A charge $Q_2 = 3\ \mu$C at *B* ($-$ 1, 3, 6) and (*c*) Both the charges of part (*a*) and (*b*) present.

13. A line charge, $\rho_l = 30$ n C/m, is located a long the line $x = 4, y = 6$, in free space (*a*) Find $\overline{E}$ at P (1, 5, $-$ 4) (*b*) If the surface $x = 4$ contains a uniform surface charge density, $\rho_s = 28$ nC/m^2, at what point in the $z = 0$ plane is $\overline{E}_{total} = 0$?

14. A Pt charge Q_1 located at the origin is free space produces a field. $\overline{E} = 20\ U_r$ V/m for $r = 4$ m (*a*) find Q_1 (*b*) using this value of Q_1, find $\overline{E}$ at *P* (2, $-$ 1, 6) in cartesian components (*c*) using the same value of Q_1, find $\overline{E}$ at *c* (2, 4, 6) in cylindrical components.

15. In free space let $Q_1 = 10$ nC be P_1 (0, – 4, 0) and $Q_2 = 20$ nC be P_2 (0, 0, 4) (*a*) find $\overline{E}$ at the origin (*b*) where should a 30 nC point change be located so that $\overline{E} = 0$ at the origin?

16. Volume charge density is given by $\rho_v = 20^{-5} e^{-500r} \sin\theta$ C/m^3 for $0 \le r \le 1$ Cm and $\rho_v = 0$ for $r > 1$ cm. Estimate $\overline{E}$ at $r = 2$ m, $\theta = 90°$, $\phi = 0$ by thinking in terms positive charge.

17. A line charge density of 44 nC/m is located is free space on the line $y = 2$, $z = 3$ (*a*) find $\overline{E}$ at P (5, –1, 3) (*b*) what point charge Q_A should be located at Q (– 3, 5,1) to cause E_y to be equal to zero at P?

18. A line charge, $\rho_l = 40$ nC/m, is located a long the line $x = 3$, $y = 4$, in free space (*a*) find $\overline{E}$ at P (1, 4, – 4) (*b*) if the surface $x = 5$ contains a uniform surface charge density, $\rho_s = 28$ nC/m^2, at what point in the $z = 0$ plane is $\overline{E}_{total} = 0$?

CHAPTER

3

Electric Flux Density, Gauss's Law and Divergence

3.1 INTRODUCTION

In this unit the concepts of electric field and electric flux density are introduced. In the previous chapter, the Coulomb's law of electrostatics was studied and electric field intensity was derived from it. Here the concept of Gauss's law is introduced which is useful in solving some problems of symmetry. Next the concept of potential gradient is also introduced and this will provide an alternative method of determining electric field intensity at a point in the field when potential is known.

3.2 ELECTRIC DISPLACEMENT OR FLUX (ψ)

Michael Faraday developed the concept of electric flux while conducting an experiment on a pair of concentric metallic spheres. It was found that a positive charge, Q placed on the inner sphere of Fig. 3.1 induced an equal negative charge on the outer sphere which was earthed, provided the spheres remained isolated from each other. The charge on the outer sphere is independent of medium between the spheres. Faraday concluded that there was some sort of 'displacement' from the inner sphere to the outer sphere which was independent of medium and so it was referred to as displacement flux (or) electric flux. If electric flux is denoted by ψ and total charge on inner sphere by Q, then from Faraday's experiment, it can be written as

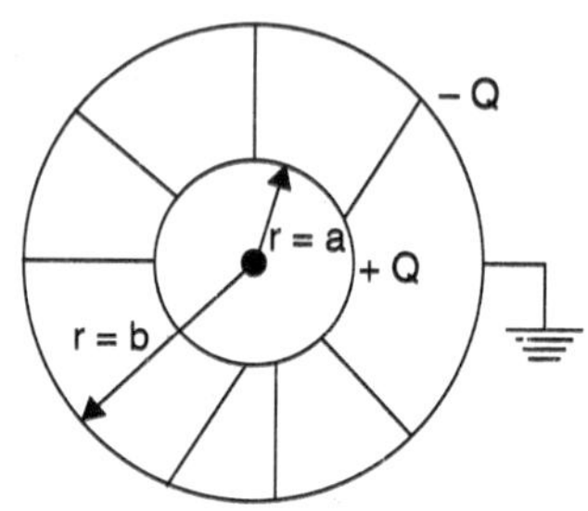

Fig. 3.1 Charged concentric spheres

$$\psi = Q$$

3.3 ELECTRIC FLUX DENSITY (D)

In free space, the electric field at any distance r from a point charge is expressed as

$$\overline{E} = \frac{Q}{4\pi\varepsilon_0 r^2} U_r \text{ N/C} \quad ...(3.1)$$

If Q is considered at center of sphere then 'r' is its radius where the field is required. So equation (3.1) can be written as

$$\varepsilon_0 E = \frac{Q}{4\pi r^2} \quad ...(3.2)$$

Dimensionally it is charge/Area which is to be defined as *surface charge density* ($4\pi r^2$ is area of spherical surface). $\varepsilon_0 E$ has the dimensions of charge density with E as vector in free space, $\varepsilon_0 E$ is shown by a symbol 'D' called *electric flux density* and is a vector quantity

$$\overline{D} = \varepsilon_0 E \quad ...(3.3)$$

The electric flux density 'D' can also be called as 'displacement flux density' or 'displacement density'. Both D and E have the same direction.

Referring to Fig. 3.1, the electric flux density is in the radial direction and has a value of

$$\overline{D}\,|_{r=a} = \frac{Q}{4\pi x^2} U_r \text{ (inner sphere)}$$

$$\overline{D}\,|_{r=b} = \frac{Q}{4\pi y^2} U_r \text{ (outer sphere)}$$

and at a radial distance 'r', where $x \le r \le y$

$$\overline{D} = \frac{Q}{4\pi r^2} U_r$$

The inner sphere resembles a point charge in the limit if it sinks and becomes smaller but the electric flux density at a point 'r' meters from the point charge is given by

$$\overline{D} = \frac{Q}{4\pi r^2} U_r \quad ...(3.4)$$

The electric flux density for a general volume charge distribution in free space is given by

$$\overline{E} = \int_{Vol} \frac{\rho_v \, dv}{4\pi\varepsilon_0 R^2} U_r \quad \text{(free space only)}$$

But $D = \varepsilon_0 E$ in free space

$$\therefore \quad \overline{D} = \varepsilon_0 \int_{Vol} \frac{\rho_v \, dv}{4\pi\varepsilon_0 R^2} U_r$$

(or) $$\overline{E} = \int_{Vol} \frac{\rho_v \, dv}{4\pi\varepsilon R^2} U_r \text{ C/m}^2 \quad (\because \varepsilon = \varepsilon_0\varepsilon_r) \quad ...(3.5)$$

Example 3.1: *Find D (in Cartesian coordinate) at P (8, 10, – 12) caused by (a) A point charge of 40 mC at the origin (b) A uniform line change ρ_L= 50 μC/m on the z-axis (c) A uniform surface charge density ρ_s = 59.2 mC/m² on plane x = 9.*

Solution:

(*a*)
$$\overline{D} = \varepsilon_0 \overline{E}$$
$$= \frac{\varepsilon_0 \times 30 \times 10^{-3}}{4\pi\varepsilon_0} \left\{ \frac{(8-0)U_x + (10-0)U_y + (-12-0)U_z}{[(8)^2 + (10)^2 + (-12)^2]^{3/2}} \right\}$$
$$= [4.71U_x + 5.89U_y - 7.07U_z]\ \mu\text{C/m}^2.$$

(*b*) Equations of *z*-axis are $x = 0$ and $y = 0$

$$\overline{D} = \varepsilon_0 \overline{E} = \varepsilon_0 \left[\frac{\rho_L}{2\pi\varepsilon_0 r} U_{12} \right] = \frac{\rho_L}{2\pi r} U_{12}$$

Point two is that at which electric flux density is desired = (8, 10, – 12). Point one is the foot of perpendicular dropped from point two on line charge = (0, 0, – 12).

$$\therefore \quad \overline{R}_{12} = (8-0)U_x + (10-0)U_y = 8U_x + 10U_y$$
$$= |\overline{R}_{12}| = \sqrt{(8)^2 + (10)^2} = 12.81$$
$$U_{12} = \frac{\overline{R}_{12}}{|\overline{R}_{12}|} = \frac{8U_x + 10U_y}{12.81}$$
$$\overline{D} = \frac{40 \times 10^{-6}}{2\pi\sqrt{69}} \left[\frac{8U_x + 10U_y}{12.81} \right] = 0.39U_x + 0.485U_y\ \mu\text{C/m}^2$$

(*c*)
$$\overline{D} = \varepsilon_0 \overline{E} = \varepsilon_0 \left[\frac{\rho_s}{2\varepsilon_0} U_N \right] = \frac{\rho_s}{2} U_N$$

The surface charge density lies on plane $x = 9$. The *x*-coordinate of point '*P*' is 8 which is less than 9. Hence $U_N = -U_x$.

$$D = \frac{59.2 \times 10^{-6}}{2}(-U_x) = -29.6\ U_x\ \mu\text{C/m}^2.$$

Example 3.2: (*a*) *Find the maximum charge that can be held on the isolated sphere 2 m diameter, the sphere being in air with dielectric strength 40 kV/cm.(b) What would be the maximum charge if this sphere were in oil of* $\varepsilon_r = 3.5$ *and dielectric strength of 75 kV/cm?*

Solution:

(*a*) In air
$$\varepsilon = \varepsilon_0 \quad |E| = \frac{Q}{4\pi\varepsilon r^2}$$
$$\therefore \quad Q = 4\pi\varepsilon r^2 |E| = \frac{1}{9 \times 10^9}(1.0 \times 10^4)\left(\frac{4 \times 10^4}{1}\right)$$
$$= \frac{1}{9 \times 10^9} \times 1.0 \times 10^4 \times 4 \times 10^4 = 0.0444 \text{ C}$$

(*b*) When in oil
$$\varepsilon = \varepsilon_0 \varepsilon_r$$
$$Q = 4\pi\varepsilon_0 \varepsilon_r r^2 |E| = \frac{1 \times 3.5}{9 \times 10^9} \times \left(\frac{2 \times 10^2}{2}\right)^2 \times 7.5 \times 10^4$$
$$= \frac{3.5 \times 10^{-9}}{9 \times 1} \times 10^4 \times 7.5 \times 10^4 = 0.29166 \text{ C}.$$

3.4 GAUSS'S LAW

Let a distribution of charge be visualized as a collection of point charges and surrounded by a closed surface bearing any shape as shown in Fig. 3.2. Let it also be assumed that the closed surface is of any material but more importantly it is a closed surface. If the total charge is assumed to be Q, then Q coulombs of electric flux will pass through the enclosing surface. At each and every point on surface the electric flux density vector D will have some value which is given by D_s, where the subscript '*s*' just shows that D must be evaluated at the surface and D_s will in general vary in magnitude and direction from one point on surface to another.

Let an incremental element on the surface be considered and the nature of the incremental element on surface be observed. An incremental element of area Δs is only a portion of a plane surface and the complete description of this surface element requires not only a statement of its magnitude Δs but also its orientation in space. In other words, it can be said that the incremental surface element is a vector quantity. The only unique direction which may be associated with Δs is the direction of normal to that plane which is tangent to surface at the point.

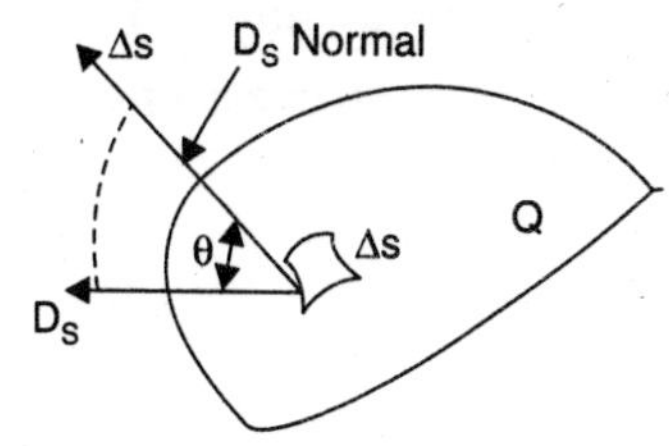

Fig. 3.2

Gauss's law states that "the electric flux passing through a closed surface is equal to the total charge enclosed by that surface".

Mathematically it can be shown as

$$\Delta\psi = \text{flux crossing } \Delta s = D_{s\,\text{normal}}\,\Delta s$$

where, $$D_{s\,normal} = D_s \cos\theta$$

$$= D_s \cdot \Delta s \qquad ...(3.6)$$

So, total flux passing through closed surface is $\psi = \int d\psi$

$$= \oint_{\text{closed surface}} D_s \cdot ds \qquad ...(3.7)$$

$$\therefore \qquad \psi = \oint_S D_s \cdot ds = \text{charge enclosed} = Q$$

But the charge which is enclosed might be the combination of several point charges by the virtue of which, it can be written as

$$Q = \sum Q_n$$

Normally all the surfaces which are taken into consideration are considered in terms of volume. As a result the final equation can be written as

$$\oint D \cdot ds = \oint_{vol} \rho_v \, dv \qquad ...(3.8)$$

Example 3.3: *The spherical surfaces r = 2, 3 and 4 carry surface charge densities of 25, – 8, and 3 nC /m² respectively (a) How much electric flux leaves the surface r = 6? (b) Find D at P (1, – 2, 3)?*

Solution:

(*a*) Sphere $r = 6$ encloses the spherical surface $r = 2$, 3 and 4. Hence charge enclosed by spherical surface ($r = 6$) = Charge present on ($r = 2$) + charge present on($r = 3$) + charge present on ($r = 4$).

i.e.,
$$Q = 25 \times 4\pi\,(2)^2 - 8 \times 4\pi\,(3)^2 + 3 \times 4\pi\,(4)^2 = 954.56 \text{ nC}.$$

Hence flux leaving through surface ($r = 6$) = $\psi = \phi = 954.56$ nC.

(*b*) r coordinate of point $P = \sqrt{1^2 + (-2)^2 + (3)^2} = \sqrt{14} = 3.741$

Point P (1, – 2, 3) lies on spherical surface $r = 3.741$. This surface encloses the spherical surface $r = 2$ and $r = 3$.

$\therefore$ Charge enclosed by ($r = 3.741$) surface = Charge present at ($r = 2$) + charge present on ($r = 3$).

$$Q = 25 \times 4\pi\,(2)^2 - 8 \times 4\pi\,(3)^2 = 112\pi \text{ nC}.$$

$$\therefore \qquad \overline{D} \text{ at } P = \frac{\psi}{area} U_r - \frac{Q}{4\pi r^2} U_r$$

$$= \frac{112\pi}{4\pi(\sqrt{14})} U_r = 2\, U_r \text{ nC/m}^2.$$

Problem 3.4: *A non uniform surface charge density of* $\frac{5\rho}{(\rho^2 + 1)}$ *nC/m² lies in the plane z = 2, where as* $\rho < 5$; $\rho_s = 0$ *for* $\rho < 5$ *(a) How much electric flux leaves the circular region* $\rho < 5$, $z = 2$*? (b) How much electric flux crosses the z = 0 plane in the –* U_z *direction? (c) How much electric flux leaves the cylinder* $\rho = 3$ *is the* U_p *direction?*

Solution:

(*a*) To the plane $z = 2$, $\pm U_z$ is perpendicular

Since circular surface is given we will use cylinder coordinate system

$$\overline{d}s = \pm(\rho\, d\phi\, d\rho)\, U_z \text{ or } ds = \rho\, d\rho\, d\phi$$

$$\therefore \qquad \psi = Q = \int dQ = \int \rho_s\, ds = \int_0^5 \int_0^{2\pi} \frac{5\rho}{\rho^2 + 1} \rho\, d\rho\, d\phi$$

$$= 5\left[\rho - \tan^{-1}\rho\right]_0^5 [\phi]_0^{2\pi}$$

$$\therefore \qquad \psi = 5 \times 3.6266 \times 2\pi = 13.933 \text{ nC}.$$

(*b*) Half of the above flux leaves the surface $z = 2$, in + U_z direction whereas as remaining half of flux is $-U_z$ direction, Hence the flux crossing the $z = 0$ plane in $-U_z$ direction will be $\frac{113.933}{2} = 56.966$ nC.

(*c*)
$$\psi = Q = \int dQ = \int \rho_s\, ds$$

$$= \int_0^3 \int_0^{2\pi} \frac{5\rho}{\rho^2 + 1} \rho\, d\rho\, d\phi$$

$$= 5\left[\rho - \tan^{-1}\rho\right]_0^3 [\phi]_0^{2\pi} = 55.008 \text{ nC}.$$

Example 3.5: *A charge distribution with spherical symmetry has density*

$$\rho_v = \begin{cases} \dfrac{2\rho_0 r}{R}, & 0 \le r \le R \\ 0, & r > R \end{cases}. \text{ Determine } \overline{E} \text{ everywhere.}$$

Solution: Applying Gauss's law *i.e.*, $\psi = Q$

$$\oint D \cdot ds = Q = \int \rho_v \, dv$$

(*a*) For $r < R$

$$D\, 4\pi r^2 = Q = \int_0^r \int_0^\pi \int_0^{2\pi} \rho_v\, r^2 \sin\theta \, d\phi\, d\theta\, dr$$

$$= \int_0^r \left[-\cos\theta\right]_0^\pi 2\pi \rho_v\, r^2\, dr = \int_0^r 4\pi r^2 \frac{2\rho_0 r}{R} dr$$

$$= \frac{8\pi \rho_0}{R} \cdot \frac{r^4}{4} = \frac{8\rho_0 \pi r^4}{R}$$

$\therefore$

$$\overline{D} = \frac{2\rho_0 r^2}{R} U_r$$

and

$$\overline{E} = \frac{\overline{D}}{\varepsilon_0} = \frac{2\rho_0 r^2}{\varepsilon_0 R} U_r$$

(*b*) For $r > R$

$$D\, 4\pi r^2 = Q = \int_0^r \int_0^\pi \int_0^{2\pi} \rho_v\, r^2 \sin\theta \, d\phi\, d\theta\, dr$$

$$= \int_0^R \frac{2\rho_0 r}{R} 4\pi r^2\, dr + 0 = 2\pi \rho_0 R^3$$

$\therefore$

$$\overline{D} = \frac{\rho_0 R^3}{2r^2} U_r$$

and

$$\overline{E} = \frac{\overline{D}}{\varepsilon_0} = \frac{\rho_0 R^3}{2\varepsilon_0 r^2} \overline{U}_r$$

Example 3.6: *Determine the variation of electric field intensity from point to point for a two concentric special shells of radii R_1 and R_2 with charges of values $3Q_1$ and $3Q_2$.*

Solution: Case (*a*): Construct a Gaussian spherical surface with $r < R_1$

Apply Gauss's law , $\psi = Q$

Charge enclosed is zero since charge lies only on the surface of the shell. Now from the relation that $D\int_S ds$ = charge enclosed by the surface.

$\therefore$

$$D \cdot 4\pi r^2 = 0$$

$$D = 0;\ E = 0$$

Therefore, electric field inside spherical shell in zero.

Case (*b*): Construct a spherical surface of radius 'r' lying between R_1 and R_2.

Apply Gauss's law, $\psi = Q$

Therefore, $D \cdot 4\pi r^2 = 3Q_1$

or
$$D = \frac{3Q_1}{4\pi r^2}$$

and
$$E = \frac{D}{\varepsilon} = \frac{3Q_1}{4\pi \varepsilon r^2}$$

Case (*c*): Apply Gauss's law to get Electric field outside both the spheres. Now the charge enclosed is the sum of charges $3Q_1$ and $3Q_2$ since both the spheres are enclosed.

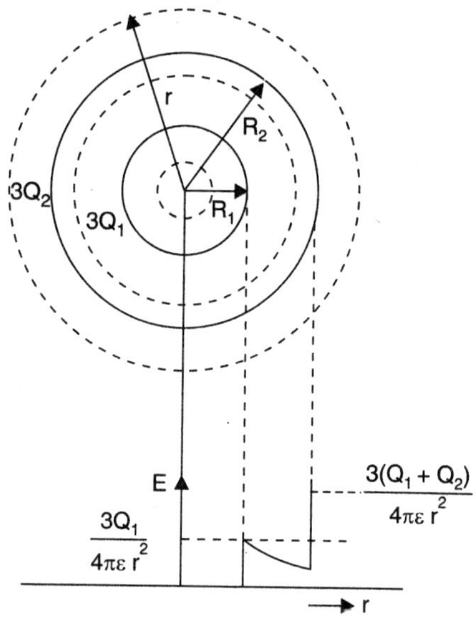

Fig. 3.3

$$D \int_S ds = 3Q_1 + 3Q_2$$

$$D \cdot 4\pi r^2 = 3Q_1 + 3Q_2$$

$$D = \frac{3Q_1 + 3Q_2}{4\pi r^2}$$

and
$$E = \frac{3Q_1 + 3Q_2}{4\pi \varepsilon r^2}$$

Example 3.7: *Given an electric flux density $D = (3\, u_x + 2\, u_y) C/m^2$, determine the net flux crossing the surface of a cube 2 m on an edge centered at the origin. (The edges of the cube are parallel to the coordinate axis).*

Solution:
$$\psi = \oint D \cdot ds = \int_{x=1} (3u_x + 2u_y) \cdot (ds\, u_z) + \int_{x=-1} (-3u_x + 2u_y) \cdot (-ds\, u_x)$$

$$+ \int_{y=1} (3u_x + 2u_y) \cdot (ds\, u_y) + \int_{y=-1} (3u_x + 2u_y) \cdot (-ds\, u_y)$$

$$+ \int_{z=1} (3u_x + 2u_y) (ds\, u_z) + \int_{z=-1} (3u_x + 2u_y) \cdot (-ds\, u_z)$$

$$= 3\int_{x=1} ds + 3\int_{x=-1} ds + 2\int_{y=1} ds - 2\int_{y=-1} ds + 0 + 0$$

$$= (3 + 3 + 2 - 2)\,(2)^2 = 24 \text{ C.}$$

3.5 APPLICATIONS OF GAUSS'S LAW

We have discussed and derived the expression for Gauss's law in the previous section. Explaining the application of Gauss's law in a nutshell we can summarize them and find the charge distribution inside any closed surface. The solution is easy if we are able to choose a closed surface called as **Gaussian surface** which satisfies these conditions:

(*i*) D_s is everywhere either normal (or) tangential to closed surface so that $D_s \cdot ds$ becomes either $D_s\, ds$ (or) zero respectively.

(*ii*) The portion of closed surface on which $D_s \cdot ds$ is not zero, D_s = constant is considered.

(*iii*) Gauss's law depends on symmetry; if the symmetry can't exist then we can't use Gauss's law.

This allows us to replace the dot product with product of scalars D_s and ds and thus bringing D_s outside the integral sign makes our job of computation easier.

3.5.1 Electric Field Due to a Point Charge

Consider a point charge Q at the origin of spherical coordinates shown in Fig. 3.4. It is easy if a spherical surface containing P is chosen which will satisfy symmetry conditions to determine E at any point 'P'. $\overline{D}$ has the same value at all points on the surface since D is everywhere normal to the Gaussian surface.

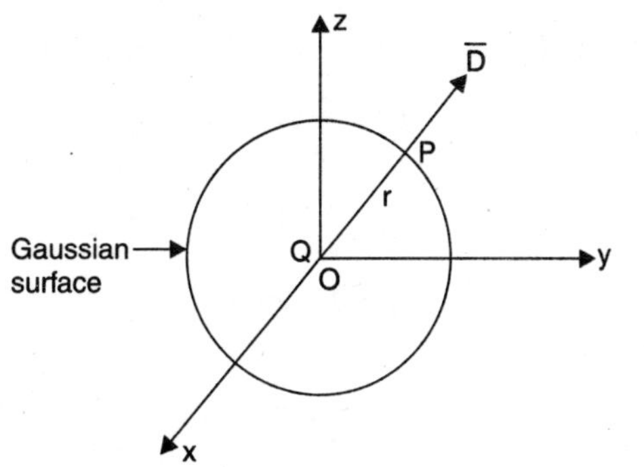

Fig. 3.4 Gaussian surface around a point charge

$$\overline{D} = D \cdot \overline{U}_r \quad ...(3.9)$$

Applying Gauss law ($\psi = Q$ enclosed)

$$Q = \oint_S \overline{D} \cdot \overline{ds} = \oint_{Sph} D \cdot ds$$

$$= D \oint_{Sph} ds = D \int_{\phi=0}^{2\pi} \int_{\theta=0}^{\pi} r^2 \sin\theta \, d\theta \, d\phi \qquad [\because \; ds = (r\, d\theta)(r \sin\theta \, d\phi)]$$

$$= D\, r^2 \int_{\theta=0}^{\pi} \sin\theta \, [\phi]_0^{2\pi} \, d\theta$$

$$= D\, r^2\, 2\pi \, [-\cos\theta]_0^{\pi} = 4\pi r^2 D$$

And hence $D = \dfrac{Q}{4\pi r^2}$

Since r may have any value and $\overline{D}$ is directed radically outward

$$\overline{D} = \frac{Q}{4\pi r^2} U_r \quad ...(3.10)$$

and

$$\overline{E} = \frac{Q}{4\pi\varepsilon_0 r^2} U_r \quad ...(3.11)$$

3.5.2 Electric Field Due to a Uniformly Charged Conductor

Let a conductor of infinite length be assumed which carries a uniform charge ρ_L C/m and lies along z-axis. On the basis of Gauss's law, in order to obtain field intensity due to infinite line charge, imagine a Gaussian surface around the conductor which is symmetrical about it. Such a surface can be seen as a circular cylindrical surface with top and bottom surfaces as shown in Fig. 3.5.

The flux leaving the top and bottom surfaces is zero since $\overline{D}$ and $\overline{ds}$ are perpendicular *i.e.*, $\overline{D} \cdot \overline{ds} = 0$. Therefore, the total flux leaving the Gaussian surfaces is equal to the flux leaving the cylindrical surface only.

Let l be the length of cylinder and ρ_l be the line charge density

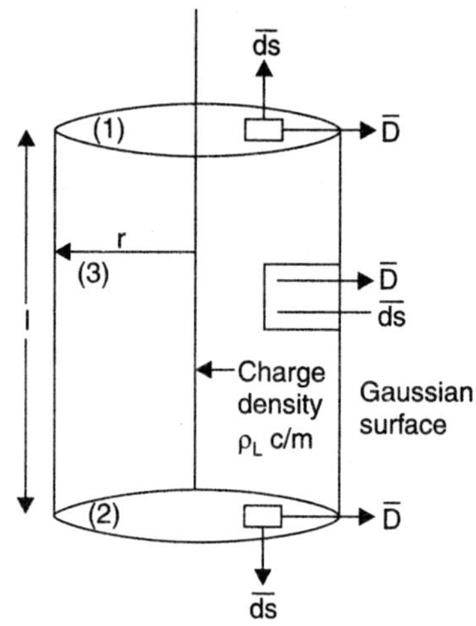

Fig. 3.5

$\therefore$ Total charge $Q = \rho_L l$ coulombs

By Gauss's law, it can be written as

$$Q = \oint_S \overline{D} \cdot \overline{ds}$$

$$Q = \oint_S \overline{D}\, ds = D \oint_S ds$$

(or) $$Q = D\,(2\pi r l) \qquad ...(3.12)$$

$$\therefore \quad D = \frac{\rho_L\, l}{2\pi\, r l}$$

$$D = \frac{\rho_L}{2\pi r} \qquad ...(3.13)$$

We have $\overline{D} = \varepsilon_0\, \overline{E}$ (or) $\overline{E} = \dfrac{\overline{D}}{\varepsilon_0}$

Therefore Electric field intensity at a distance 'r' from the charge line is given by

$$\overline{E} = \frac{\rho_L}{2\pi\, \varepsilon_0\, r}\, U_r \qquad ...(3.14)$$

So, it can be seen that the electric field due to a line charge is inversely proportional to the distance r. Therefore, it can be stated as electric field intensity decreases as the distance 'r' increases.

Example 3.8: *Two infinite line charges, one having line charge density ρ_L and the other $-\rho_L$ are separated by a distance l. Determine the electrical field intensity at point P of Fig.3.6 using Gauss law.*

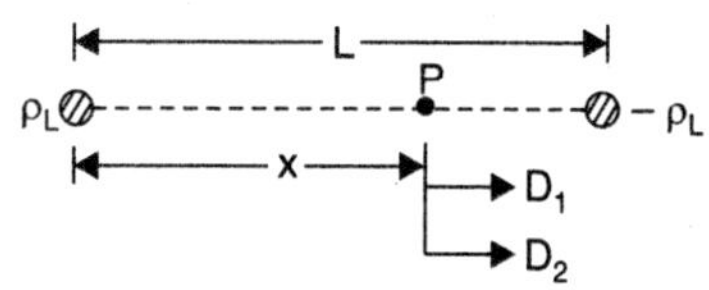

Fig. 3.6

Solution: By Gauss law, for cylindrical surface of unit height going through point P, $2\pi x D_1 = \rho_L$ and $-2\pi x D_1 = -\rho_L$.

$$\therefore \quad D_1 = \frac{\rho_L}{2\pi} \text{ and } D_2 = \frac{\rho_L}{2 \times \pi(1-x)}$$

$$D = D_1 + D_2 = \frac{\rho_L}{2 \times \pi}\left(\frac{1}{x} + \frac{1}{1-x}\right) = \frac{\rho_L}{2 \times \pi \times x(1-x)}$$

Field intensity, $E = \dfrac{\rho_L}{2\varepsilon_0 \times \pi \times x(1-x)}$

3.5.3 Electric Field Due to an Infinite Plane Sheet of Charge

Consider a thin, non-conducting, infinite plane sheet of charge shown in Fig. 3.7. Let the surface charge density be ρ_s C/m^2. Let P_1 be a point at a distance r from the sheet and it is required to calculate E at P_1. For this purpose let a point P_2 which is symmetrical with P_1 on the other side of sheet be considered and a cylindrical Gaussian surface with one end on one side and other end on the other side be constructed. Let A be the area of cross-section of each plane end. By symmetry E is everywhere perpendicular to plane.

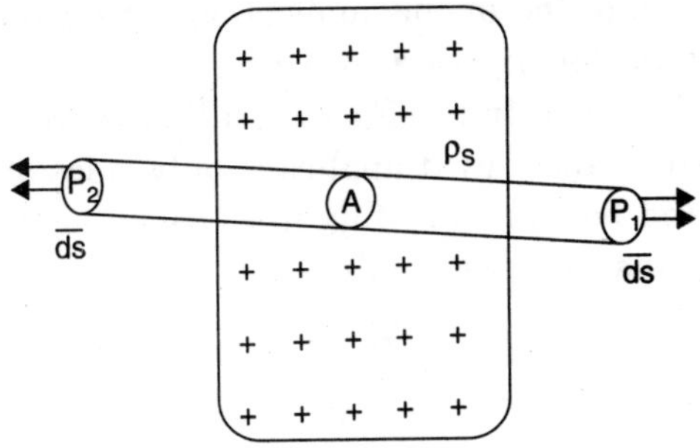

Fig. 3.7

Since $\overline{E}$ and $\overline{ds}$ are at right angles everywhere on curved surface, the flux through the curved surfaces of Gaussian cylinder is zero. As $\overline{D}$ and $\overline{ds}$ are parallel to each other at end caps of Gaussian surface and hence they contribute to electric flux. The electric flux through these end caps is given by

$$\psi = \oint \overline{D} \cdot \overline{ds} + \oint \overline{D} \cdot \overline{ds}$$

$$= \oint D\, ds + \oint D\, ds \qquad [\because \theta = 0°]$$

$$= DA + DA = 2DA$$

$\therefore$ The total flux through the Gaussian cylinder

$$\psi = 2DA \qquad ...(3.15)$$

Net charge enclosed by Gaussian cylinder

$$Q = \rho_s A$$

By Gauss's law,

$$2DA = \rho_s A$$

$$D = \frac{\rho_s}{2} \qquad ...(3.16)$$

And

$$E = \frac{D}{\varepsilon_0} = \frac{\rho_s}{2\varepsilon_0}$$

$\therefore$

$$E = \frac{\rho_s}{2\varepsilon_0} \qquad ...(3.17)$$

From the above equation it can be seen that magnitude of electric field is independent of distance from the sheet. Thus E is same for all points on each side of plane.

Example 3.9: *Charge lies in the $z = -4$ m plane in the form of a square sheet defined by $-2 \le x \le 2$ m, $-2 \le y \le 2$ m with charge density $\rho_s = 2(x^2 + y^2 + a)$ n C/m^2. Find E at the origin.*

Solution:

$$R = -xu_x - yu_y + 4u_z$$

$$d\phi = \rho_s\, dx\, dy = 2(x^2 + y^2 + a)^{3/2} \times 10^{-9}\, dx\, dy$$

$\therefore$

$$dE = \frac{2(s^2 + y^2 + a)^{3/2} \times 10^{-9}\, dx\, dy}{4\pi\varepsilon_0 (x^2 + y^2 + 16)} \times \frac{(-xu_y - yu_y + 4u_z)}{\sqrt{x^2 + y^2 + 16}}$$

Due to symmetry, only the z-component of E exists.

$$\therefore \qquad E = \int_{-2}^{2}\int_{-2}^{2} \frac{8\times10^{-9}\,dx\,dy}{4\pi\varepsilon_0}\,u_z = 1152\,u_z \text{ V/m.}$$

3.5.4 Electric Flux Density in a Cylinder Due to Line Charge

One of the major applications of Gauss's law is the electric flux density due to a line charge in a cylindrical shape container. In the Fig. 3.8, if line charge is considered. As per the Gauss law it applies to only sides of figure but not to top and bottom since flux lines come out radically. By applying Gauss law, we obtain

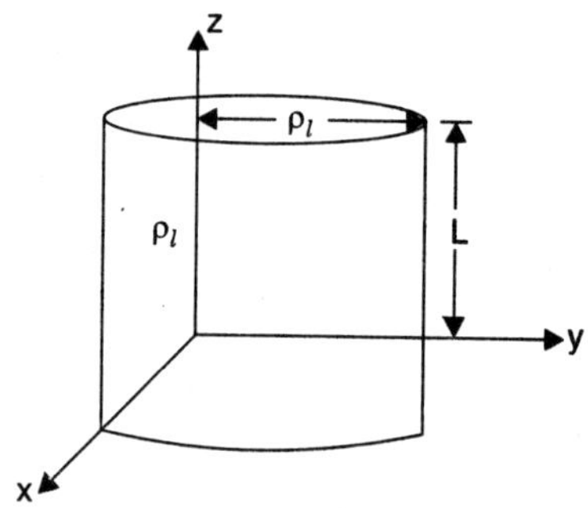

Fig. 3.8

$$Q = \oint D_s\cdot ds = D_s \int_{sides} ds + 0\int_{top} + 0\int_{bottom} ds$$

$$= D_s \int_{z=0}^{l}\int_{\phi=0}^{2\pi} \rho\, d\phi\, dz = D_s\, 2\pi\,\rho_L$$

$$\therefore\ D_s = \frac{Q}{2\pi\rho_l} \qquad \text{...(3.18)}$$

Example 3.10: *Two identical uniform line charges lie along the x and y-axis with charge density* $\rho_L = 30\ \mu C/m$. *Obtain D at (3, 3, 3).*

Solution: The distance from the observation point to either line charge is $3\sqrt{2}$ m, considering first, the line charge on the x-axis

$$D_1 = \frac{\rho_l}{2\pi r_1}u_{r_1} = \frac{30\times10^{-6}}{2\pi\,(3\sqrt{2})}\left(\frac{u_y+u_z}{\sqrt{2}}\right)$$

Considering the y-axis line charge

$$D_2 = \frac{\rho_l}{2\pi r_2}u_{r_2} = \frac{30\times10^{-6}}{2\pi\,(3\sqrt{2})}\left(\frac{u_x+u_z}{\sqrt{2}}\right)$$

The total density is the vector sum

$$D = \frac{30}{2\pi\,(3\sqrt{2})}\left(\frac{u_x+u_y+2u_z}{\sqrt{2}}\right) = (1.126)\left(\frac{u_x+u_y+2u_z}{\sqrt{2}}\right)\ \mu\text{C/m}^2.$$

3.5.5 Differential Volume Element

Let us now apply the methods of Gauss law to a different type of problem which does not have any symmetry. Just by our sense it might seen that our mode is wrong since without symmetry a simple Gaussian surface can't be chosen such that the normal component of D is constant or zero every where on the surface. The integral can't be evaluated without such a surface. The only way to prevent these difficulties is to choose a very small closed surface that D is almost constant over the surface and the small charge in D can be adequately represented by using first two terms of Taylor's series expansion for D.

Let us consider any point P as shown in Fig. 3.9 located by Cartesian coordinate system. Let the value of D at the point P be expressed in Cartesian components $D_0 = D_{x_0} U_x + U_{y_0} U_y + D_{z_0} U_z$. Let a closed surface be chosen, a small rectangular box centered at P, having side of length Δx, Δy and Δz

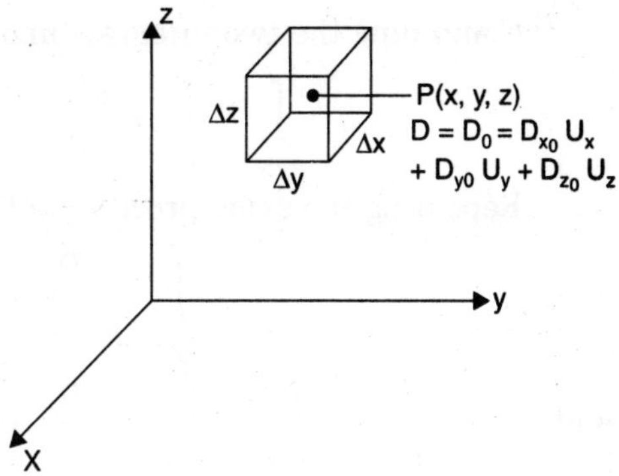

Fig. 3.9

Applying Gauss law we get

$$\oint_S D \cdot ds \qquad \text{...(3.19)}$$

Since integral has to be evaluated over closed surface, the integral should be broken into six integrals like

$$\oint D \cdot ds = \int_{front} + \int_{back} + \int_{left} + \int_{right} + \int_{top} + \int_{bottom}$$

Let us choose the first one in detail. Since the surface element is very small, D is essentially constant and as a result

$$\int_{front} = D_{front} \cdot \Delta s_{front}$$

$$= D_{front} \cdot \Delta y, \Delta z\, U_x$$

$$= D_{x\,front}\, \Delta y, \Delta z$$

The front face is at a distance of $\frac{\Delta x}{2}$ from P and hence

$$D_{x\,front} = D_{x_0} + \frac{\Delta x}{2} \text{ (rate of charge of } D_x \text{ with } x)$$

$$= D_{x_0} + \frac{\Delta x}{2}\frac{\partial D_x}{\partial x} \qquad \text{...(3.20)}$$

Where D_{x_0} is value of D_x at P and here a partial derivative must be used to express the rate of charge of D_x with x, since D_x in general also varies with y and z. So here partial derivative must be used to express rate of charge of D_x with x.

$$\int_{front} = \left(D_{x_0} + \frac{\Delta x}{2}\frac{\partial D_x}{\partial x}\right)\Delta y\, \Delta z \qquad \text{...(3.21)}$$

Consider integral over back surface

$$\int_{back} = D_{back} \cdot \Delta s_{back}$$

$$= D_{back} \cdot (-\Delta y\, \Delta z\, U_x)$$

$$= -D_{x\,back}\, \Delta y\, \Delta z$$

$$\therefore \quad D_{x\,back} = \left(-D_{x_0} + \frac{\Delta x}{2}\frac{\partial D_x}{\partial x}\right)\Delta y\, \Delta z \qquad \text{...(3.22)}$$

Combining the two integrals in equations (3.21) and (3.22)

$$\int_{front} + \int_{back} = \frac{\partial D_x}{\partial x} \Delta x \, \Delta y \, \Delta z \qquad ...(3.23)$$

Repeating the same process we find that

$$\int_{right} + \int_{left} = \frac{\partial D_y}{\partial y} \Delta x \, \Delta y \, \Delta z$$

and

$$\int_{top} + \int_{bottom} = \frac{\partial D_z}{\partial z} \Delta x \, \Delta y \, \Delta z$$

Summing up all the results it can be written as

$$\oint_S D \cdot ds = \left(\frac{\partial D_x}{\partial x} + \frac{\partial D_y}{\partial y} + \frac{\partial D_z}{\partial z} \right) \Delta x \, \Delta y \, \Delta z \qquad ...(3.24)$$

(or)

$$\oint_S D \cdot ds = Q = \left(\frac{\partial D_x}{\partial x} + \frac{\partial D_y}{\partial y} + \frac{\partial D_z}{\partial z} \right) \Delta V \qquad ...(3.25)$$

So, charge enclosed in volume ΔV.

$$\oint_S D \cdot ds = \left(\frac{\partial D_x}{\partial x} + \frac{\partial D_y}{\partial y} + \frac{\partial D_z}{\partial z} \right) (\text{volume } \Delta V) \qquad ...(3.26)$$

3.6 MAXWELL'S FIRST EQUATION

It is required to combine the information in last two sections and to provide an interpretation of the divergence operation as it relates to electric flux density. The equation (3.26) can be written as

$$\left(\frac{\partial D_x}{\partial x} + \frac{\partial D_y}{\partial y} + \frac{\partial D_z}{\partial z} \right) = \frac{\oint D \cdot ds}{\Delta V} = \frac{Q}{\Delta V} \qquad ...(3.27)$$

(or) as a limit

$$\left(\frac{\partial D_x}{\partial x} + \frac{\partial D_y}{\partial y} + \frac{\partial D_z}{\partial z} \right) = \lim_{\Delta V \to 0} \frac{\oint_S D \cdot ds}{\Delta V} = \lim_{\Delta V \to 0} \frac{Q}{\Delta V}$$

It is evident that the last term in the above expression is the volume change density ρ_v and hence

$$\left(\frac{\partial D_x}{\partial x} + \frac{\partial D_y}{\partial y} + \frac{\partial D_y}{\partial y} \right) = \lim_{\Delta V \to 0} \frac{\oint_S D \cdot ds}{\Delta V} = \rho_v \qquad ...(3.28)$$

Now we know that

$$\text{Div } D = \frac{\partial D_x}{\partial x} + \frac{\partial D_y}{\partial y} + \frac{\partial D_y}{\partial y}$$

From the equation (3.28)

$$\text{div } D = \lim_{\Delta V \to 0} \frac{\oint_S D \cdot ds}{\Delta V} \quad \ldots(3.29)$$

and

$$\text{div } D = \frac{\partial D_x}{\partial x} + \frac{\partial D_y}{\partial y} + \frac{\partial D_y}{\partial y} \quad \ldots(3.30)$$

$\therefore$

$$\text{div } D = \rho_v \quad \ldots(3.31)$$

The first equation *i.e.*, equation (3.29) is definition of divergence, the second equation (3.30) is result of applying the definition to a differential volume element in Cartesian coordinates giving us an equation by which divergence of vector expressed in Cartesian coordinates may be evaluated and third is merely written using new term div D.

This can be derived even by using Gauss law

i.e.,

$$\oint_S D \cdot ds = Q \quad \ldots(3.32)$$

per unit volume,

$$\frac{\oint_S D \cdot ds}{\Delta V} = \frac{Q}{\Delta V}$$

As volume approaches to zero

$$\lim_{\Delta V \to 0} \frac{\oint_S D \cdot ds}{\Delta V} = \lim_{\Delta V \to 0} \frac{Q}{\Delta V} \quad \ldots(3.33)$$

Div D can be taken on L.H.S. and volume change density on R.H.S.

$$\text{div } D = \rho_v \quad \ldots(3.34)$$

Example 3.11: *Find the volume charge density that is associated with each of the following fields. (a)* $\overline{D} = (2xy^2\overline{U}_x + 2\,yx^2\overline{U}_y + 3z\overline{U}_z) C/m^2$ *(b)* $D = (2\rho z^2 \sin^2\phi\, \overline{U}_\rho + \rho z^2 \sin\phi \cos\phi \overline{U}_\phi + 3\rho^2 z \sin^2\phi \overline{U}_z)\, C/m^2$ *(c)* $4\overline{U}_r\, C/m^2$.

Solution:

(*a*)

$$\rho_v = \nabla \cdot \overline{D} = \frac{\partial}{\partial x}(2xy^2) + \frac{\partial}{\partial y}(2yx^2) + \frac{\partial}{\partial y}(3z)$$

$$= 2y^2 + 2x^2 + 3 \text{ C/m}^2.$$

(*b*)

$$\rho_v = \nabla \cdot \overline{D} = \frac{1}{\rho}\frac{\partial}{\partial \rho}(2\rho z^2 \sin^2\phi) + \frac{1}{\rho}\frac{\partial}{\partial \phi}(\rho z^2 \sin\phi \cos\phi) + \frac{\partial}{\partial z}(3\rho^2 z \sin^2\phi)$$

$$= 4z^2 \sin^2\phi + z^2 \cos^2\phi + 3\rho^2 \sin^2\phi$$

$$= 2z^2 + 3\rho^2 \sin^2\phi \text{ C/m}^3.$$

(*c*)

$$\rho_v = \nabla \cdot \overline{D} = \frac{1}{r^2}\frac{\partial}{\partial r}(r^2 \times 4)$$

$$= \frac{8}{r} \text{ C/m}^3.$$

Example 3.12: *Given that D = z ρ cos² ϕ U_z C/m², calculate the charge density at* $\left(2, \frac{\pi}{4}, 4\right)$ *and total charge enclosed by cylinder of radius with* $-2 \le z \le 2m$.

Solution: $$\rho_v = \nabla \cdot \bar{D} = \frac{\partial D_t}{\partial z} = \rho \cos^2 \phi$$

At $\left(2, \frac{\pi}{4}, 4\right)$, $\rho_v = 2 \cdot \cos^2\left(\frac{\pi}{4}\right) = 0.999 \text{ C/m}^2$

Applying Gauss's law

$$Q = \psi = \oint \bar{D} \cdot \bar{d}s = \int_s \bar{D} \cdot \bar{d}s + \int_t \bar{D} \cdot \bar{d}s + \int_b \bar{D} \cdot \bar{d}s$$

$$= \psi_s + \psi_t + \psi_b$$

Where, ψ_s, ψ_t, ψ_b are the fluxes through the sides, top surface and bottom surface of cylinder respectively. Since $\bar{D}$ does not have components along $U_\rho \cdot \psi_s = 0$

For ψ_t, $d_s = \rho \, d\phi \, d\rho \, U_z$,

So,

$$\psi_t = \int_{\rho=0}^{2} \int_{\phi=0}^{2\pi} 2\rho \cos^2 \phi \rho \, d\phi \, d\rho \Big|_{z=2}$$

$$= 2 \int_0^2 \rho^2 \, d\rho \int_0^{2\pi} \cos^2 \phi \, d\phi$$

$$= 2\left(\frac{8}{3}\right)(\pi) = \frac{16\pi}{3}$$

And for ψ_b, $\overline{ds} = -\rho \, d\phi \, U_z$,

So,

$$\psi_b = -\int_{\rho=0}^{2} \int_{\phi=0}^{2\pi} 2\rho \cos^2 \phi \, \rho \, d\phi \, d\rho \Big|_{z=2}$$

$$= 2 \int_0^2 \rho^2 \, d\rho \int_0^{2\pi} \cos^2 \phi \, d\phi$$

$$= 2\left(\frac{8}{3}\right)\pi = \frac{16\pi}{3}$$

Thus, $$Q = \psi = 0 + \frac{16\pi}{3} + \frac{16\pi}{3} = \frac{32\pi}{3}$$

Example 3.13: *If D = 2e^{-x} sin y* U_x *+ 3e^{-x} cos y* U_y *+ 7z* U_z*. Find div D at point P (0.3, 0.5, 0.6).*

Solution: Given $D = 2e^{-x} \sin y \, U_x + 3e^{-x} \cos y U_y + 7zU_z$

We know that, $$\text{div } D = \frac{\partial D_x}{\partial x} + \frac{\partial D_y}{\partial y} + \frac{\partial D_z}{\partial z}$$

We have

$$D_x = 2e^{-x} \sin y$$

$$D_y = 3e^{-x} \cos y$$

$$D_z = 7z$$

$$\frac{\partial D_x}{\partial x} = -2e^{-x} \sin y; \quad \frac{\partial D_y}{\partial y} = -3e^{-x} \sin y; \quad \frac{\partial D_z}{\partial z} = 7$$

$\therefore$ $\quad$ $\text{div } D = -2e^{-x} \sin y - 3e^{-x} \sin y + 7$

At point P (0.3, 0.5, 0.6)

$$\text{div } D = -2e^{-0.3} \sin(0.5) - 3e^{-0.3} \sin(0.5) + 7 = 6.967.$$

Example 3.14: *Calculate the divergence of $\overline{G}$ at P (2, – 3, 4) if $\overline{G}$ = (a) $2xU_x + 3yU_y + 4zU_z$ (b) $\rho\, U_\rho$ (c) $r\, U_r$ (d) $6r^2 \sin\theta\, U_r + 2\, r^2 \cos\theta\, U_\theta$.*

Solution:

(*a*) Using the expressions

$$\text{div}\,\overline{D} = \frac{\partial D_x}{\partial x} + \frac{\partial D_y}{\partial y} + \frac{\partial D_z}{\partial z}$$

We have $\frac{\partial}{\partial x}(2x) + \frac{\partial}{\partial y}(3y) + \frac{\partial}{\partial z}(4z) = 2 + 3 + 4 = 9$

(*b*) Using equation

$$\text{div } D = \frac{1}{\rho}\frac{\partial}{\partial \rho}(D_\rho) + \frac{1}{\rho}\frac{\partial D_\phi}{\partial \phi} + \frac{\partial D_z}{\partial z}$$

Divergence of $\quad \overline{G} = \frac{1}{\rho}\frac{\partial}{\partial \rho}(\rho \times \rho)$

(*c*) Using the equation

$$\text{div } D = \frac{1}{\rho}\frac{\partial}{\partial \rho}(\rho\, D_\rho) + \frac{1}{\rho}\frac{\partial D_\phi}{\partial \phi} + \frac{\partial D_z}{\partial z} \text{ we have}$$

Divergence of $\quad \overline{G} = \frac{1}{r^2}\frac{\partial}{\partial r}(r^2 \times r)$

(*d*) Using the above equations as shown in (*c*) again we get div of

$$\overline{G} = \frac{1}{r^2}\frac{\partial}{\partial r}[r^2 (6r^2 \sin\theta)] + \frac{1}{r \sin\theta}\frac{\partial}{\partial \theta}[\sin\theta\,(2r^2 \cos\theta)]$$

$$= 24\, r \sin\theta + \frac{2r\cos^2\theta}{\sin\theta}$$

for point P, $\quad r = 5.385$ and $\theta = 42.03°$. Hence at point P

$$\text{div of } \overline{G} = 24\,(5.385) \sin(42.03)° + \frac{2(5.385)\cos(84.062°)}{\sin(42.03°)} = 8.193$$

3.7 VECTOR OPERATOR ∇ AND DIVERGENCE THEOREM

Here we define the del operator ∇ as a vector operator

$$\nabla = \frac{\partial}{\partial x} U_x + \frac{\partial}{\partial y} U_y + \frac{\partial}{\partial z} U_z \qquad ...(3.35)$$

Similar such scalar operators appear in several methods of solving differential equation where we often let D replace $\frac{\partial}{\partial x}$, D^2 replace $\frac{\partial^2}{\partial x^2}$ and so consider $\nabla \cdot D$ signifying.

$$\nabla \cdot D = \left(\frac{\partial}{\partial x}U_x + \frac{\partial}{\partial y}U_y + \frac{\partial}{\partial z}U_z\right) \cdot (D_x U_x + D_y U_y + D_z U_z) \qquad ...(3.36)$$

We consider dot product of unit vector, discarding the six zero terms and having left with

$$\nabla \cdot D = \frac{\partial}{\partial x}(D_x) + \frac{\partial}{\partial y}(D_y) + \frac{\partial}{\partial z}(D_z)$$

Where the parenthesis can be removed by differentiating

$$\nabla \cdot D = \frac{\partial D_x}{\partial x} + \frac{\partial D_y}{\partial y} + \frac{\partial D_z}{\partial z}$$

This can be seen as divergence of D, so we have

$$\therefore \qquad \text{div } D = \nabla \cdot D = \frac{\partial D_x}{\partial x} + \frac{\partial D_y}{\partial y} + \frac{\partial D_z}{\partial z} \qquad ...(3.37)$$

The use of $\nabla \cdot D$ in various cases is much more than that of div D, although both have their advantages. Solving with $\nabla \cdot D$ indeed allows us to obtain simply and quickly the correct partial derivatives whereas on the other hand div D is good reminder of physical interpretation to divergence.

If we consider D in cylindrical coordinates, then $\nabla \cdot D$ still indicates the divergence of D (or)

$$\nabla \cdot D = \frac{1}{\rho}\frac{\partial}{\partial \rho}(\rho D_\rho) + \frac{1}{\rho}\frac{\partial D_\phi}{\partial \phi} + \frac{\partial D_z}{\partial z} \qquad ...(3.38)$$

If we take the concept of divergence, it applies to any vector field for which the appropriate partial derivative exists. Starting from Gauss law,

$$\oint_S D \cdot ds = Q \quad \text{and} \quad \text{letting } Q = \int_{vol} \rho_v \, dv \qquad ...(3.39)$$

and then replacing ρ_v by it equal

$$\nabla \cdot D = \rho_v \qquad ...(3.40)$$

We have $\oint_S D \cdot ds = Q = \int_{vol} \rho_v \, dv = \int_{vol} \nabla \cdot D \, dv \qquad ...(3.41)$

Divergence theorem can be shown as $\oint_S D \cdot ds = \int_{vol} \nabla \cdot D \, dv \qquad ...(3.42)$

It can be stated as the integral of normal component of any vector field over a closed surface is equal to integral of divergence of that vector field throughout the volume enclosed by closed surface.

Example 3.15: *Find a numerical value for the divergence of $\overline{D}$ at the point indicated if (a) $\overline{D} = (30xy^2(z+1)\overline{U}_x + 10x^2y(z+1)\overline{U}_y + 20x^2y^2\overline{U}_z)$ C/m² at P_A (0.1, 0.2, 0.3) (b) $\overline{D} = (2\rho z \sin\phi\, \overline{U}_\rho + 3\rho z \cos\phi\, \overline{U}_\phi + 5\rho^2 \sin\phi\, \overline{U}_z)$ C/m² at $P_B\left(2, \frac{\pi}{4}, 3\right)$; (c) $\overline{D} = (\sin\theta\cos\phi\overline{U}_r + \cos\theta\cos\phi\,\overline{U}_\theta - \sin\phi\,\overline{U}_\phi)$ C/m² at $P_C\left(3, \theta=\frac{\pi}{2}, \phi=\frac{\pi}{3}\right)$.*

Solution:

(*a*)
$$\text{div } \overline{D} = \frac{\partial}{\partial x}[30xy^2(z+1)] + \frac{\partial}{\partial y}[10x^2y(z+1)] + \frac{\partial}{\partial z}[20x^2y^2]$$
$$= [30y^2(z+1) + 10x^2(z+1) + 0] \text{ C/m}^2$$

$$\therefore \quad \text{div } \overline{D} \text{ at } P_A = 30(0.20)^2(0.3+1) + 10(0.1)^2(0.3+1) = 1.69 \text{ C/m}^2$$

(*b*)
$$\text{div } \overline{D} = \frac{1}{\rho}\frac{\partial}{\partial\rho}(\rho \times 2\rho z \sin\phi) + \frac{1}{\rho}\frac{\partial}{\delta\phi}(3\rho z\cos\phi) + \frac{\partial}{\partial z}(5\rho^2\sin\phi)$$
$$= \frac{1}{\rho}(4\rho z\sin\phi) + \frac{1}{\rho}(-3\rho z\sin\phi) + 0$$

$$\therefore \quad \text{div}\overline{D} \text{ at } P_B = 4\times 3\times\sin\frac{\pi}{4} - 2\times 3\times \sin\frac{\pi}{4} = 4.242 \text{ C/m}^3$$

(*c*)
$$\text{div } \overline{D} = \frac{1}{r^2}\frac{\partial}{\partial\rho}(r^2 \times \sin\theta\cos\phi) + \frac{1}{r\sin\theta}\frac{\partial}{\partial\theta}(\sin\theta\times\cos\theta\cos\phi)$$
$$+ \frac{1}{r\sin\theta}\frac{\partial}{\partial\phi}(-\sin\phi) = \frac{2\sin\theta\cos\phi}{r} + \frac{\cos^2\theta\cos\phi}{r\sin\theta} - \frac{\cos\phi}{r\sin\theta}$$

$$\therefore \quad \text{div}\overline{D} \text{ at } P_c = \frac{2\sin\left(\frac{\pi}{2}\right)\cos\left(\frac{\pi}{3}\right)}{3} + \frac{\cos\left(\frac{\pi}{2}\right)\cos\left(\frac{\pi}{3}\right)}{3\sin\left(\frac{\pi}{2}\right)} - \frac{\cos\left(\frac{\pi}{3}\right)}{3\sin\left(\frac{\pi}{2}\right)}$$
$$= 0.333 - 0.166 - 0.166 = 0.001 \text{ C/m}^3.$$

Example 3.16: *Given the flux density $\overline{D} = \left[\left(\frac{2\cos\theta}{r^3}\right)U_r + \left(\frac{\sin\theta}{r^3}\right)\overline{U}_\theta\right]$ C/m³. Evaluate both sides of divergence theorem for region defined by $1 < r < 2$, $0 < \theta < \frac{\pi}{2}$, $0 < \phi < \frac{\pi}{2}$.*

Solution: The divergence theorem is $\oint_S D\cdot ds = \int_{vol} \nabla\cdot D\, dv$

$$\text{L.H.S.} = \oint_S D\cdot ds = \int_{r=1} D\cdot ds + \int_{r=2} D\cdot ds + \int_{\theta=0} D\cdot ds + \int_{\theta=\frac{\pi}{2}} D\cdot ds$$
$$= \int_0^{\pi/2}\int_0^{\pi/2} \overline{D}\underset{(r=1)}{(-r^2\sin\theta\, d\theta\, d\phi\, U_r)} + \int_0^{\pi/2}\int_0^{\pi/2}\overline{D}\underset{(r=2)}{(r^2\sin\theta\, d\theta\, d\phi\, U_r)} +$$

$$\int_1^2 \int_0^{\pi/2} \overline{D}\,(-r \sin\theta\, d\theta\, d\phi\, U_\theta)_{(\theta=0)} + \int_1^2 \int_0^{\pi/2} \overline{D}\,(r \sin\theta\, d\theta\, d\phi\, U_\theta)_{(\theta=\pi/2}$$

$$= \left[\frac{\cos 2\theta}{2}\right]^{\frac{\pi}{2}} [\phi]_0^{\pi/2} - \frac{1}{2}\left[\frac{\cos 2\theta}{2}\right]^{\frac{\pi}{2}} [\phi]_0^{\pi/2} + 0 + \left[\frac{-1}{r}\right]_1^2 [\phi]_0^{\pi/2}$$

$$= -\frac{\pi}{2} + \frac{\pi}{4} + \frac{\pi}{4} = 0$$

$$\text{R.H.S.} = \int_{vol} \overline{\nabla} \cdot \overline{D}\, dv$$

$$\nabla \cdot D = \frac{1}{r^2}\frac{d}{dr}\left(r^2 \times \frac{2\cos\theta}{r^3}\right) + \frac{1}{r \sin\theta}\frac{d}{d\theta}\left(\sin\theta \times \frac{\sin\theta}{r^3}\right)$$

$$= \frac{-2\cos\theta}{r^4} + \frac{2\cos\theta}{r^4} = 0$$

$\therefore$ R.H.S. = 0

Hence, L.H.S. = R.H.S.

Example 3.17: *Uniform surface charge densities of 150, – 50, and 40 $\mu C/m^2$ lie in the $x = 0$, $y = 0$ and $z = 0$ plane respectively. Find $|\overline{D}|$ as points 3.5 m from every coordinate plane in each of eight octants.*

Solution: Eight points in eight octanes are given below

1 (3.5, 3.5, 3.5), 2 (– 3.5, 3.5, 3.5), 3 (– 3.5, – 3.5, 3.5), 4 (3.5, – 3.5, 3.5),
5 (3.5, 3.5, – 3.5), 6 (– 3.5, 3.5, – 3.5), 7 (– 3.5, – 3.5, – 3.5), 8 (3.5, – 3.5, – 3.5)

At point 1

$$\overline{D} = \frac{150}{2} U_x \text{ (x-coordinate of point 1 is greater than } x = 0)$$

$$+ \frac{-50}{2} U_y \text{ (– y-coordinate of point 1 is greater than } y = 0)$$

$$+ \frac{40}{2} U_z \text{ (z-coordinate of point 1 is greater than } z = 0)$$

$$\overline{D} = 75U_x - 25U_y + 20U_z\ \mu C/m^2$$

(or) $$|\overline{D}| = \sqrt{(75)^2 + (25)^2 + (20)^2} = 81.547\ \mu C/m^2$$

The magnitude of $\overline{D}$ remains the same at all above points.

Example 3.18: *The electric flux density $\overline{D} = 20r^3\, U_r\, C/m^2$ in the region $0 \le r < 1$ m.*

(a) Find ρ_v at $r = 0.5$ m (b) find D_r at $r = 0.5$ m (c) What total charge is contained in the region $0 \le r \le 0.6$ m (d) If $\rho_v = 0$ for $r > 1$m, find D_r for $r > 1$m?

Solution:

(*a*) $$\rho_v = \overline{\nabla} \cdot \overline{D} = \frac{1}{r^2}\frac{\partial}{\partial r}(r^2 \times 20r^3) = \frac{1}{r^2}(20 \times 5r^4)$$

$$= 100\, r^2\ C/m^3$$

$\therefore$ ρ_v at $r = 0.4$ M is equal to $100\,(0.5)^2 = 25\ C/m^3$.

(*b*) $D_r = 20\,r^3$, Hence at $r = 0.5$ M, $D_r = 20\,(0.5)^3$

$$= 2.5 \text{ C/m}^2.$$

(*c*) $Q = \int \rho_v \, dV = \int_0^{0.6} \int_0^{\pi} \int_0^{2\pi} 100\, r^2 \cdot r^2 \sin\theta \, dr\, d\theta\, d\phi$

$$= 100 \left[\frac{r^5}{5}\right]_0^{0.6} [-\cos\theta]_0^{\pi} [\phi]_0^{2\pi} = 19.49 \text{ C}.$$

(*d*) $Q = \int_0^{1} \int_0^{\pi} \int_0^{2\pi} 100\, r^2 \times r^2 \sin\theta \, dr\, d\theta\, d\phi$

$$= 100 \left[\frac{r^5}{5}\right]_0^{1} [-\cos\theta]_0^{\pi} [\phi]_0^{2\pi} = 251.2 \text{ C}.$$

Area of sphere $= 4\pi r^2$ m^2

$$D_r = \frac{Flux}{area} = \frac{251.2}{4\pi r^2} = \frac{20}{r^2} \text{ C/m}^2 \text{ for } r > 1\text{m}.$$

Example 3.19: *Find the flux density at a point A (6, 4, – 5) caused by: (a) a point charge of 20 mC at the origin, (b) a uniform line charge* $\rho_l = 20\ \mu C/m$ *on the z-axis, (c) a uniform charge density* $\rho_s = 60\ \mu C/m^2$ *at a plane x = 8.*

Solution:

The flux density can be obtained from the expression of field intensity at a point due to different charge distributions.

(*a*) The field at a point due to a point charge is given by

$E = \dfrac{Q}{4\pi\varepsilon_0 r^2} U_r$ and flux density at a point is given by

$$D = \varepsilon_0 E = \frac{Q}{4\pi r^2} U_r$$

$$= \frac{20 \times 10^{-3}}{4\pi \times 77} \times \frac{6U_x + 4U_y - 5U_z}{\sqrt{77}}$$

$$D = 14.13U_x + 9.42U_y - 11.78U_z \ \mu\text{C/m}^2.$$

(*b*) The field at a point due to a line charge in *z*-axis is $E = \dfrac{\rho_l}{2\pi\varepsilon_0 r} U_r$

or flux density $D = \dfrac{\rho_l}{2\pi r} U_r$

For this case $r = 6U_x + 4U_y$

$$D = \frac{20 \times 10^{-6}}{2\pi\sqrt{52}} \times \frac{6U_x + 4U_y}{\sqrt{52}}$$

$$= 0.37U_x + 0.25U_y \,\mu\text{C/m}^2.$$

(*c*) The field due to infinite plane of charge density ρ_s is $E = \pm \frac{\rho_s}{2\varepsilon_0} U_n$

The flux density $D = \varepsilon_0 E = \pm \frac{\rho_s}{2} U_n$

The plane is at $x = 8$ and the point is at $x = 6$, *i.e.*, it is below the plane so flux density at point (6, 4, – 5) can be written as

$$D = -\frac{\rho_s}{2} U_n = -\frac{\rho_s}{2} U_n \qquad (U_n \text{ in this case is } U_x)$$

$$= -\frac{60}{2} U_x$$

$$= -30U_x \,\mu\text{C/m}^2.$$

Example 3.20: *Let $D = 2y^2z^2U_x + 3xy^2z^2U_y + 2xyzU_z\ pC/m^2$ in free space. Find (a) the total electrical flux passing through the surface $x = 2$, $0 \le y \le 2$ and $0 \le z \le 2$ in a direction away from the origin, (b) the total charge contained in an incremental sphere of a radius 1μm centered at P (2, 2, 2).*

Solution:

Given $D = 2y^2z^2U_x + 3xy^2z^2U_y + 2xyzU_z$ pC/m²

(*a*) The elemental surface of the plane at $x = 2$, $0 \le y \le 2$ and $0 \le z \le 2$ is $ds = dy\, dz\, U_x$

So $$\int_S D \cdot ds = \int_S 2y^2z^2\, dy\, dz$$

$$= \int_{y=0}^{2} \int_{z=0}^{2} 2y^2z^2\, dy\, dz = 14.22 \text{ pC}.$$

(*b*) $$\nabla \cdot D = \frac{\partial D_x}{\partial x} + \frac{\partial D_y}{\partial y} + \frac{\partial D_z}{\partial z}$$

$$= 0 + 6xyz^2 + 2xy \text{ pC/m}^3.$$

This value at point P (2, 2, 2) is given by 104 pC/m³.

As per Gauss's law in differential form $\nabla \cdot D = \rho_v$

Now $\rho_v = 104$ pC/m³.

ΔQ = charge contained in the incremental sphere of radius 1 μm

$= \rho_v \times$ volume

$= \rho_v \times \frac{4}{3}\pi r^3 = 104 \times 10^{-12} \times \frac{4}{3}\pi \times 1 \times 10^{-18} = 4.36 \times 10^{-28}$ C.

Example 3.21: *Using the Gauss's law in differential form, obtain the electric field intensity at different points due to the following charge distribution in spherical coordinates.*

$$\rho(r, \theta, \phi) = \begin{bmatrix} \rho_0\left(\frac{r}{a}\right) \\ 0 \end{bmatrix} \qquad 0 < r < a$$

Solution:

The Gauss's law in differential form states

$$\nabla \cdot D = \rho_v$$

In spherical coordinates, in general

$$\nabla \cdot D = \frac{1}{r^2}\frac{\partial}{\partial r}(r^2 D_r) + \frac{1}{r \sin\theta}\frac{\partial}{\partial \theta}(D_\theta \sin\theta) + \frac{1}{r\sin\theta}\frac{\partial D_\phi}{\partial \phi}$$

With variation only in r direction $\nabla \cdot D = \frac{1}{r^2}\frac{\partial}{\partial r}(r^2 D_r)$

For $0 < r < a$, at any radius r (variable)

$$\frac{1}{r^2}\frac{\partial}{\partial r}(r^2 D_r) = \rho_0 \frac{r}{a}$$

$$\frac{\partial}{\partial r}(r^2 D_r) = \rho_0 \frac{r^3}{a}$$

Integrating it $r^2 D_r = \int_0^r \rho_0 \frac{r^3}{a} dr$ (constant of integration being zero by symmetry of the problem)

$$D_r = \frac{\rho_0 r^2}{4a}$$

$$\therefore \qquad E = \frac{\rho_0 r^2}{4\varepsilon_0 a} U_r$$

Considering the case $a < r < \infty$, where charge is enclosed only up to radius 'a'.

$$\frac{\partial}{\partial r}(r^2 D_r) = \rho_0 \frac{r^3}{a}$$

$$r^2 D_r = \int_{r=0}^{a} \rho_0 \frac{r^3}{a} dr$$

On simplication $$E = \frac{\rho_0 a^3}{4\varepsilon_0 r^2} U_r$$

At $r = a$ $$E = \frac{\rho_0 a}{4\varepsilon_0} U_r$$

Example 3.22: *Determine D at (4, 0, 3) if there is a point charge – 5π mC at (4, 0, 0) and a line charge 3π mC/m along the y-axis.*

Solution:

Let $D = D_Q + D_L$ where D_Q and D_L are flux densities due to the point charge and line charge, respectively.

$$D_Q = \varepsilon_0 E = \frac{Q}{4\pi R^2} u_R = \frac{Q(r - r')}{4\pi |r - r'|^3}$$

where $r - r' = (4, 0, 3) - (4, 0, 0) = (0, 0, 3)$.

Hence, $$D_Q = \frac{-5\pi \cdot 10^{-3} (0, 0, 3)}{4\pi |(0, 0, 3)|^3} = -0.138\, U_z \text{ mC/m}^2$$

Also $$D_L = \frac{\rho_L}{2\pi\rho} U_\rho$$

In this case $$u_\rho = \frac{(4, 0, 3) - (0, 0, 0)}{|(4, 0, 3) - (0, 0, 0)|} = \frac{(4, 0, 3)}{5}$$

$$\rho = |(4, 0, 3) - (0, 0, 0)| = 5$$

Hence, $$D_L = \frac{3\pi}{2\pi(25)}(4u_x + 3u_z) = 0.2\, u_x + 0.18 u_z \text{ mC/m}^2$$

Thus $D = D_Q + D_L = 240U_x + 42U_y\ \mu\text{C/m}^2$.

SHORT QUESTIONS AND ANSWERS

1. **What are the different types of charge configurations and name their units?**

   The different types of charge configurations are

   (*i*) Line charge, its unit is C/m.

   (*ii*) Surface charge, its unit is C/m^2.

   (*iii*) Volume charge, its unit is C/m^3.

2. **Indicate the applications of Gauss's law in electric field.**

   Gauss's law is useful to find the field due to (*a*) point charge (*b*) line charge (*c*) surface charge.

3. **Write the two conditions of the field of symmetrical line charge.**

   The conditions of symmetrical line charge are as follows:

   (*a*) All the vertical components of the field are getting cancelled.

   (*b*) All the radial components of the field are additive.

4. What are equipotential surfaces?

Equipotential surfaces are those on which the potential is same through out the surface.

5. What is the principle of linear superposition?

The linear superposition principle states that the total response of system of charges is always equal to the response of individual charges.

6. How will you find the direction of intensity of electric field in a place?

It is always opposite to the negative of gradient of potential in the field. So by observing the direction potential rise or fall, the direction of the electric field can be easily found.

7. Why is electric field zero inside a conductor?

Always the charge tend to distribute over the surface a conductor, so the charge inside the conductor is zero and hence the electric field inside the conductor is also zero.

8. State Gauss's law.

The net electric flux passing through any closed surface is equal to the total charge enclosed by that surface.

9. What is point charge?

Point charge is one whose maximum dimension is very small in comparison with any other length.

10. Define irrotational and solenoidal fields.

If curl of a vector field is zero, then that field is called as irrotational field. And if divergence of a vector field is zero, then that field is called as solenoidal.

11. State the condition that the vector $\overline{F}$ to be solenoidal.

If $\overline{F}$ to be solenoidal, then the condition is $\nabla \cdot \overline{F} = 0$

12. What is Maxwell's first equation?

The point from of Gauss's law is called as 'Maxwell's first equation'.

i.e., $$\nabla \cdot \overline{D} = \rho$$

13. Define electric flux and flux density.

The lines of electric force is electric flux. It is denoted by 'ψ'

$$\psi = Q \text{ Coulomb}$$

Electric flux density is defined as electric flux per unit area.

It is denoted by D. And is a vector quantity.

$$D = \frac{Q}{A} \text{ C/m}^2$$

14. What is the physical significance of divergence of $\overline{D}$?

Divergence of $\overline{D}$ shows that whether the field is continuous or discontinuous, as per Gauss's law divergence of $\overline{D}$ is equal to ρ, so we can say that electric field is discontinuous.

15. Should the potential at a point always have a reference? Justify.

Yes. As the potential is defined as the total work done in bringing a unit positive charge from infinite distance to the point, where the potential is to be found, so always to measure the total work done there should be one reference.

MULTIPLE CHOICES

1. Identify the correct relation

   (a) $\nabla \cdot \overline{D} = \rho_v$ (b) div $D = \rho_v$

   (c) $\oint_S \overline{D} \cdot \overline{ds} = Q$ (d) all the above

2. Which of the following statements is not true?

   (a) The divergence of $\overline{D}$ is equal to the volume charge density

   (b) Divergence theorem converts a line integral to a volume integral

   (c) $\nabla \cdot \overline{D} = \rho_v$ is the Maxwell's first equation

   (d) $\nabla \cdot \overline{D} = 0$ for an insulator

3. For electrostatic fields $\oint_S \overline{D} \cdot \overline{ds}$ = ...

   (a) 0 (b) ρ_v

   (c) Q (d) $\overline{E}$

4. D = $10x^2y\ U_x$, ρ_v at a point $P(2, 1, -1)$ is ....

   (a) 40 C/m^3 (b) 4 C/m^2

   (c) 0 (d) 40

5. Potential V = ...

   (a) $-\int \overline{E} \cdot dl$ (b) $-\int \overline{Q} \cdot d\overline{l}$

   (c) $-\int \overline{E}^2 \cdot d\overline{l}$ (d) none of these.

6. $-\oint \overline{E} \cdot d\overline{l}$ = ...

   (a) infinite (b) one

   (c) zero (d) ρ_v

7. Identify the Maxwell's equation from the following:

   (a) $\nabla \cdot \overline{D} = \rho_v$ (b) $\nabla \times \overline{E} = 0$

   (c) $\oint \overline{E} \cdot d\overline{l} = 0$ (d) all the above

8. V = $10xy^2$, $\overline{E}$ = ... at $p(1, 2, 3)$.

   (a) $40u_x + 40u_z$ (b) $40u_x + 40u_y$

   (c) $-40u_x - 40u_v$ (d) $-40u_x - 40u_z$

9. D = $10u_x + 5u_y + 20u_z$, $\overline{E}$ = ...

(*a*) $10\varepsilon u_x + 5\varepsilon u_y + 20\varepsilon u_z$

(*b*) $10/\varepsilon u_x + 5/\varepsilon u_v + 20\varepsilon u_z$

(*c*) 0

(*d*) $10/\varepsilon u_x + 5\varepsilon u_y + 20\varepsilon u_z$

10. The electric flux Ψ = ...

(*a*) $\oint \overline{E} \cdot d\overline{s}$

(*b*) Q

(*c*) $\int_S \overline{D} \cdot d\overline{s}$

(*d*) both (*b*) & (*c*)

11. Maxwell's first equation integral form is ....

(*a*) $\oint \overline{E} \cdot d\overline{s} = \int \rho_v \, d_v$

(*b*) $\nabla \cdot \overline{D} = \rho_v$

(*c*) $\oint \overline{E} \cdot d\overline{l} = 0$

(*d*) none of these

12. $\overline{D} = 40xyz^2u_v + 20u_v$; find at ρ_v a point $P(1, 2, -1)$ ....

(*a*) 10

(*b*) – 80

(*c*) – 40

(*d*) 80

13. $V = 10xz^2$; find ∇V at $P(1, 1, -1)$ ....

(*a*) $10u_x + 20u_y$

(*b*) $10u_x - 20u_z$

(*c*) $10u_x$

(*d*) $10u_y$

14. $V = -10x^2yz$, find $\overline{E}$ at a point $(-1, 2, 0)$

(*a*) $10u_x + 20u_y$

(*b*) $-20u_z$

(*c*) $20u_z$

(*d*) $20u_x$

15. The point form of Gauss's law is

(*a*) $\nabla \cdot \overline{D} = Q_{enc}$

(*b*) $\nabla \times \overline{D} = \rho_v$

(*c*) $\nabla \cdot \overline{D} = \rho_v$

(*d*) $\nabla \cdot \overline{D} = \rho_s$

16. The total electric flux passing through any closed surface is equal to

(*a*) 0

(*b*) ρ_1

(*c*) ρ_s

(*d*) Q_{enc}

17. Inside a conductor ....

(*a*) V_{ab} = infinite

(*b*) $V_{ab} = 0$

(*c*) electrons will not be there

(*d*) none of these

18. Magnitude of electric field intensity at any point between infinitely extended surface with charge density + ρ_s C/m^2 and – ρ_s C/m^2 is

(*a*) $\dfrac{\rho_s}{2\varepsilon_0}$

(*b*) $2\dfrac{\rho_s}{2\varepsilon_0}$

(*c*) $\dfrac{\rho_s}{\varepsilon_0}$

(*d*) none of these

19. I = ...

(a) dQ/dt (b) $\int_S \bar{J} \cdot d\bar{s}$

(c) both (a) & (b) (d) none of these

20. The electric field intensity $\overline{E}$ at 'P' due to an infinite line charge kept along one axis of the coordinate system is ....

(a) inversely proportional to the distance between them

(b) inversely proportional to the square root of the distance

(c) independent of the distance

(d) none of these

21. V is ...., where V is electric potential.

(a) a vector (b) a scalar

(c) not but E (d) none of these

22. Electric potential is

(a) a vector (b) a scalar

(c) not but E (d) none of these

23. The work in moving a point charge around a closed path an electric field $\overline{E}$ is

(a) zero (b) not possible to find

(c) infinite (d) none of these

24. The units of electric flux is ...., C stands for coulombs.

(a) C/m^2 (b) C/m^3

(c) C (d) none of these

25. The relaxation time 'T_r' is ....

(a) independent of conductance

(b) inversely proportional to the square of conductance

(c) inversely proportional to the conductance

(d) none of these

26. The unit of electric flux density is

(a) V/m (b) C

(c) C/m^3 (d) C/m^2

27. (1) If there is no net charge inside a Gaussian surface, E is zero all around the surface.

(2) If E is zero, everywhere on the surface, there is no net charge.

(a) 1, 2 are true (b) 1 true, 2 false

(c) 1, 2 false (d) 1 false, 2 true

28. The electric flux of a hemisphere of radius R and electric field E is

(a) $4\pi R^2 E$ (b) $\pi R^2 E$

(c) $2\pi R^2 E$ (d) $\frac{1}{2} \pi R^2 E$

KEY

1. (*d*)	**2.** (*b*)	**3.** (*c*)	**4.** (*c*)	**5.** (*a*)	**6.** (*a*)
7. (*d*)	**8.** (*c*)	**9.** (*b*)	**10.** (*d*)	**11.** (*a*)	**12.** (*d*)
13. (*b*)	**14.** (*c*)	**15.** (*c*)	**16.** (*d*)	**17.** (*b*)	**18.** (*c*)
19. (*c*)	**20.** (*a*)	**21.** (*b*)	**22.** (*b*)	**23.** (*a*)	**24.** (*c*)
25. (*c*)	**26.** (*d*)	**27.** (*d*)	**28.** (*b*)		

REVIEW QUESTIONS

1. What is the significance of the negative sign in the equation $\vec{E} = -\nabla V$?
2. State Gauss's law.
3. What are the limitations of Gauss's law?
4. Can we use Gauss's law in each of the following cases:

   (*a*) $\rho_v = k\rho^2$ (*b*) $\rho_v = k\rho \cos \phi$

   (*c*) $\rho_v = k\rho$ and $\rho \neq 0$ (*d*) $\rho_v = kr$

   (*e*) $\rho_v = kr \cos \theta$ (*f*) $\rho_v = kr \cos \theta$?
5. Explain briefly about electric displacement.
6. A hollow conductor in the form of a matchbox is placed in a static electric field. What is the electric field inside the conductor? Sketch the charge distribution on the inner and the outer surfaces of the conductor.
7. Derive the expression for electric flux density.
8. Is it necessary to find a Gaussian surface on which the normal component of the $\vec{D}$ field is constant in order to apply Gauss's law? Give reasons to justify your answer.
9. Determine the electric field due to a point charge using Gauss's law.
10. A charge of 10 mC is placed inside a conduction shell. What is the induced charge on the inner surface? What is the charge on the outer surface? Does it matter where the charge is placed inside the shell?
11. What is the physical significance of $\nabla \cdot \vec{D}$?
12. Show that Gauss's law depends on the inverse square law *i.e.*, $E \propto \frac{1}{r^2}$ for a point charge.
13. If the charge is distributed over *a* thin spherical shell of radius *b*, what is $\vec{E}$ inside the shell?
14. Show that div $D = \rho_v$ using Gauss's law.
15. When we apply Gauss's law in a dielectric medium, should we take into account the bound charge densities?
16. State and prove Gauss's law and determine the field due to an infinite line charge using this.

17. State Gauss's law, derive the same from Coulomb's law and explain the significance of equipotential surfaces.
18. Using Gauss's law find an expression for electric field intensity at any point in the electric field produced by an infinite line charge of uniform density.

EXERCISE PROBLEMS

1. Find $\overline{D}$ and plot D_r V_s r for each of following charge distribution

(*a*) 1 C at origin (*b*) $\frac{1}{4\pi}$ C/m^2 at $r = 1$ m

(*c*) $\frac{4}{4\pi}$ C/m^2 from $0 < r < 1$, 0 elsewhere

2. Let $\overline{D} = [(8x + 4x^3)U_x - 2yU_y + 2zU_z]$ C/m^2 (*a*) use Gauss's law to determine charge enclosed is cubical region, $-a < x, y, z < a$ (*b*) obtain as approximate value for above charge. Evaluate the derivative at center of volume.
3. A cubical region of space defined by surface $x = 1.8, y = 1.8, z = 1.8, x = 2, y = 2$ and $z = 2$. If $\overline{D} = (3y^2U_x + 3x^2yU_y)$ C/m^3 (*a*) Find the exact value of total charge enclosed within the cube by surface integration (*b*) Find an approximate value for enclosed charge of evaluation for derivative at center of cube.
4. If $G = 5r \sin^2\theta \cos^2\phi\, U_r$ evaluate both sides of divergence theorem for region $r \le 2$.
5. Given the field $D = \left(\frac{20}{\rho^2}\right) - (-\sin^2\phi\, U_\rho + \sin^2\phi\, U_\phi)$ C/m^2. Find the total charge lying with in the volume $1 < \rho < 2$, $0 < \phi < \frac{\pi}{2}$, $0 < z < 1$ by a different methods.
6. Let $D = xU_x$ and find value of $\oint_S D \cdot ds$ over the surface of sphere $r = 1$.
7. Let $\overline{D} = (6xy\, z^2U_x + 3x^2z^2U_y + 6x^2yzU_z)$ C/m^2. Find the total charge lying within the region bounded by $x = 1$ and 3, $y = 0$ and 1, and $z = -1$ and 1 by separately evaluating each side of divergence theorem.
8. Uniform line changes of 20 nC/M each lie in the $z = 0$ plane at $x = 0, \pm 1, \pm 2 - \pm 5$. (*a*) Find the total electric flux leaving the spherical surface $r = 2.5$ (*b*) Find $\overline{D}$ at P (0, 2.5, 4).
9. A point change of 90 nC is located at origin and there are also three uniform surface change distribution: – 8 nC/m^2 at $r = 2$ and 1.2 nC/m^2 at $r = 3$ (*a*) Find $\overline{D}$ everywhere.
10. Cylindrical surface at $P = 2$, 4 and 6 m carry uniform change densities of 20 nC/m^2, – 4 nC/m^2 and ρ so respectively (*a*) Find $\overline{D}$ at $\rho = 1$, 3 and 5m. (*b*) Determine ρ so such that $\overline{D} = 0$ at $\rho = 7$ m.
11. Let $\rho_V = 8z(1 - z)$ C/m^3 for $0 < z <$ 1m, $8z(1 + z)$ for $-1 < z < 0$ and 0 for $|z| > l$ (*a*) Find $\overline{D}$ everywhere.

12. A point charge Q lies at the origin. Show that $\nabla \cdot \overline{D} = 0$ everywhere except at the origin (*b*) Replace the point charge with a uniform volume charge density ρ_v for $0 < r < a$. Relate ρ_v, Q and a so that the total charge is the same. Find $\overline{\nabla} \cdot \overline{D}$ everywhere.

13. Let $\overline{D} = x\overline{U}_x$ and find the value of $\oint \overline{D} \cdot \overline{ds}$ over the surface of sphere $r = 1$?

14. A cubical region of space (1) defined by surface $x = 1.8$, $y = 1.8$, $z = 1.8$, $x = 2$, $y = 2$ and $z = 2$. If $\overline{D} = 3y^2U_x + 3x^2\, yU_y$ C/m² (*a*) Find the exact value of total change enclosed within the cube by surface integration (*b*) Find approximate value for the enclosed charge by evaluation of the derivatives at the center of cube?

15. Let $\overline{D} = (8x + 4x^3)U_x - 2yU_y + 2zU_z$ C/m² (*a*) Use Gauss's law to determine the change enclosed in the cubical region $-a < x, y, z < a$ (*b*) Obtain as approximate value for the above change. Evaluate the derivatives at the center of volume (*c*) Show that the answers agree in the limit as '*a*' becomes very small.

16. A point change of 6 C is located at the origin, a uniform line charge density of 180 nC/m lies along the x-axis and a uniform sheet of charge equals to 25 nC/m² lies in the $z = 0$ plane. (*a*) Find $\overline{D}$ at A (0, 0, 4); (*b*) Find $\overline{D}$ at B (1, 2, 4); (*c*) Calculate the total electric flux leaving the surface of a sphere of 4 m radius centered at the origin.

17. The spherical surfaces $r = 1$, 2 and 3 carry surface charge densities of 25, – 9, and 3 nC/m² respectively (*a*) How much electric flux leaves the surface $r = 6$? (*b*) Find D at P (1, – 1, 4).

18. Point charges $Q_1 = 5 \times 10^{-9}$ C and $Q_2 = 3 \times 10^{-9}$ C are located at (0, 3, 0) and (0.8, 0, 0) respectively. Determine the electric field at all points in $x - y$ plane and also at P(0.5, 0.3) by first finding the potential due to the charges and then taking the gradient of the potential.

19. Consider an interface between free space and a dielectric medium with dielectric constant of 5. The normal component of the $\vec{D}$ field in free space is 10 C/m², and the tangential component of the $\vec{E}$ field in the dielectric medium is 100 V/m. Find their counterparts.

CHAPTER

4

Energy and Potential, Conductors, Dielectrics and Capacitors

4.1 INTRODUCTION

From the previous chapters, we have become acquainted with Coulomb's law in finding the electric field about several simple charge configurations and also with Gauss law and its application in determining the field about some symmetrical charge arrangement. The use of Gauss law was invariably easier for these highly symmetrical distributions; because the problem of integration always disappeared when proposed charged surface was chosen.

The application of Coulomb's law is laborious, detailed and often quite complex. The reason for this being precisely the fact that electric field intensity, a vector field must be found directly from the charge distribution. Three different integrations are needed in general, one for each component. Certainly it would be desirable if we could find some as yet undefined scalar function with single integration and then determine the electric field from this scalar by some straight forward procedure.

4.2 ENERGY EXPENDED (OR) WORK DONE IN MOVING A POINT CHARGE IN ELECTRIC FIELD

The electric field intensity is defined as force on a unit test charge at that point at which we wish to find the value of this vector field. Suppose a charge Q is moved a distance dL in an electric field E. The force on charge Q due to electric field is given by

$$F_E = QE \quad \text{...(4.1)}$$

Where the subscript reminds the fact that this force is due to field.

The component of this force in the direction dL which must be overcome is

$$F_{EL} = F_E \cdot U_L = QE \cdot U_L \quad \text{...(4.2)}$$

Where U_L is a unit vector in direction of dL.

The force which must be applied should be equal and opposite to force due to field

$$F_{appl} = -QE \cdot U_L \quad \text{...(4.3)}$$

and our expenditure of energy is product of force and distance.

Differential work done by external source moving

$$dW = -QE \cdot U_L\, dL$$
$$= -QE \cdot dL$$

(or) $$dW = -QE \cdot dL \qquad ...(4.4)$$

Where $U_L\, dL$ is replaced by dL, so work required to move a charge a finite distance must be determined as

$$W = -Q \int_{initial}^{final} E \cdot dL \qquad ...(4.5)$$

4.3 POTENTIAL DIFFERENCE AND ELECTRIC POTENTIAL

We need to define a new concept from the expression for work done by an external source in moving a charge Q from one point to another in an electric field E.

$$W = -Q \int_{initial}^{final} \overline{E} \cdot \overline{dL} \qquad ...(4.6)$$

Potential difference V is defined as work done by an external source in moving a unit positive charge from one point to another in an electric field

Potential difference, $$V = \int_{initial}^{final} \overline{E} \cdot \overline{dL} \qquad ...(4.7)$$

For reference, the potential difference between the two points P and Q is written as V_{PQ} and is the work done in moving the unit charge from Q to P.

Referring Fig. 4.1, in a uniform field the '+ve' test charge is moved from point Q to point P opposite to the field direction. An external source has to perform work in order to move a test charge from 'Q' to 'P' against the field and the $P.D.$ between the points P and Q is

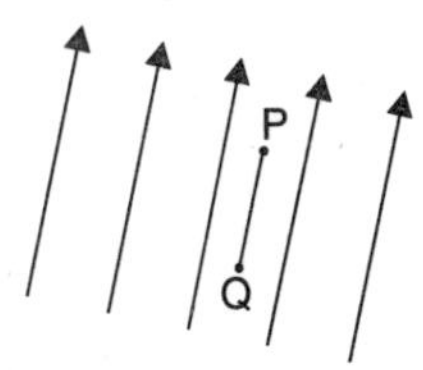

Fig. 4.1 Movement of test charge from Q to P

$$V_{PQ} = -\int_{Q}^{P} \vec{E} \cdot \vec{dl} \qquad ...(4.8)$$

The unit of potential difference is J/C or volt. It is easy to speak of the potential, or absolute potential of a point, rather than potential difference between two points. Another widely used reference 'point' is infinity. In actual case, the ground has zero potential; but for our case we shall assume a point at infinity to have zero potential.

With this concept in the frame of mind we can write potential at point 'P' is 'V_P' and potential at Q is V_Q, then

$$V_{PQ} = V_P - V_Q \qquad ...(4.9)$$

Both V_P and V_Q can be assumed to have same zero reference points.

It can be summarized as

(*i*) P is the initial point and Q is the final point in determining V_{PQ}.

(*ii*) There is a loss of potential energy in moving charge 'q' from P to Q if V_{PQ} is negative. This implies that the work is being done by the field. However if V_{PQ} is positive, there is a gain in potential energy in the movement as an external agent performs the work.

(*iii*) V_{PQ} is independent of the path taken.

(*iv*) V_{PQ} is measured in joules per coulomb, commonly referred to as volts (V).

4.4 POTENTIAL FIELD OF A POINT CHARGE

Consider the electric field due to a point charge. Let two points 'A' and 'B' in the electric field be situated at distances r_A and r_B respectively from the point charge.

The field intensity at a distance '*r*' from a point charge can be written as

$$\vec{E} = \frac{Q}{4\pi\varepsilon\, r^2} \cdot u_r \qquad ...(4.10)$$

The potential difference between 'A' and 'B' can be found out for any specified path between 'A' and 'B'. Considering spherical coordinates, the differential length vector is given by

$$\vec{dl} = dr\, u_r + r\, d\theta\, u_\theta + r \sin\phi\, d\phi\, u_\phi$$

We have

$$V_{AB} = -\int_B^A \bar{E}\, \bar{dl}$$

$$= -\int_{r-r_B}^{r-r_A} \left[\frac{Q}{4\pi\varepsilon_0 r^2}\right] [dr\, u_r + r\, d\theta\, u_\theta + r \sin\phi\, d\phi\, u_\phi]$$

$$= -\int_{r_B}^{r_A} \frac{Q}{4\pi\varepsilon_0 r^2}\, dr = \frac{Q}{4\pi\varepsilon_0 \left[\frac{1}{r}\right]_{r_B}^{r_A}}$$

$$= \frac{Q}{4\pi\varepsilon_0\, r_A} - \frac{Q}{4\pi\varepsilon_0\, r_B} \qquad ...(4.11)$$

It is clear from equation (4.11) that potential difference between two points is dependent upon their distances from the point charges (r_A and r_B) and not on the path from B to A.

Since, $V_{AB} = V_A - V_B$, it can be written as

$$V_A = \frac{Q}{4\pi\varepsilon_0\, r_A} \quad \text{and} \quad V_B = \frac{Q}{4\pi\varepsilon_0\, r_B}$$

$$V_{AB} = V_A - V_B = \frac{Q}{4\pi\varepsilon_0}\left[\frac{1}{r_A} - \frac{1}{r_B}\right] \qquad ...(4.12)$$

If $r_B \to \infty$, then

$$V_{AB} = V_A = \frac{Q}{4\pi\varepsilon_0\, r_A} \qquad ...(4.13)$$

This is called absolute potential at point A. In general for any point '*P*' the absolute potential at a distance '*r*' from the point charge Q, is given as

$$V_r = \frac{Q}{4\pi\varepsilon_0\, r}$$

Potential due to a point charge is inversely proportional to 'r'.

or $$V_r = -\int_{\infty}^{r} \vec{E}\,\vec{dl} \qquad ...(4.14)$$

This is the work done per coulomb to bring positive test charge from infinity to the point against the field.

Example 4.1: *Given a field* $E = \left[\frac{-5y}{x^2}u_x + \frac{5}{x}u_y + 6u_z\right]$ *V/m. Calculate the potential difference between A (– 7, 2, 1) and B (4, 1, 2).*

Solution: $$V_{AB} = -\int_B^A \vec{E}\,\vec{dl}$$

$$= -\int_B^A \left\{\left(\frac{-5y}{x^2}\right)\bar{u}_x + \left(\frac{5}{x}\right)\bar{u}_y + 6\bar{u}_z\right\}\cdot(dx\,\bar{u}_x + dy\,\bar{u}_y + dz\,\bar{u}_z)$$

$$= -\int_B^A \left(\frac{-5y}{x^2}dx + \frac{5}{x}dy + 6dz\right)$$

$$= +\int_{\substack{x=4\\y=1}}^{x=-7} \frac{5y}{x^2}dx + \int_{\substack{y=1\\x=4}}^{y=2} \frac{-5}{x}dy + \int_{z=2}^{1} -6dz$$

$$= \int_4^{-7} \frac{5}{x^2}dx - \frac{5}{4}\int_1^2 dy - 6\int_2^1 dz = -\left[\frac{5}{x}\right]_4^{-7} - \frac{5}{4}[y]_1^2 - 6[z]_2^1$$

$$= -5\left[-\frac{1}{7} - \frac{1}{4}\right] - \frac{5}{4}[2-1] - 6[1-2]$$

$$= \frac{1.964}{1} - \frac{5}{4} + 6 = 6.714 \text{ V.}$$

Example 4.2: *A point charge of 5 nC is located in free space. Find potential if point P is located at (0.2, 0.3, – 0.3) and (a) V = 0 at infinite, (b) V = 0 at (3, 0, 0) and (c) V = 35 v at (– 0.4, 2, – 2).*

Solution:

(*a*) $$V = \frac{Q}{4\pi\varepsilon_0 r} + C, \text{ where } C \text{ is a constant}$$

Since at $r = \infty$; $V = 0$ and thus $C = 0$

Hence, $$V = \frac{Q}{4\pi\varepsilon_0 r}$$

Where, $$r = \sqrt{x^2 + y^2 + z^2} \text{ at } P\,(0.2, 0.3, -0.3)$$

$$= \sqrt{(0.2)^2 + (0.3)^2 + (-0.3)^2} = 0.46$$

$\therefore$ $$V = \frac{5\times10^{-9}}{4\pi\times8.854\times10^{-12}\times0.46} = 97.7 \text{ V.}$$

(*b*) $V = \dfrac{Q}{4\pi\varepsilon_0 r} + C$ and given that at (3, 0, 0), $V = 0$; where $r = 3$

$$0 = \frac{5 \times 10^{-9}}{4\pi \times 8.854 \times 10^{-12} \times 3} + C$$

$$= 14.9 + C \Rightarrow C \simeq 15$$

$$\therefore \quad V = \left[\frac{Q}{4\pi\varepsilon_0 r} - 15\right] = \frac{5 \times 10^{-9}}{4\pi \times 8.854 \times 10^{-12} \times 0.46} - 15$$

or $\quad V = 97.7 - 15 = 82.7$ V

(*c*) $\quad V = \dfrac{Q}{4\pi\varepsilon_0 r} + C$ and at (– 0.4, 2, – 2); $V = 35$ V

where, $\quad r = \sqrt{(0.4)^2 + (2)^2 + (-2)^2} = 2.856$

$$35 = \frac{5 \times 10^{-9}}{4\pi \times 8.854 \times 10^{-12} \times 2.856} + C$$

$\Rightarrow \quad C = 19.2$

At P (0.2, 0.3, – 0.3);

$$V = \frac{Q}{4\pi\varepsilon_0 r} + C$$

$$= 97.7 + 19.2 = 116.9 \text{ V}.$$

Example 4.3: *A negative point charge of magnitude 5 μC is situated in air at origin and two positive point charges of 3 μC each at point y = ± 2 m. Determine the electric field strength and electric potential at a point 5 m from the origin on the x-axis.*

Solution: From the Fig. 4.2, the electric field E_1 due to point charge at (0, 0) is given by

$$\vec{E}_1 = \frac{Q}{4\pi\varepsilon_0 r^2}\bar{u}_x = \frac{-5 \times 10^{-6}}{4\pi\varepsilon_0 \times 5^2}$$

$$= -\frac{9 \times 10^9 \times 5 \times 10^{-6}}{25}$$

$$= -1.8 \times 10^3\, \bar{u}_x$$

$$\vec{E}_1 = -1800\, \bar{u}_x \text{ V/m}$$

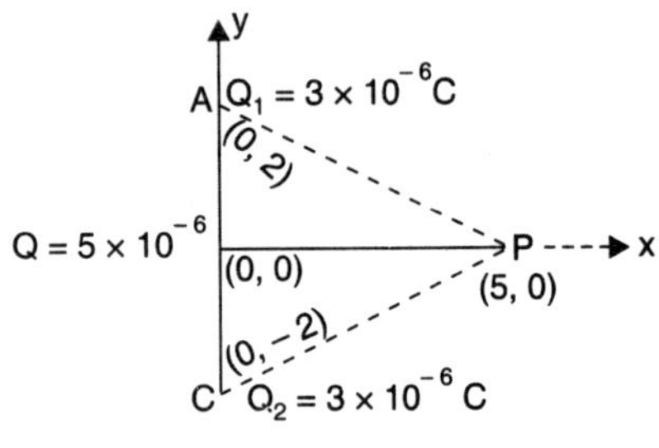

Fig. 4.2

Now, the resultant electric field strength at point P are contributions from electric field strengths due to the two point charges at points (y = ± 2) is given by

$$\vec{E}_2 = \frac{3 \times 10^{-6}}{4\pi\varepsilon_0 (AP)^2}\cos\theta \cdot \bar{u}_x = \frac{3 \times 10^{-6}}{4\pi\varepsilon_0 (\sqrt{29})^2}\cos\theta\, \bar{u}_x$$

$$\left[\because \cos\theta = \frac{5}{\sqrt{29}} \text{ and } AP = CP = \sqrt{29}\right]$$

$$= \frac{9\times10^9 \times 3\times10^{-6}}{29} \times \frac{5}{\sqrt{29}} \cdot \bar{u}_x = 864.443\ \bar{u}_x \text{ V/m}$$

$\therefore$ $$\bar{E}_2 = 864.443\ \bar{u}_x$$

$\therefore$ $$\bar{E} = \bar{E}_1 + \bar{E}_2 = -1800\bar{u}_x + 864.443\bar{u}_x = -935.557\bar{u}_x \text{ V/m}$$

or $$|\ \bar{E}\ | = 935.557 \text{ V/m.}$$

Electric potential V_1 due to point charge at origin (0, 0) is given by

$$V_1 = \frac{Q}{4\pi\varepsilon_0 \times r} = \frac{-5\times10^{-6}}{4\pi\varepsilon_0 \times 5} = -9000 \text{ volts}$$

Similarly, electric potentials due to point charges at points (0, ± 2) are

$$V_2 = \frac{3\times10^{-6}}{4\pi\varepsilon_0 \times \sqrt{29}} = \frac{9\times10^9 \times 3\times10^{-6}}{5.385} = 5.013 \times 10^3 \text{ volts}$$

$$= 5013 \text{ volts}$$

and $$V_3 = V_2 = 5013 \text{ volts.}$$

Therefore, resultant electric potentials at point P is given by

$$V = V_1 + V_2 + V_3 = -9000 + 5013 + 5013$$
$$= 1026$$

$\therefore$ $$V = 1026 \text{ volts.}$$

Example 4.4: *Determine the potential at a point P from a charged line of 1 m length whose charge is $\rho_L = 2\ \mu C/m$. The point P is at a distance 50 cm from mid-point of the charged line.*

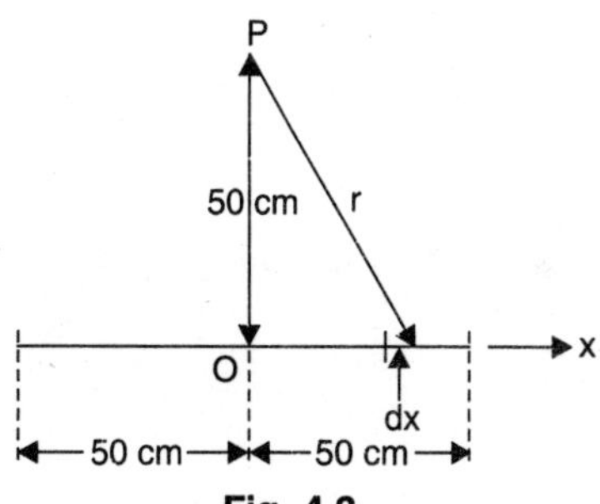

Fig. 4.3

Solution: Potential at P due to a concentrated charged $Q_L\ dx$ at a distance r from P and x from O.

$$dV = \frac{1}{4\pi\varepsilon_0} \frac{Q_L\ dx}{r}$$

$\therefore$ $$V_{(P)} = \int dV = \frac{Q_L}{4\pi\varepsilon_0} \int \frac{dx}{r}$$

$\therefore$ $$V_{(P)} = 9 \times 10^9 \times 2 \times 10^{-6} \int_{-0.5}^{0.5} \frac{dx}{\sqrt{x^2 + (0.5)^2}} = 18\times10^3 \int_{-0.5}^{0.5} \frac{dx}{\sqrt{x^2+(0.5)^2}}$$

Replacing $$x = 0.5 \tan\theta$$

$$dx = 0.5 \sec^2\theta\ d\theta$$

$$x^2 + 0.5^2 = 0.5^2 \sec^2\theta$$

$\therefore$ $$V_{(P)} = 18 \times 10^3 \int_{-\pi/4}^{\pi/4} \sec\theta\ d\theta = 18 \times 10^3\ [\ln\ |\ \sec\theta + \tan\theta\ |\ ]_{-\pi/4}^{\pi/4}$$

$$= 18\times10^3 \left[\ln\left| \sec\left(\frac{\pi}{4}\right) + \tan\left(\frac{\pi}{4}\right)\right| - \left(\ln\left|\sec\left(-\frac{\pi}{4}\right) + \tan\left(-\frac{\pi}{4}\right)\right|\right)\right]$$

$$= 18 \times 10^3 \ln \left[\frac{\left| \sec\left(\frac{\pi}{4}\right) + \tan\left(\frac{\pi}{4}\right) \right|}{\left| \sec\left(-\frac{\pi}{4}\right) + \tan\left(-\frac{\pi}{4}\right) \right|} \right]$$

$$= 18 \times 10^3 \ln \left(\frac{\sqrt{2}+1}{\sqrt{2}-1} \right) = 18 \times 10^3 \ln \frac{2.41}{0.41}$$

$$= 18 \times 10^3 \times 1.77 = 3.186 \times 10^4 \text{ V.}$$

Example 4.5: *Find the electric potential at the point at which* **E** *is zero, when point charges of + 4 μC and + 8 μC are located at (0, 0) and (0.5, 0).*

Solution:

Let the field be zero, P at a distance x from origin and lying between the charges.

$$\therefore \quad \overline{E}_{(P)} = \frac{4 \times 10^{-6}}{4\pi\varepsilon_0 x^2} U_x - \frac{8 \times 10^{-6}}{4\pi\varepsilon_0 (0.5 - x)^2} U_{(x)} = 0$$

$$\therefore \quad \frac{4 \times 10^{-6}}{4\pi\varepsilon_0 x^2} = \frac{8 \times 10^{-6}}{4\pi\varepsilon_0 (0.5 - x)^2}$$

i.e.,

$$4(0.5 - x)^2 = 8x^2$$

$$4(0.25 - x + x^2) = 8x^2$$

$$1.0 - 4x + 4x^2 = 8x^2$$

$$4x^2 + 4x - 1.0 = 0$$

$$\therefore \quad x = \frac{-4 \pm \sqrt{16+16}}{8} = 1.656 \text{ m}$$

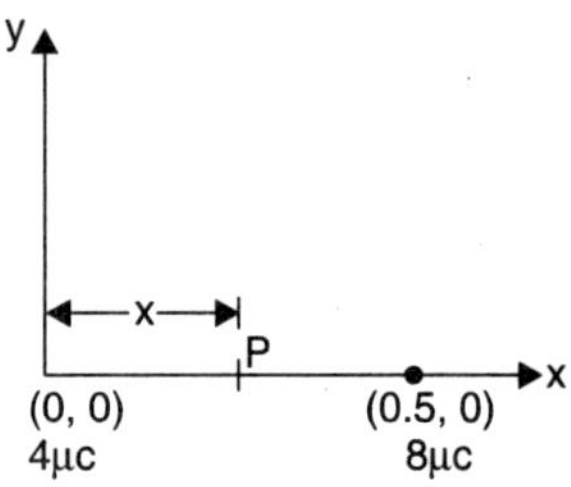

Fig. 4.4

Now potential at P

$$V_{(P)} = V_{1P} + V_{2P}$$

$$= \frac{4 \times 10^{-6}}{4\pi\varepsilon_0 (1.656)} - \frac{8 \times 10^{-6}}{4\pi\varepsilon_0 (0.5 - 1.656)}$$

$$= 9 \times 10^9 \times 10^{-6} \left[\frac{4}{1.656} - \frac{8}{1.156} \right] = 40.5 \times 10^3 \text{ V.}$$

4.5 EQUIPOTENTIAL SURFACES

Equipotential surfaces are the surfaces on which the potential is same at every point. In other words, the locus of all points which have the same electric potential is called equipotential surface. Potential field is a scalar field.

No work is done in taking a charge from one point to another on an equipotential surface as the potential difference between any two points surface is zero. This is only possible when the charge taken is perpendicular to the field. In this way equipotential surface at every point is perpendicular to the field.

The direction of field at different points in an electric field is represented by lines of force. Hence the lines of force at every point of equipotential surface are perpendicular to the surface. On the other hand the equipotential surfaces are perpendicular to the lines of force. The electric potential is a linear function of the value of its charge. So, the potential at a point P, at distances $r_1, r_2, r_3, ...$ from point charges $Q_1, Q_2, Q_3, ...$ respectively is

$$V_P = \frac{Q}{4\pi\varepsilon_0}\left[\frac{Q_1}{r_1} + \frac{Q_2}{r_2} + \frac{Q_3}{r_3} + ...\right]$$

$$V_P = \sum_{i=1}^{n} \frac{Q_i}{4\pi\varepsilon_0 r_i} \qquad ...(4.15)$$

This is also true for different charge distributions. Line charge, surface charge and volume charge.

4.6 POTENTIAL GRADIENT (RELATIONSHIP BETWEEN E AND V)

Here Electric field E is calculated with the help of potential V. Consider two surfaces having potentials V and $V + dV$ in an electric field. Let a positive test charge 'Q' at point 'A' on the surface 'V' be moved to the point B on the surface $V + dV$ along the vector displacement $\overline{dl}$ by an external source.

The force experienced by the test charge 'Q' at 'A' due to electric field will be $Q\overline{E}$. This force is in the direction of $\overline{E}$ which is at right angle to surface V.

The external source must apply an equal and opposite force $\overline{F}$ as shown in Fig. 4.5 in order to move the test charge Q without acceleration. Thus

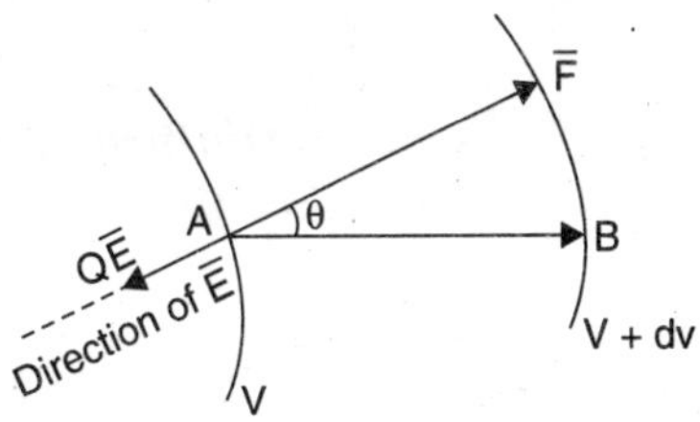

Fig. 4.5

$$\overline{F} = -Q\overline{E} \qquad ...(4.16)$$

The work done by the external source to move the test charge from A to B along $\overline{dL}$ is given by

$$dw = \overline{F} \cdot \overline{dL}$$

(or)

$$dw = -Q\overline{E} \cdot \overline{dL}$$

$$\therefore \quad \frac{dw}{Q} = -\overline{E} \cdot \overline{dL} \qquad ...(4.17)$$

We know that $\frac{dw}{Q}$ = is the potential difference dV between points A and B. Thus

$$dV = -\overline{E} \cdot \overline{dL} \qquad ...(4.18)$$

Let the coordinate of A and B be (x, y, z) and $(x + dx, y + dy, z + dz)$ respectively.

The potential difference dV between A and B can be expressed as

$$dV = \frac{\partial V}{\partial x} dx + \frac{\partial V}{\partial y} dy + \frac{\partial V}{\partial z} dz$$

$$= \left(\frac{\partial V}{\partial x} U_x + \frac{\partial V}{\partial y} U_y + \frac{\partial V}{\partial z} U_z \right) \cdot (dx\, U_x + dy\, U_y + dz\, U_z)$$

$$= (\text{grad } V) \cdot \overline{dL} \qquad ...(4.19)$$

$$= \nabla V \cdot \overline{dL} \qquad \left[\because \; \nabla = \frac{\partial}{\partial x} U_x + \frac{\partial}{\partial y} U_y + \frac{\partial}{\partial z} U_z \right]$$

Comparing equations (4.18) and (4.19), we get

$$\overline{E} = -\text{ grad } V = -\nabla V \qquad ...(4.20)$$

$$dV = -\overline{E} \cdot \vec{dL} = -\overline{E}_x\, dx - \overline{E}_y\, dy - \overline{E}_z\, dz \qquad ...(4.21)$$

Where, $\overline{E}_x, \overline{E}_y$ and $\overline{E}_z$ are components of $\overline{E}$, then equation (4.21) can also be written as

$$E_x U_x + E_y U_y + E_z U_z = \left[-\frac{\partial}{\partial x} U_x + \frac{\partial}{\partial y} U_y + \frac{\partial}{\partial z} U_z \right]$$

Thus we have $\quad E_x = -\dfrac{\partial V}{\partial x};\; E_y = -\dfrac{\partial V}{\partial y}$ and $E_z = -\dfrac{\partial V}{\partial z}$

Thus the electric field intensity at a point in the electric field is equal to negative potential gradient in that direction.

4.7 POISSON'S AND LAPLACE'S EQUATIONS

From the point form of Gauss law it is extremely easy to obtain Poisson's equation

$$\nabla \cdot D = \rho_v \qquad ...(4.22)$$

From definition of D, $\quad D = \varepsilon E \qquad ...(4.23)$

It can also be written as $\quad E = -\nabla V$

By substitution we have $\nabla \cdot D = \nabla \cdot (\varepsilon E) = -\nabla \cdot (\varepsilon \nabla V) = \rho_v$

$$\nabla \cdot (\nabla V) = -\frac{\rho_v}{\varepsilon}$$

$$\therefore \qquad \nabla^2 V = -\frac{\rho_v}{\varepsilon} \qquad ...(4.24)$$

This equation is called Poisson's equation. But the double ∇ operator must be understood and expanded at least in Cartesian coordinates, before the equation can be used for deriving other equations.

$$\nabla \cdot A = \frac{\partial A_x}{\partial x} + \frac{\partial A_y}{\partial y} + \frac{\partial A_z}{\partial z}$$

$$\nabla V = \frac{\partial V}{\partial x} U_x + \frac{\partial V}{\partial y} U_y + \frac{\partial V}{\partial z} U_z$$

$$\therefore \qquad \nabla \cdot \nabla V = \frac{\partial}{\partial x}\left(\frac{\partial V}{\partial x}\right) + \frac{\partial}{\partial y}\left(\frac{\partial V}{\partial y}\right) + \frac{\partial}{\partial y}\left(\frac{\partial V}{\partial y}\right)$$

$$= \frac{\partial^2 V}{\partial x^2} + \frac{\partial^2 V}{\partial y^2} + \frac{\partial^2 V}{\partial z^2} \qquad ...(4.25)$$

Usually, the operation $\nabla \cdot \nabla = \nabla^2$

$$\therefore \qquad \nabla^2 V = \frac{\partial^2 V}{\partial x^2} + \frac{\partial^2 V}{\partial y^2} + \frac{\partial^2 V}{\partial z^2} \qquad ...(4.26)$$

In Cartesian coordinates

$$\therefore \qquad \nabla^2 V = -\frac{\rho_v}{\varepsilon}$$

If $\rho_v = 0$, it indicates zero volume charge density and it is known that for a charge free space $\rho_v = 0$. Since there are no free charges in dielectrics, equation (4.24) becomes

$$\nabla^2 V = 0 \qquad ...(4.27)$$

This is called Laplace's equation. The ∇^2 operation of V is called laplacian of V.

In Cartesian coordinates Laplace's equation is given by

$$\nabla^2 V = \frac{\partial^2 V}{\partial x^2} + \frac{\partial^2 V}{\partial y^2} + \frac{\partial^2 V}{\partial z^2} = 0 \text{ (Cartesian)} \qquad ...(4.28)$$

In Cylindrical coordinate Laplace's equation can be written as

$$\nabla^2 V = \frac{1}{\rho}\frac{\partial}{\partial \rho}\left(\rho \frac{\partial V}{\partial \rho}\right) + \frac{1}{\rho^2}\left(\frac{\partial^2 V}{\partial \phi^2}\right) + \left(\frac{\partial^2 V}{\partial z^2}\right) \qquad ...(4.29)$$

In spherical coordinates, Laplace's equation is given by

$$\nabla^2 V = \frac{1}{r^2}\frac{\partial}{\partial r}\left(r^2 \frac{\partial V}{\partial r}\right) + \frac{1}{r^2 \sin\theta}\frac{\partial}{\partial \theta}\left(\sin\theta \frac{\partial V}{d\theta}\right) + \frac{1}{r^2 \sin^2\theta}\frac{\partial^2 V}{\partial \phi^2}$$

...(4.30)

4.8 SOLUTION OF LAPLACE'S EQUATION

The equation which is used for solving the second order partial differential equation is known as Laplace's equation. The first and simplest method is the direct integration method. The other method is the product method which is used only in difficult problems. The direct integration method is applicable, however, only to problems of one dimension or in which potential V is a function of only one variable of three coordinates.

4.8.1 Solution of Laplace's Equation in Rectangular Coordinates

Let a parallel plate capacitor be considered with one plate kept at $x = 0$ and other plate kept at $x = d$ as shown in Fig. 4.6. Assume that the space between plates is filled with dielectric material. The left plate is at 'zero potential and that of right plate is at potential 'V_0'. Laplace's equation is used to calculate the potential distribution between the plates.

As there is no variation in 'y' and 'z' direction, the problem is one dimensional and Laplace's equation can be written as

$$\frac{\partial^2 V}{\partial x^2} + \frac{\partial^2 V}{\partial y^2} + \frac{\partial^2 V}{\partial z^2} \qquad ...(4.31)$$

Substituting $\frac{\partial^2 V}{\partial y^2} = 0$ and $\frac{\partial^2 V}{\partial z^2} = 0$ in equation (4.31).

$$\therefore \qquad \frac{\partial^2 V}{\partial x^2} = 0$$

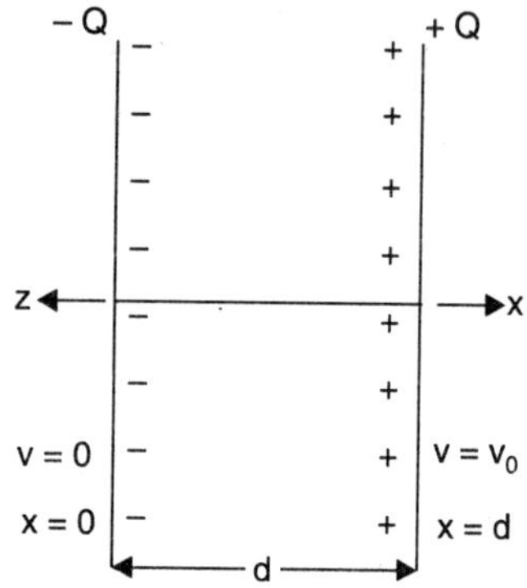

Fig. 4.6 Parallel plate capacitor

The partial derivative can be replaced by ordinary derivative as 'V' is not a function of y and z.

$$\therefore \qquad \frac{d^2 V}{dx^2} = 0$$

On integration we obtain

$$\int \frac{d^2 V}{dx^2} = 0 \quad \text{(or)} \quad \frac{dV}{dx} = \text{constant} = A \text{ (say)} \qquad ...(4.32)$$

Again integrating $\int \frac{dV}{dx} = \int A$

$$\int dV = \int A\, dx$$

(or) $$V = Ax + B \qquad ...(4.33)$$

This is the solution of Laplace's equation in Cartesian coordinate system while considering variation along x-direction only.

In equation (4.33) A and B are constants of integration which can be determined by the following boundary conditions.

At $x = 0$; $\quad V = 0$

Substituting in equation (4.33) we get

$B = 0$, *i.e.*, $V = Ax$

When $x = d$; $V = V_0$

Substituting in equation (4.33), we get

$$V_0 = Ad \quad \text{(or)} \quad A = \frac{V_0}{d}$$

Substitute $A = \frac{V_0}{d}$ and $B = 0$ in equation (4.33)

$$\therefore \qquad V = \left(\frac{V_0}{d}\right) x \text{ volts.} \qquad ...(4.34)$$

4.8.2 Solution of Laplace's Equation in Cylindrical Coordinates

Consider a cylindrical capacitor with inner radius 'a' and outer radius 'b' as shown in Fig. 4.7. It is assumed that variation of field is only w.r.t. 'ρ' and not with 'ϕ' (or) 'z'.

Laplacian of V in cylindrical coordinates is given as

$$\nabla^2 V = \frac{1}{\rho}\frac{\partial}{\partial \rho}\left(\rho \frac{\partial V}{\partial \rho}\right) + \frac{1}{\rho^2}\frac{\partial^2 V}{\partial \phi^2} + \frac{\partial^2 V}{\partial z^2} \quad \text{...(4.35)}$$

According to the assumption taken above, equation (4.35) in cylindrical coordinate can be reduced to

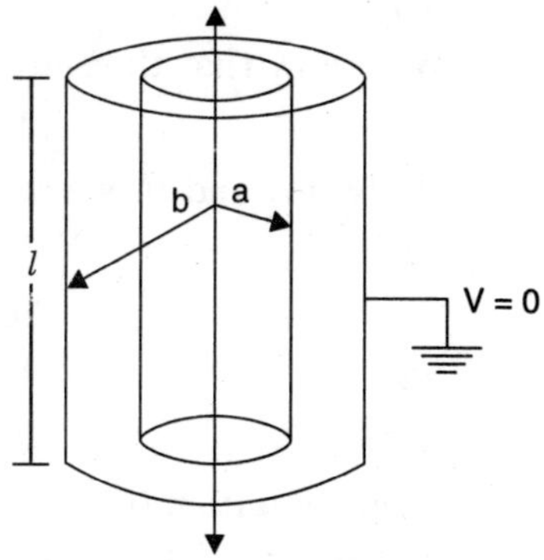

Fig. 4.7 Cylindrical capacitor

$$\nabla^2 V = \frac{1}{\rho}\frac{\partial}{\partial \rho}\left(\rho \frac{\partial V}{\partial \rho}\right) = 0$$

(or) $$\frac{1}{\rho}\frac{\partial}{\partial \rho}\left(\rho \frac{\partial V}{\partial \rho}\right) = 0$$

$$\rho \frac{\partial V}{\partial \rho} = A$$

$$\frac{\partial V}{\partial \rho} = \frac{A}{\rho}$$

$$\int dv = A \int \frac{\partial \rho}{\rho}$$

(or) $$V = A \ln \rho + B \quad \text{...(4.36)}$$

It is needed to determine potential in the region between inner cylinder and outer cylinder. 'ρ' is the radius of cylinder in cylindrical coordinate system.

When $\rho = b$; $V = 0$

$\therefore \quad 0 = A \ln b + B \Rightarrow B = -A \ln b$

Substituting 'B' in equation (4.36)

$$V = A \ln \rho - A \ln b$$

$$V = A \ln\left(\frac{\rho}{b}\right) \quad \text{...(4.37)}$$

Where, $\rho = a$, $\quad V = V_0$

$$\therefore \quad V_0 = A \ln\left(\frac{a}{b}\right) \quad \text{...(4.38)}$$

(or) $$A = \frac{V_0}{\ln\left(\frac{a}{b}\right)}$$

From equations (4.37) and (4.38)

$$\frac{V}{V_0} = \frac{A \ln\left(\frac{\rho}{b}\right)}{A \ln\left(\frac{a}{b}\right)}$$

$$V = V_0 \frac{\ln\left(\frac{\rho}{b}\right)}{\ln\left(\frac{a}{b}\right)} \qquad ...(4.39)$$

Example 4.6: *In cylindrical coordinates, V = 50 volts at ρ = 5 mm and V = 0 at ρ = 65 mm. Find the voltage at ρ = 120 mm, if the potential depends only on ρ.*

Solution:

Given $V = 50$ V at $\rho = 5$ mm; and $V = 0$ at $\rho = 65$ mm

We know that the solution of Laplace's equation to find V is

$$V = A \ln \rho + B$$

To find A and B, we can apply boundary conditions specified in the problem.

(1) $V = 50$ at $\rho = 5$ mm

$\therefore$ $50 = A \ln 5 + B$ ...(a)

(2) $V = 0$ V at $\rho = 65$ mm

$\therefore$ $0 = A \ln 65 + B$ ...(b)

From equation (b), $B = -\ A \ln 65$

Substituting the value of B in equation (a), we get

$$50 = A \ln 5 - A \ln 65$$
$$= A (\ln 5 - \ln 65)$$
$$= A \ln\left(\frac{5}{65}\right) = -2.5770\,A$$

or $A = \frac{50}{-2.577} = -19.40$

We know that $B = -A \ln 65$

$$= -(-19.40) \ln 65$$
$$= 80.98$$

$\therefore$ Potential field $V = -19.40 \ln \rho + 80.98$

To find V at $\rho = 140$ mm, put $\rho = 140$ mm in the above expression V

$$V = -19.40 \ln 140 + 80.98$$
$$= -14.88 \text{ V.}$$

4.8.3 Solution of Laplace's Equation in Spherical Coordinates

Consider two concentrically placed spherical capacitors of radii 'a' and 'b' as shown in Fig. 4.8. Let the potential of inner sphere be V_0 and let the outer sphere be grounded so that its potential is zero.

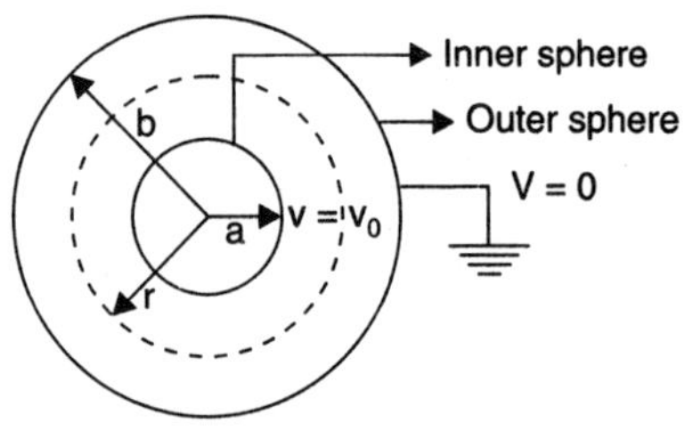

Fig. 4.8 Spherical capacitor

Let it be required to obtain potential field on basis of Laplace's equation. Laplacian of V in spherical coordinate is given as follows:

$$\nabla^2 V = \frac{1}{r^2}\frac{\partial}{\partial r}\left(r^2 \frac{\partial V}{\partial r}\right) + \frac{1}{r^2 \sin\theta}\frac{\partial}{\partial \theta}\left(\sin\theta \frac{\partial v}{\partial \theta}\right) + \frac{1}{r^2 \sin^2\theta}\frac{\partial^2 V}{\partial \phi^2} = 0 \quad ...(4.40)$$

Consider variation along direction of '*r*' alone. The second and third terms of equations (4.40) vanish.

By Laplace's equation

$$\nabla^2 V = 0$$

$$\frac{1}{r^2}\frac{\partial}{\partial r}\left(r^2 \frac{\partial V}{\partial r}\right) = 0$$

$$\frac{\partial}{\partial r}\left(r^2 \frac{\partial V}{\partial r}\right) = 0$$

The partial derivative is also replaced by ordinary derivative as V is not a function of θ and ϕ.

$$\therefore \quad \frac{\partial}{\partial r}\left(r^2 \frac{\partial V}{\partial r}\right) = 0$$

Integrating w.r.t r

$$r^2 = \frac{\partial V}{\partial r} = A$$

$$\frac{\partial V}{\partial r} = \frac{A}{r^2}$$

Again integrating $\quad V = -\frac{A}{r} + B \quad ...(4.41)$

To evaluate the constants A and B, boundary conditions are applied

We have $V = 0$ at $r = b$

$$0 = -\frac{A}{b} + B \; ; \; B = \frac{A}{b} \quad ...(4.42)$$

Substituting equation (4.42) in (4.41)

$$V = -\frac{A}{r} + \frac{A}{b} \quad ...(4.43)$$

$$V = A\left[\frac{1}{b} - \frac{1}{r}\right]$$

When, $\quad V = V_0$ at $r = a$

$$V_0 = A\left[\frac{1}{b} - \frac{1}{a}\right] \quad ...(4.44)$$

Let $\quad V_0 = A\,k$; where $K = \left[\frac{1}{b} - \frac{1}{a}\right]$

(or) $$A = \frac{V_0}{K}$$

Substituting $A = \frac{V_0}{K}$ in equation (4.43).

$$V = \frac{V_0}{K}\left[\frac{1}{b} - \frac{1}{r}\right] \quad ...(4.45)$$

4.9 ELECTRIC DIPOLE AND DIPOLE MOMENT

4.9.1 Electric Dipole

A combination of two charges + Q and – Q separated by a small distance 'l' constitute an 'Electric dipole' (or) simply a dipole.

4.9.2 Dipole Moment

It is defined as product of charge 'Q' and the distance 'd' between the charges. The electric dipole moment of this combination is defined as vector.

$$\bar{p} = Q\,\bar{d} \quad ...(4.46)$$

Where $\bar{d}$ is vector joining the negative charge to the positive charge. The line along direction of dipole moment is called axis of dipole.

4.10 ELECTRIC POTENTIAL DUE TO DIPOLE

Let the negative charge '– Q' be placed at point 'A' and positive charge '+ Q' be placed at point 'B'.

The separation $AB = d$ and the middle of AB is 'O'. The potential is to be evaluated at a point p where

$$OP = r \quad \text{and} \quad \angle POB = \theta$$

Also let $r >> d$

Let AA' be the perpendicular from A to PO and BB' be the perpendicular from B to PO.

As d is very small compared to 'r'.

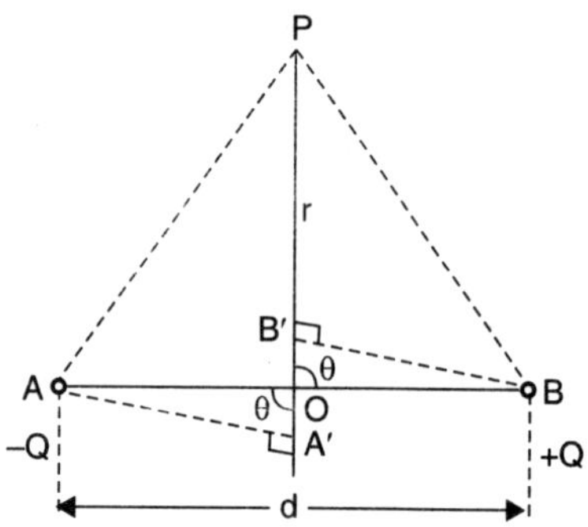

Fig. 4.9

$$\therefore \quad AP \approx A'P = OP + OA'$$

$$= OP + AO\cos\theta$$

$$= r + \frac{d}{2}\cos\theta \quad ...(4.47)$$

Similarly $\quad BP \approx B'P = OP - OB'$

$$= r - \frac{d}{2}\cos\theta \quad ...(4.48)$$

The potential at P due to charge – Q is

$$V_1 = -\frac{1}{4\pi\varepsilon_0}\frac{Q}{AP}$$

$$= -\frac{1}{4\pi\varepsilon_0} \times \frac{Q}{\left(r + \frac{d}{2}\cos\theta\right)} \quad ...(4.49)$$

and that due to charge + Q is

$$V_2 = \frac{1}{4\pi\varepsilon_0}\frac{Q}{BP}$$

$$= \frac{1}{4\pi\varepsilon_0} \times \frac{Q}{\left(r - \frac{d}{2}\cos\theta\right)} \quad \text{...(4.50)}$$

The net potential at P due to dipole is

$$V = V_1 + V_2$$

$$= \frac{1}{4\pi\varepsilon_0}\frac{Qd\cos\theta}{r^2}$$

or

$$V = \frac{p\cos\theta}{4\pi\varepsilon_0 r^2} \quad \text{...(4.51)}$$

Generalized definition of electric dipole

In general any charge distribution that produces electric potential given by

$$V = \frac{p\cos\theta}{4\pi\varepsilon_0 r^2}$$

is called an electric dipole. The constant P is called dipole moment and direction from which the angle θ is measured to get above equation is called direction of dipole moment.

Example 4.7: *Show that the potential due to a electric dipole satisfies Laplace's equation.*

Solution: Potential due to electric dipole, $V = \dfrac{Ql\cos\theta}{4\pi\varepsilon r^2}$

And, Laplace's equation is $\nabla^2 V = 0$

Now

$$\nabla^2 V = \frac{1}{r^2}\frac{\partial}{\partial r}\left(r^2\frac{\partial V}{\partial r}\right) + \frac{1}{r^2\sin\theta}\frac{\partial}{\partial\theta}\left(\sin\theta\frac{\partial v}{\partial\theta}\right) + \frac{1}{r^2\sin^2\theta}\frac{\partial^2 V}{\partial\phi^2}$$

As V for dipole depends only on θ and r; its variation with ϕ is zero.

$\therefore$

$$\nabla^2 V = \frac{1}{r^2}\frac{\partial}{\partial r}\left[r^2\frac{\partial}{\partial r}\left(\frac{Ql\cos\theta}{4\pi\varepsilon r^2}\right)\right] + \frac{1}{r^2\sin\theta}\frac{\partial}{\partial\theta}\left[\sin\theta\left(\frac{Ql\cos\theta}{4\pi\varepsilon r^2}\right)\right] + 0$$

$$= \frac{1}{r^2}\frac{\partial}{\partial r}\left[-\frac{Ql\cos\theta}{2\pi\varepsilon r}\right] + \frac{1}{r^2\sin\theta}\frac{\partial}{\partial\theta}\left(-\frac{Ql\sin^2\theta}{4\pi\varepsilon r^2}\right)$$

$$= \frac{1}{r^2}\left(\frac{Ql\cos\theta}{2\pi\varepsilon r^2}\right) + \frac{1}{r^2\sin\theta}\left(\frac{-Ql}{4\pi\varepsilon r^2}2\sin\theta\cos\theta\right)$$

$$= \frac{Ql\cos\theta}{2\pi\varepsilon r^4} - \frac{Ql\cos\theta}{2\pi\varepsilon r^4} = 0$$

Hence Laplace's equation is satisfied by the potential of a dipole.

4.11 ELECTRIC FIELD DUE TO A DIPOLE

We know that potential due to dipole is

$$V = \frac{p\cos\theta}{4\pi\varepsilon_0 r^2} \quad ...(4.52)$$

Using gradient relationship in spherical coordinates, we find electric field intensity $\overline{E}$ at point P is

i.e., $\overline{E} = -\Delta V$

$$= -\left[\frac{\partial V}{\partial r}U_r + \frac{1}{r}\frac{\partial V}{\partial \theta}U_\theta + \frac{1}{r\sin\theta}\frac{\partial V}{\partial \phi}U_\phi\right] \quad ...(4.53)$$

Potential V is independent of ϕ. Thus $\frac{\partial V}{\partial \phi} = 0$

$$\overline{E} = -\left[\frac{\partial V}{\partial r}U_r + \frac{1}{r}\frac{\partial V}{\partial \theta}U_\theta\right] \quad ...(4.54)$$

$$\frac{\partial V}{\partial r} = \frac{-2p\cos\theta}{4\pi\varepsilon_0 r^3} \quad ...(4.55)$$

and

$$\frac{\partial V}{\partial \theta} = \frac{-p\sin\theta}{4\pi\varepsilon_0 r^2}$$

$$\therefore \quad \frac{1}{r}\frac{\partial V}{\partial \theta} = -\frac{p\sin\theta}{4\pi\varepsilon_0 r^3} \quad ...(4.56)$$

Substituting equations (4.55) and (4.56) in equation (4.54)

$$\overline{E} = \frac{p}{4\pi\varepsilon_0 r^3}[2\cos\theta\, U_r + \sin\theta\, U_\theta] \quad ...(4.57)$$

Example 4.8: *A dipole at the origin in free space has* $\overline{P} = 95\pi\,\varepsilon_0\, U_Z$ *C-m. (a) Find V at P (x, y, z) in Cartesian coordinated (b) Find* $\overline{E}$ *at P (x, y, z) in Cartesian coordinate.*

Solution:

(*a*)

$$V = \frac{\overline{P}\cdot(\overline{r}_2 - \overline{r}_1)}{4\pi\varepsilon_0 |\overline{r}_2 - \overline{r}_1|^3}\,; \text{ where } \overline{P} = 95\,\pi\varepsilon_0\, U_z \text{ C-m.}$$

Point 1 is the point at which dipole is located = (0, 0, 0)

Point 2 is the point at which potential is desired = (x, y, z)

$$\overline{r}_2 - \overline{r}_1 = (xU_x + yU_y + zU_z) - (0U_x + 0U_y + 0U_z)$$
$$= xU_x + yU_y + zU_z$$

$$\overline{P}\cdot(\overline{r}_2 - \overline{r}_1) = 95\,\pi\,\varepsilon_0\, z$$

$$|\,\overline{r}_2 - \overline{r}_1\,| = \sqrt{x^2 + y^2 + z^2}$$

$$\therefore \quad V = \frac{95\pi\varepsilon_0 z}{4\pi\varepsilon_0(x^2+y^2+z^2)^{3/2}} = \frac{22.5z}{(x^2+y^2+z^2)^{1.5}}$$

(*b*) We have

$$E = -\frac{\partial}{\partial x}\left[\frac{22.5z}{(x^2+y^2+z^2)^{1.5}}\right]U_x + \frac{\partial}{\partial y}\left[\frac{22.5z}{(x^2+y^2+z^2)^{1.5}}\right]U_y + \frac{\partial}{\partial z}\left[\frac{22.5z}{(x^2+y^2+z^2)^{1.5}}\right]U_z$$

$$= \frac{67.5\,xzU_x + 67.5\,zyU_y + 22.5\,(x^2+y^2-2z^2)U_z}{(x^2+y^2+z^2)^{2.5}}.$$

Example 4.9: *Consider a dipole shown in Fig. 4.10. It is located at the origin of x – y system and dipole pointing along the positive y-axis. Calculate E at point P(x, y).*

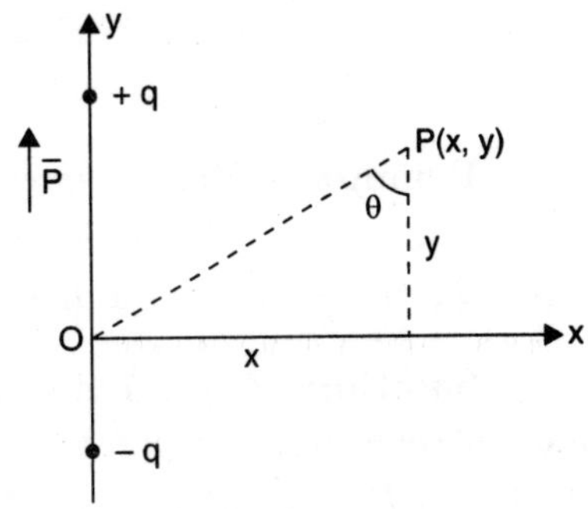

Fig. 4.10

Solution: The potential of a dipole is given by

$$V = \frac{p\cos\theta}{4\pi\varepsilon_0 r^2}$$

From the Fig. 4.10, it can be written as

$$r = \sqrt{x^2+y^2}$$

and $$\cos\theta = \frac{y}{r} = \frac{y}{(x^2+y^2)^{1/2}}$$

$$\therefore \qquad V = \frac{1}{4\pi\varepsilon_0}\frac{py}{(x^2+y^2)^{3/2}}$$

The component of field along *x*-axis is given by

$$E_x = -\frac{\partial V}{\partial x} = -\frac{\partial}{\partial x}\left[\frac{p}{4\pi\varepsilon_0}\times\frac{y}{(x^2+y^2)^{3/2}}\right]$$

$$= -\frac{py}{4\pi\varepsilon_0}\times\left[-\frac{3}{2}(x^2+y^2)^{-5/2}(2x)\right] = \frac{3p}{4\pi\varepsilon_0}\times\frac{xy}{(x^2+y^2)^{5/2}}$$

$$= \frac{3p(r\sin\theta)(r\cos\theta)}{4\pi\varepsilon\, r^3} = \frac{3p\sin\theta\cos\theta}{4\pi\varepsilon\, r^3}$$

Similarly the component of field along *y*-axis can be written as

$$E_y = -\frac{\partial V}{\partial y} = -\frac{\partial}{\partial y}\left[\frac{p}{4\pi\varepsilon_0}\times\frac{y}{(x^2+y^2)^{3/2}}\right]$$

$$= -\frac{p}{4\pi\varepsilon_0}\times\frac{\partial}{\partial y}\,[y\times(x^2+y^2)^{-3/2}]$$

$$= \frac{-p}{4\pi\varepsilon_0}\left[y - \frac{3}{2}(x^2+y^2)^{-5/2}\,2y + (x^2+y^2)^{-3/2}\right]$$

$$= \frac{p}{4\pi\varepsilon_0}\left[\frac{3y^2}{(x^2+y^2)^{5/2}} - \frac{1}{(x^2+y^2)^{3/2}}\right]$$

$$E_y = \frac{p}{4\pi\varepsilon_0}\times\frac{1}{(x^2+y^2)^{3/2}}\left[\frac{3y^2}{x^2+y^2} - 1\right]$$

$$\therefore \qquad E_y = \frac{p}{4\pi\varepsilon_0} \times \frac{1}{r^3} [3\cos^2\theta - 1]$$

If we put $\theta = 0$ *i.e.*, point '*P*' is on *y*-axis

Then $$E_x = 0 \quad \text{and} \quad E_y = \frac{p}{4\pi\varepsilon_0} \times \frac{1}{r^3} (2)$$

$$= \frac{p}{2\pi\varepsilon_0 r^3}$$

If point '*P*' is on *x*-axis, then $\theta = 90°$

$$\therefore \qquad E_x = 0 \quad \text{and} \quad E_y = \frac{-p}{4\pi\varepsilon_0 r^3}.$$

Example 4.10: *Find the potential due to an electric dipole consisting of* $+Q$ *at* $\left(\frac{a}{2}, 0, 0\right)$ *and* $-Q$ *at* $\left(\frac{-a}{2}, 0, 0\right)$ *at a distance point* (r, θ, ϕ) *in spherical coordinates.*

Solution: Here the dipole consists of charges $+Q$ and $-Q$ separated through a small distance '*a*'. The dipole moment is $P = Q \cdot dU_x$, where U_x is the unit vector along the positive *x*-direction, *x*-direction (the direction of *P* is from $-Q$ to $+Q$).

Referring to Fig. 4.11, the *x*-coordinate of point '2' is

$$X = (r \sin\theta) \cos\phi$$

And d = distance between the two charges $+Q$ and $-Q$ constituting an electric dipole

$$= 2x = 2r\sin\theta\cos\phi$$

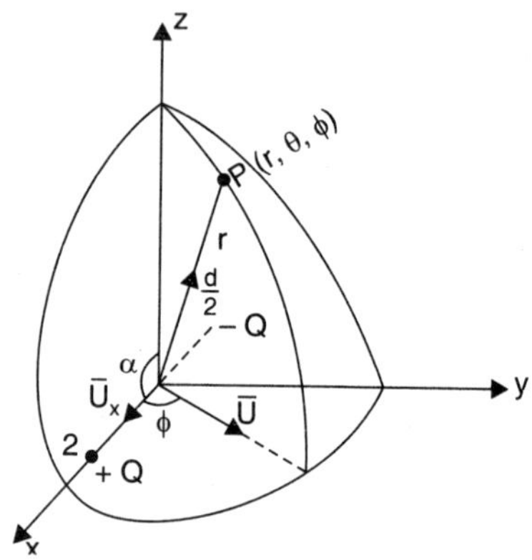

Fig. 4.11

Now, the potential at $P(r, \theta, \phi)$ due to the electric dipole is

$$V = \frac{Qa\cos\alpha}{4\pi\varepsilon_0 r^2} = \frac{Qa(u_x \cdot u_r)}{4\pi\varepsilon_0 r^2} \qquad ...(a)$$

But, $$u_x \cdot u_r = \cos\alpha \qquad ...(b)$$

Where, $$u_r = u_\rho \sin\theta \qquad ...(c)$$

From equations (*b*) and (*c*)

$$u_x \cdot u_r = u_x \cdot u_\rho \sin\theta = (u_x \cdot u_\rho)\sin\theta$$

$$= \cos\phi\sin\theta \qquad ...(d)$$

Introducing equation (*d*) in equation (*a*), we have

$$V = \frac{Qa\sin\theta\cos\theta}{4\pi\varepsilon_0 r^2}.$$

Example 4.11: *Show that the family of equipotential for a dipole may be given by* $r^2 = d_1 \cos\theta$. *Further show that* $r = d_2 \sin^2\theta$ *represents the line of force of the dipole, where* d_1 *and* d_2 *are constants.*

Solution: Potential at for point $P(r, \theta, \phi)$ for a dipole, $\phi = \dfrac{QL\cos\theta}{4\pi\varepsilon_0 r^2}$. For equi-potentials V should be constant.

Now $V = P'\dfrac{\cos\theta}{r^2}$, where, $P' = \dfrac{QL}{4\pi\varepsilon_0}$ which is constant for a particular dipole.

$$r^2 = \frac{P'}{V}\cos\theta = d_1\cos\theta$$

where, $$d_1 = \frac{P'}{V} = \text{constant}$$

We know that $$\frac{dx}{E_x} = \frac{dy}{E_y} = \frac{dz}{E_z}$$

$\therefore$ $$\frac{dr}{E_r} = \frac{r\,d\theta}{E_0}$$

Now $$E_r = \frac{2QL\cos\theta}{4\pi\varepsilon_0 r^3} \text{ and } E_0 = \frac{QL\cos\theta}{4\pi\varepsilon_0 r^3}$$

$\therefore$ $$dr = E_r\frac{r\,d\theta}{E_\theta} = \left(\frac{2QL\cos\theta}{4\pi\varepsilon_0 r^3}\right)\left(\frac{4\pi\varepsilon r^3}{QL\sin\theta}\right) r\,d\theta$$

$$= \frac{2r\cos\theta}{\sin\theta}d\theta$$

$\therefore$ $$\frac{dr}{r} = 2\frac{\cos\theta}{\sin\theta}d\theta$$

or $$\ln r = 2\ln\sin\theta + d = \ln\sin^2\theta + \ln d_2$$

$$= \ln d_2\sin^2\theta$$

$\therefore$ $$r = d_2\sin^2\theta.$$

Example 4.12: *A dipole at the origin in free space has $\overline{P} = 70\pi\varepsilon_0 u_z$ C-m. (a) Find V at P(x, y, z) in Cartesian coordinates (b)Find E at P (x, y, z) in Cartesian coordinates.*

Solution:

(*a*) $$V = \frac{\overline{P}\cdot(\overline{r}_2 - \overline{r}_1)}{4\pi\varepsilon_0 |\overline{r}_2 - \overline{r}_1|^3}\text{ ; where } P = 70\pi\varepsilon_0 U_z \text{ C-m.}$$

Point 1 is the point at which dipole is located = (0,0,0) and Point 2 is the point at which potential is desired = (x, y, z)

$$\overline{r}_2 - \overline{r}_1 = (xU_x + yU_y + zU_z) - (0U_x + 0U_y + 0U_z)$$
$$= xU_x + yU_y + zU_z$$

$$P.(\overline{r}_2 - \overline{r}_1) = 70\,\pi\,\varepsilon_0\,z$$

$$|\,\overline{r}_2 - \overline{r}_1\,| = \sqrt{x^2 + y^2 + z^2}$$

$\therefore$ $$V = \frac{70\pi\varepsilon_0 z}{4\pi\varepsilon_0 (x^2 + y^2 + z^2)^{3/2}} = \frac{17.5z}{(x^2 + y^2 + z^2)^{1.5}}$$

(b) We have
$$E = -\left\{\frac{\partial}{\partial x}\left[\frac{17.5z}{(x^2+y^2+z^2)^{1.5}}\right]U_x + \frac{\partial}{\partial y}\left[\frac{17.5z}{(x^2+y^2+z^2)^{1.5}}\right]U_y + \frac{\partial}{\partial z}\left[\frac{17.5z}{(x^2+y^2+z^2)^{1.5}}\right]U_z\right\}$$
$$= \frac{52.5\,xzU_x + 52.5\,zyU_y + 17.5\,(x^2+y^2-2z^2)U_z}{(x^2+y^2+z^2)^{1.5}}.$$

4.12 TORQUE ON AN ELECTRIC DIPOLE PLACED IN ELECTRIC FIELD

Consider as electric dipole placed in a uniform electric field $\overline{E}$.

The charge separation is at $2d$ meters. The force acting at A is $\overline{F} = \overline{E}Q$ and the force at B is $\overline{F} = -\overline{E}Q$. Let θ be the orientation of the dipole with the uniform electric field (E). Resolving forces at A and B along and perpendicular to AB. The component of $\overline{F}$ along AB are neutralized and the components perpendicular at AB are $F \sin \theta$.

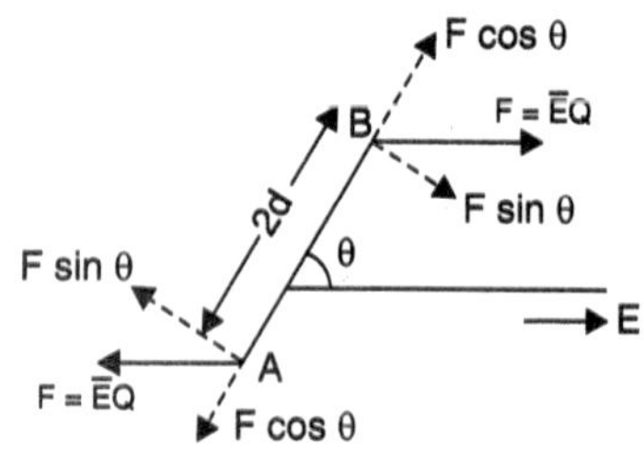

Fig. 4.12

The torque on the dipole is given by

$$T = F \sin\theta \times 2d = EQ\,2d \sin\theta \quad ...(4.58)$$
$$= 2dQE \sin\theta \quad ...(4.59)$$
$$= PE \sin\theta \quad ...(4.60)$$

where, $P = 2dQ$ = dipole moment, acts from B to A

The torque on the dipole is a vector and is denoted as $T = \overline{P} \times \overline{E}$. ...(4.61)

The torque on the dipole is only $\overline{P} \times \overline{E}$ but not $\overline{E} \times \overline{P}$ because the cross product $\overline{P} \times \overline{E}$ rotates $\overline{P}$ towards $\overline{E}$ in defining the cross product and is in the clockwise direction. This rotation of $\overline{P}$ in the clockwise direction coincides with the direction of the couple produced by the forces $F \sin\theta$ acting at the ends A and B.

4.13 WORK DONE IN ELECTROSTATIC FIELD

In order to bring a positive charge from infinity into the field of another positive charge it requires work to be done, the work being done by the external source moving the charge. Imagine that the external source carries the charge up to a point near the fixed charge and holds it there. Energy must be conserved and energy expended in bringing this charge into position now represents potential energy.

In order to find potential energy present in a system of charges work done by external source in positioning the charges should be found out first. Consider an empty universe. Bringing a charge Q_1 from infinity to any position requires no work, as there is no field present. The positioning of Q_2 at a point in field of Q_1 requires an amount of work given by product of charge Q_2 and potential at that point due to Q_1. If this potential is represented as V_{21} where the first subscript indicates the location and second subscript indicates the source, then

Work to position $Q_2 = Q_2 V_{21}$ ...(4.62)

Similarly the work required to position each additional charge in field of all these already present can be represented as

Work to position $Q_3 = Q_3 V_{31} + Q_3 V_{32}$

Similarly on $Q_4 = Q_4 V_{41} + Q_4 V_{42} + Q_4 V_{43}$ and so on ...(4.63)

So total positioning work = potential energy of field = W_E

$$= Q_2 V_{21} + Q_3 V_{31} + Q_3 V_{32} + Q_4 V_{41} + Q_4 V_{42} \quad ...(4.64)$$

Noting the form of representative term is above equation

$$Q_3 V_{31} = Q_3 \cdot \frac{Q_1}{4\pi\varepsilon_0 R_{13}} = Q_1 \frac{Q_3}{4\pi\varepsilon_0 R_{31}}$$

where R_{13} and R_{31} each represent the scalar distance between Q_1 and Q_3. So if each term is replaced by its equal

$$W_E = Q_1 V_{12} + Q_1 V_{13} + Q_2 V_{23} + Q_1 V_{14} + Q_2 V_{24} + Q_3 V_{34} + \ ...(4.65)$$

Combining the eqns.(4.64) and (4.63), we can write as

$$2W_E = Q_1 (V_{12} + V_{13} + V_{14} + ...) + Q_2 (V_{21} + V_{23} + V_{24} + ...) + Q_3 (V_{31} + V_{32} + V_{34} + ...) \quad ...(4.66)$$

In other words $V_{12} + V_{13} + V_{14} + ... = V_1$ ...(4.67)

The potential at location of Q_1 due to presence of Q_2, Q_3 can write as

$$W_E = \frac{1}{2} (Q_1 V_1 + Q_2 V_2 + Q_3 V_3 + \ ...) \quad ...(4.68)$$

$$W_E = \frac{1}{2} \sum_{m=1}^{N} Q_m V_m \quad ...(4.69)$$

Example 4.13: *Find the work done in moving a point charge 6 μC from (5, π, 0) to (7, π,0) in the field* $E = \frac{10^5}{\rho} u_\rho + 10^4 z\, u_z$.

Solution:

We know that $\vec{E} = E_\rho \bar{u}_\rho + E_\theta \bar{u}_\theta + E_z \bar{u}_z$

But, $E_\rho = \frac{10^5}{\rho}$; $E_z = 10^4 z$

And $\vec{dl} = d\rho \bar{u}_\rho + \rho\, d\theta\, \bar{u}_\theta + dz\, \bar{u}_z$

$$\vec{E} \cdot \vec{dl} = E_\rho\, d\rho$$

Work done $= -Q \int \vec{E} \cdot \vec{dl}$

$$= -6 \times 10^{-6} \int E_\rho \, d\rho = -6 \times 10^{-6} \int_5^7 \frac{10^5}{\rho} \, d\rho$$

$$= -6 \times 10^{-6} \times 10^5 \left[\ln \rho\right]_4^6$$

$$= -0.2422 \text{ Joules.}$$

Example 4.14: *Find the work done in moving a 5 μC charge from the origin (0,0,0) to point P (4, – 2, 6) through the field E = 5xyz u_x + 3x^2z u_y +2 xy^2 u_z V/m, via a straight line x = – 5y, z = 3x.*

Solution: Given that

$$\vec{E} = 5xyz\,\overline{u}_x + 3x^2 z\overline{u}_y + 2xy^2\overline{u}_z$$

$$\vec{dl} = dx\,\overline{u}_x + dy\,\overline{u}_y + dz\,\overline{u}_z$$

$$\vec{E} \cdot \vec{dl} = 5xyz\,dx + 3x^2 z dy + 2xy^2 dz$$

Also $x = -5y$; $z = 3x$

and $dx = -5dy$; $dz = 3dx$

$$\text{Total work done} = -Q \int \vec{E} \cdot \vec{dl}$$

$$= Q \int_{(0,0,0)}^{(4,-2,6)} (5xyz\,dx + 3x^2 z\,dy + 2xy^2 dz)$$

$$= -Q \int_0^4 \left(-3x^3\,dx - \frac{9}{5}x^3\,dx + \frac{6}{25}x^3\,dx\right)$$

$$= -Q \int_0^4 x^3\,dx \left[\frac{-114}{25}\right] = +5 \times 10^{-6} \times \frac{114}{25} \times 64 = 1459.2 \ \mu \text{ joules.}$$

4.14 BEHAVIOUR OF CONDUCTORS IN ELECTRIC FIELD

A conductor has its valence electrons (or) conduction electrons in its outer shell. Whenever sufficient energy is given they proceed into empty conduction band and start conducting current. Now if the conductor is kept under influence of an electric field, the electron will experience a force over it. Now the electron will have a charge Q. Because of this force the electron would accelerate and continuously increase its velocity and make collisions with other electrons and a constant velocity will soon be obtained. This velocity V_d is called as drift velocity and is linearly related to electric field intensity E by mobility of electron in given material

$$V_d = -\mu_e E \qquad \text{...(4.70)}$$

The mobility is positive by definition. But since the electron velocity is in a direction opposite to direction of E, so the negative sign.

Current density 'J' is given by

$$J = \rho_v V_d \qquad \text{...(4.71)}$$

So, substituting value of V_d, we get

$$= -\rho_v \mu_e E$$

$$= -\rho_\rho \mu_e E \quad ...(4.72)$$

The relationship between J and E can be given by the following equation which can be written as $J = \sigma E$.

where, $\sigma = -\rho_\rho \mu_e$ is defined as conductivity

The conductivity of material depends on temperature and frequency. A material with high conductivity is called as conductor ($\sigma >> 1$) and low conductivity ($\sigma << 1$) is called as insulator.

4.14.1 Convection and Conduction Currents

Electric current is generally caused by motion of electric charges. The current through a given area is the electric charge passing through the area per unit time.

Which implies $$I = \frac{dQ}{dt} \quad ...(4.73)$$

So, the current is transferred at rate of one coulomb/sec. If ΔI flow through the surface Δs the current density is

$$J = \frac{\Delta I}{\Delta s}$$

$$\Delta I = J\,\Delta s \text{ (assuming angle between } J \text{ and } \Delta s \text{ to be 0)} \quad ...(4.74)$$

So, total current flowing through surface S is

$$I = \int_S J \cdot ds \quad ...(4.75)$$

Depending on how current is produced, there are three different kinds of current densities (*a*) convection current density (*b*) conduction current density and (*c*) displacement current density. Convention current does not involve conductors and thus does not satisfy Ohm's law. It occurs when current flows through an insulting medium such as liquid (or) vaccum. A beam of electrons in a vaccum tube is convection current consider a filament shown in Fig. 4.13.

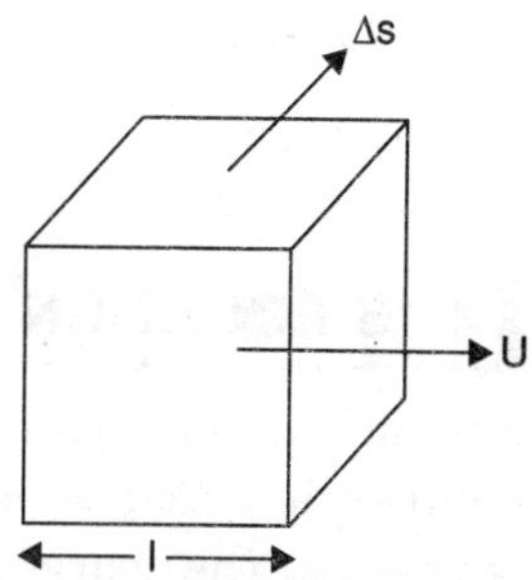

Fig. 4.13

If there is flow of charge of density ρ_v at a velocity U, the current through the filament

$$\Delta J = \frac{\Delta Q}{\Delta t} = \rho_v\,\Delta s\,\frac{\Delta l}{\Delta t} = \rho_v\,\Delta s\,U_y \quad ...(4.76)$$

$$\left(\text{Since velocity } U = \frac{\text{distance}}{\text{time taken}}\right)$$

Now $$J_y = \frac{\Delta I}{\Delta s}$$

$$= \rho_v U_y$$

Hence is general, $$J = \rho_v U \quad ...(4.77)$$

Current I is conduction current and J is convection current density.

Conduction current requires a conductor. A conductor is characterized by large amount of free electrons that provide conduction current due to impressed electric field. When an electric field E is applied the force on electron with charge $-e$ is

$$F = -e\,E \quad ...(4.78)$$

Since electron is not in free space, it suffers collision with other atomic lattice and drifts from one atoms to another. If electron with mass M is moving in an electric field E with an average drift velocity U handling to Newton's law, the average change in momentum of free electrons must match the applied force.

$$\frac{MU}{T} = -eE \quad ...(4.79)$$

$$U = \frac{-e\tau}{M}E$$

Where, τ = average time interval between the collisions.

This indicated drift velocity is proportional to applied field. If there are n electrons per unit volume, the electron charge density is given by

$$\rho_v = -ne \quad ...(4.80)$$

Thus, conduction current density is

$$J = \rho_v\,U$$

$$= \frac{ne^2\tau}{M}E = \sigma E \quad ...(4.81)$$

Where, $\sigma = \frac{ne^2\tau}{M}$ is the conductivity of conductor and also J = σE is called as point form of Gauss law.

4.15 EQUATION OF CONTINUITY

The differential equation relating the current density J and charge density ρ at each point in a circuit is called as equation of continuity. This equation is based on law of conservation of charge *i.e.*, the charge can neither be created nor destroyed. Same charge keen flowing in the circuit. But according to law of conversation of charge

$$I_{cs} = -\frac{dI}{dt} \quad ...(4.82)$$

Where I_{cs} is the current due to flows of charges out of a closed surface S and Q is the charge enclosed by S. In terms of current density J and charge density ρ, I_{cs} and Q are given by

$$I_{cs} = \oint_S \bar{J}\cdot\bar{ds} \quad ...(4.83)$$

And

$$Q = \int_V \rho\cdot ds \quad ...(4.84)$$

Where, V is volume bounded by S.

Substitute equations (4.84) and (4.83) in equation (4.82)

$$\oint_S \bar{J} \cdot \overline{ds} = -\frac{d}{dt}\int_V \rho \cdot dv \qquad ...(4.85)$$

According to Gauss divergence theorem

$$\oint_S \bar{J} \cdot \overline{ds} = \int_V \bar{J}\, dv \qquad ...(4.86)$$

From equations (4.85) and (4.86), we get

$$\int_V \nabla \cdot J\, dv = -\int_V \frac{d\rho}{dt}\, dv \qquad ...(4.87)$$

(or) $$\int_V \left(\nabla \cdot J + \frac{d\rho}{dt}\right) dv = 0 \qquad ...(4.88)$$

This integral must be zero for any arbitrary volume. This is possible when integral is zero. Thus,

$$\left(\nabla \cdot J + \frac{d\rho}{dt}\right) dv = 0 \qquad ...(4.89)$$

This equation is called equation of continuity. For static fields (or) for steady currents $\frac{d\rho}{dt} = 0$ and equation (4.88) reduces to

$$\nabla \cdot J = 0 \qquad ...(4.90)$$

(or) $$\text{div}\, J = 0$$

4.15.1 Ohm's Law in Point Form

Current flowing through a conductor is directly proportional to the potential difference across it provided temperature is kept constant.

The proportionality constant is conductance, G.

$$I = GV; I = \frac{V}{R}$$

$$I = \frac{V}{\frac{\rho l}{a}} = \frac{V_a}{\rho l} \qquad \left[\because\ R = \frac{\rho l}{a}\right]$$

$\Rightarrow$ $$\frac{1}{a} = \frac{V}{l}\frac{1}{\rho}$$

$\therefore$ $$\bar{J} = \sigma \bar{E} \qquad ...(4.91)$$

The reciprocal of resistivity ρ is conductivity σ.

In equation (4.91), σ is conductivity and E is electric field intensity. This equation is called **point form of ohm's law.** So, it can be seen that direction of current density vector J is same as electric field vector E. It relates the current density at a point to the total field E at the point and the conductivity σ of the material.

4.16 DIELECTRICS

These are the substances which do not contain free electrons (or) the number of such electrons is too low to constitute the electric current. In dielectrics, the electrons are tightly bound to the nucleus of atom. Mica, glass, plastic etc. are examples of dielectrics. The dielectric helps in the following three ways:

(*i*) It helps in maintaining two large metal plats at very small separation.

(*ii*) It increases the potential difference which a capacitor can withstand without breakdown.

(*iii*) It increases capacitance of capacitor.

4.16.1 Non-polar Dielectric in an Electric Field

The dielectrics which are polarized only when they are placed in an electric field are called non polar dielectrics.

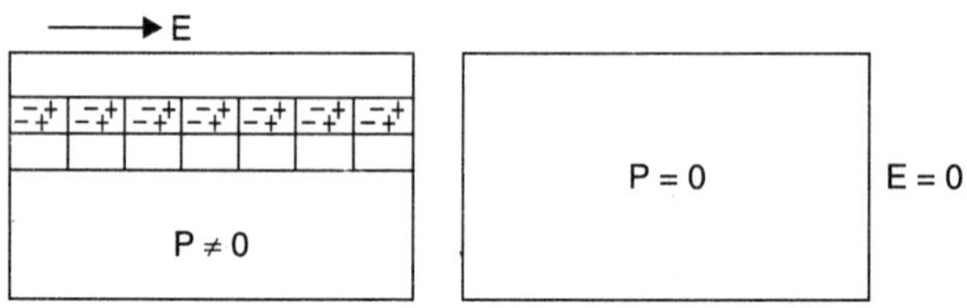

Fig 4.14 (*a*) **Fig 4.14 (*b*)**

In these substances there is no separation between the center of gravity of positive and negative charges of atoms/molecules resulting zero dipole moment. But when the substances are exposed to electric fields, the negative charges of an atom of the substances move towards the field and positive charges move away from the field. This results in formation of dipoles in the direction of field as shown in Fig. 4.14 (*a*) and such dipoles are called as induced dipoles. But when the external field is removed, the formation of above disappears as shown in Fig. 4.14 (*b*). The number of dipoles formed per unit volume is more when the applied electric field intensity is maximum. The dipole moment per unit volume is defined as the number of dipoles formed per unit volume which is also known as polarization.

4.16.2 Polar Dielectric in an Electric Field

In the polar materials there is formation of permanent dipoles in the absence of external electric field.

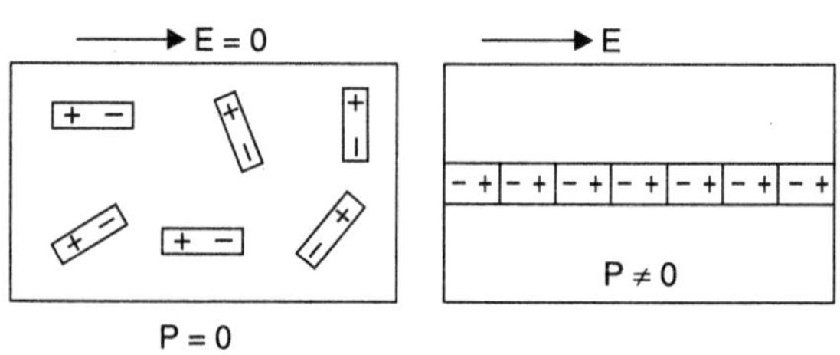

Fig. 4.15 (*a*) **Fig. 4.15 (*b*)**

But these dipoles are arranged in random manner as shown in Fig. 4.15 (*a*) so that the average dipole moment is zero. When these materials are subjected to electric fields, the dipoles align themselves in the direction of field. When the strength of the electric field is more, additional dipoles may be formed. The alignment of dipoles is shown in Fig. 4.15 (*b*) which is due to torque acting on the dipoles.

4.16.3 Gauss's Law in Dielectrics

The Gauss's law states that electric flux ϕ through any closed surface is $\frac{1}{\varepsilon_0}$ times the charge enclosed by the surface. Hence

$$\phi = \int_S \overline{E} \cdot \overline{ds} = \frac{Q}{\varepsilon_0} \qquad \text{...(4.92)}$$

Where E is electric field intensity and ds is the surface area.

Figure 4.16 shows a parallel plate condenser (*a*) without (*b*) with dielectric constant K. The charge Q on plates is same. The Gaussian surfaces are also shown in same figure.

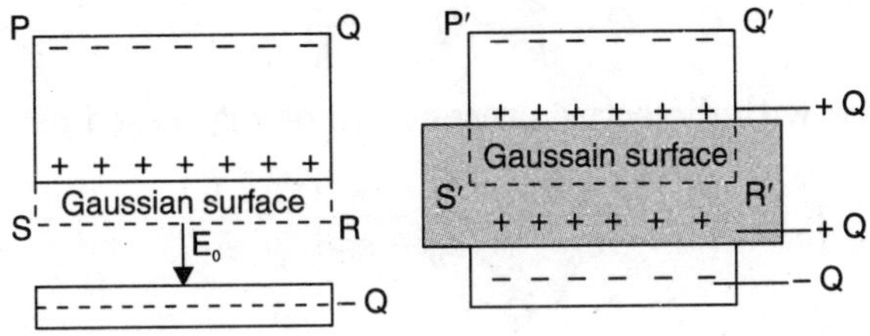

Fig. 4.16 (*a*) Fig. 4.16 (*b*)

(*i*) When no dielectric is present, then Gauss law gives

$$\int_S \overline{E_0} \cdot \overline{ds} = \frac{Q}{\varepsilon_0},$$

where, Q is charge enclosed

(or)
$$E_0 \int_S ds = E_0 A = \frac{Q}{\varepsilon_0}$$

$$\Rightarrow \qquad E_0 = \frac{Q}{\varepsilon_0 A} \qquad \text{...(4.93)}$$

Where, A is area of plate

(*ii*) When dielectric is placed between the plates of condenser the net charge within Gaussian surface is $(Q - Q')$.

where, Q' is induced surface charge

Let E' be resultant field within dielectric, then by Gauss's law

$$\int_S \overline{E} \cdot \overline{ds} = \left(\frac{Q - Q'}{\varepsilon_0}\right)$$

(or)
$$EA = \left(\frac{Q - Q'}{\varepsilon_0}\right)$$

$$\therefore \qquad E = \left(\frac{Q - Q'}{\varepsilon_0 A}\right) = \frac{Q}{\varepsilon_0 A} - \frac{Q'}{\varepsilon_0 A} \qquad ...(4.94)$$

Here it should be remembered that E is less than E_0 because induced charges produce their own field which opposes external field E_0.

By definition of dielectric constant K,

$$K = \frac{E_0}{E} \text{ and hence } E = \frac{E_0}{K}$$

So, we can write equation (4.94) as

$$\frac{E_0}{K} = \frac{Q}{\varepsilon_0 A} - \frac{Q'}{\varepsilon_0 A}$$

Substituting value of E_0 from equation (4.93), we have

$$\frac{Q}{K\varepsilon_0 A} = \frac{Q}{\varepsilon_0 A} - \frac{Q'}{\varepsilon_0 A}$$

(or) $$Q' = Q - \frac{Q}{K} = Q\left[1 - \frac{1}{K}\right] \qquad ...(4.95)$$

Now Gauss law with dielectric present can be expressed as

$$\int_S \overline{E} \cdot \overline{ds} = \frac{Q - Q'}{\varepsilon_0} = \frac{Q}{\varepsilon_0} - \frac{Q}{\varepsilon_0}\left(1 - \frac{1}{K}\right)$$

(or) $$\int_S \overline{E} \cdot \overline{ds} = \frac{Q}{\varepsilon_0} - \frac{Q}{\varepsilon_0} + \frac{Q}{\varepsilon_0 K} = \frac{Q}{\varepsilon_0 K}$$

(or) $$K \int_S \overline{E} \cdot \overline{ds} = \frac{Q}{\varepsilon_0} \qquad ...(4.96)$$

This is Gauss law in dielectrics.

4.17 POLARIZATION

Dielectrics have few electrons available for conduction of current in contrast to conductors which have electron in abundance. So, the main difference between a conductor and dielectric lies in availability of free electron in the outer most shells to conduct current. Although charge in a dielectric is not able to more freely, we can expect a displacement if fore is applied.

Consider an atom of dielectric consisting of negative charge ($-Q$) and a positive charge ($+Q$). When an electric field E is applied, the positive charge is displaced from its position in direction of E by force $F^+ = QE$ while negative charge is displaced in the opposite direction with a force $F^- = QE$. A dipole results and the dielectric is said to be polarized. Now

$$P = Qd \qquad ...(4.97)$$

If there are n dipoles in volume ΔV of dielectric.

$$\text{Total dipole moment} = \sum_{k=1}^{n} Q_k\, d_k$$

Now polarization P is dipole moment/unit volume of dielectric.

$$P = \lim_{\Delta V \to 0} \sum_{k=1}^{n} \frac{Q_k d_k}{\Delta V} \qquad ...(4.98)$$

4.18 ELECTRIC FIELD INSIDE A DIELECTRIC MATERIAL

A dielectric is an electric field can be viewed as free space arrangement of microscopic. In the dielectric material, no molecule has dipole moment and $P = 0$ throughout the material. Somewhere in the interior of dielectric we select an incremental surface element Δs and apply an electric field.

The electric field produces a moment $P = Qd$ in each molecule such that P and d make an angle θ with Δs. If there are n molecules/m^3 then net charge that crosses the element surface in an upward direction is $nQd \cos \theta \, \Delta s$.

(or) $$\Delta Q_b = n\, Q\, d \cdot \Delta s \qquad ...(4.99)$$

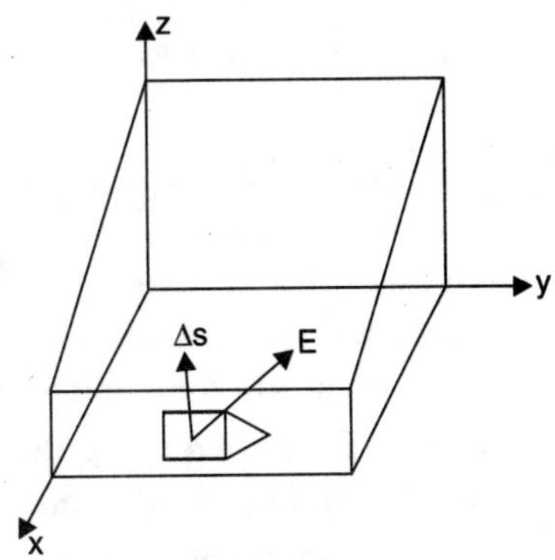

Fig. 4.17

Where subscript Q_b reminds us that we are dealing with a bound charge.

Interms of polarization, we have

$$\Delta Q_b = P \cdot \Delta s$$

If we interpret Δs as an element of closed surface inside dielectric material, the direction of Δs is outward and net increase in bound charge closed surface is obtained through integral.

$$Q_b = -\int_S P \cdot ds \qquad ...(4.100)$$

As Δs is going outward, the bound charge will be decreasing. So, the negative sign is present. The above relation is like Gauss law. The total charge is written as bound charge + free charge. It can be given as follows:

$$Q_T = Q_b + Q \qquad ...(4.101)$$

$$Q_T = \oint_S \varepsilon_0\, E \cdot ds$$

It can also be written as

$$Q = Q_T - Q_b$$

$$= \oint \varepsilon_0\, E \cdot ds - (-) \oint P \cdot ds$$

$$= \oint_S (\varepsilon_0\, E + P) \cdot ds \qquad ...(4.102)$$

Now define D as $\quad D = \varepsilon_0 E + P$

Thus there is an added term to D which appears when a polarization material is present.

Utilizing several volume charge distribution

$$Q_b = \int_V \rho_b \, dv$$

Where ρ_b = bound volume charge density.

$$Q = \int_V \rho_v \, dv$$

Where ρ_v = volume charge density.

$$Q_T = \int_V \rho_T \, dv$$

Where ρ_T = Total volume charge density.

4.19 DIELECTRIC BOUNDARY CONDITIONS

When flux lines are passing through a single medium, they are continuous. The flux lines get refracted while passing through medium 1 to medium 2 of two different dielectric materials. This can be studied by using boundary conditions.

The rules governing the behaviour of E and D at the boundary between two dielectrics are known as boundary conditions. Following are the two boundary conditions:

(*i*) The normal component of electric displacement, D is same on both sides of boundary of two media of different dielectrics. The normal component of displacement vector is continuous across the charge free boundary between two dielectrics.

(*ii*) The tangential components of electrical intensities are same on both sides of dielectrics.

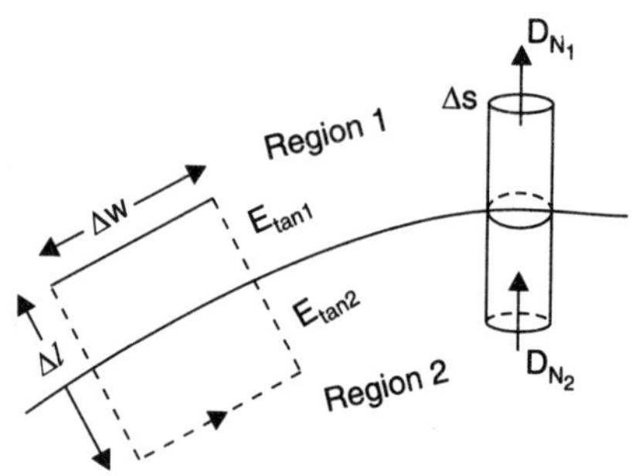

Fig. 4.18

Assuming that there are two different dielectrics and examining the tangential component by the equation

$$\int E \cdot dL = 0$$

It can be written as

$$E_{\tan 1} \, \Delta l - E_{\tan 2} \, \Delta l = 0$$

$$\therefore \quad E_{\tan 1} = E_{\tan 2}$$

If tangential field is continuous across the boundary then intensity is given as

$$E_{\tan 1} = \frac{D_{\tan 1}}{\varepsilon_1} \quad \text{and} \quad E_{\tan 2} = \frac{D_{\tan 2}}{\varepsilon_2}$$

$$\frac{D_{\tan 1}}{D_{\tan 1}} = \frac{\varepsilon_1}{\varepsilon_2} \qquad ...(4.103)$$

Now the boundary condition on normal components are found by apply the Gauss law to the diagram shown in Fig. 4.18. The sides are again short and flux leaving top and button surface is

$$D_{N_1} \Delta s - D_{N_2} \Delta s = \Delta Q = \rho_s \Delta s$$

from which $$D_{N_1} - D_{N_2} = \rho_s \quad ...(4.104)$$

This surface charge density can't be a bound surface charge density because we are taking polarization of dielectric into effect by using dielectric constant different from unity.

Assuming $\rho_s = 0$; we get $D_{N_1} = D_{N_2}$

i.e., $$\varepsilon_1 E_{N_1} = \varepsilon_2 E_{N_2} \quad ...(4.105)$$

These conditions may be combined to show change in vectors D and E at the surface. Let D_1 and E_1 make an angle θ_1 with normal to the surface.

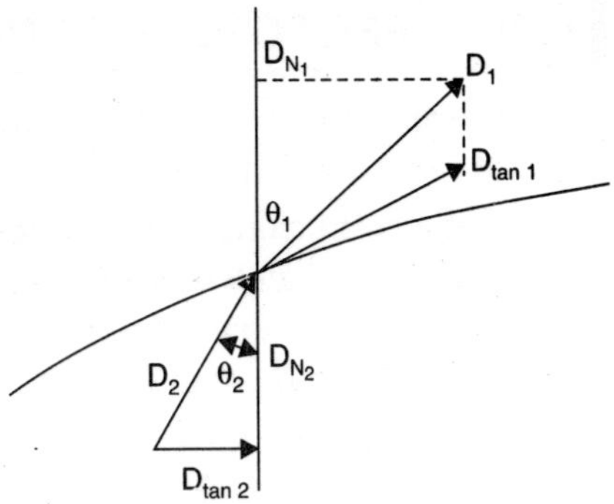

Fig. 4.19

$$D_{N_1} = D_1 \cos\theta_1; \; D_2 \cos\theta_2 = D_{N_2} \quad ...(4.106)$$

Ratio of tangential component is

$$\frac{D_{\tan 1}}{D_{\tan 2}} = \frac{D_1 \sin\theta_1}{D_2 \sin\theta_2} = \frac{\varepsilon_1}{\varepsilon_2}$$

$$\varepsilon_2 D_1 \sin\theta_1 = \varepsilon_1 D_2 \sin\theta_2 \quad ...(4.107)$$

Dividing equation (4.107) with equation (4.106) we get

$$\frac{\tan\theta_1}{\tan\theta_2} = \frac{\varepsilon_1}{\varepsilon_2}$$

We know the $\varepsilon_2 D_1 \sin\theta_1 = \varepsilon_1 D_2 \sin\theta_2$

i.e., $$D_2 \sin\theta_2 = \frac{\varepsilon_2}{\varepsilon_1} D_1 \sin\theta_1 \quad ...(4.108)$$

and also, $$D_2 \cos\theta_2 = D_1 \cos\theta_1 \quad ...(4.109)$$

For magnitude of D_1 and D_2, squaring on both sides and adding equations (4.108) and (4.109)

$$D_2^2 \sin^2\theta_2 + D_2^2 \cos^2\theta_2 = D_1^2 \left[\cos^2\theta_1 + \left(\frac{\varepsilon_2}{\varepsilon_1}\right)^2 \sin^2\theta_1 \right]$$

$$D_2 = D_1 \sqrt{\cos^2\theta_1 + \left(\frac{\varepsilon_2}{\varepsilon_1}\right)^2 \sin^2\theta_1} \quad ...(4.110)$$

Similarly, $D_1 \sin\theta_1 = D_2 \frac{\varepsilon_1}{\varepsilon_2} \sin\theta_2$ and $D_1 \cos\theta_1 = D_2 \cos\theta_2$

$$D_1 = D_2 \sqrt{\cos^2\theta_2 + \left(\frac{\varepsilon_1}{\varepsilon_2}\right)^2 \sin^2\theta_2} \quad ...(4.111)$$

4.20 EFFECT OF DIELECTRIC DISCONTINUITIES ON THE ELECTRIC FIELD

The nature of electric field in space containing dielectrics of various permitivities may be illustrated by an example as shown in Fig. 4.20.

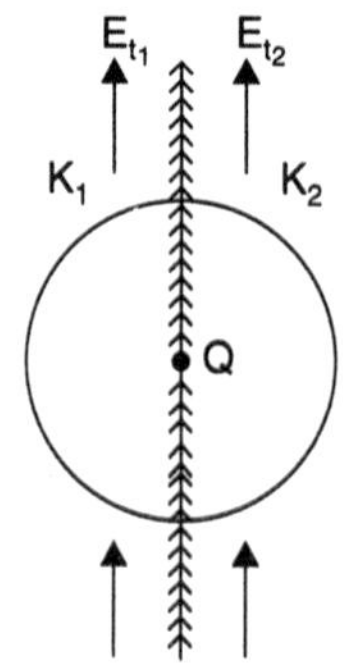

Fig. 4.20

Let a point charge 'Q' be embedded in the interface between two large dielectric blocks of dielectric constants k_1 and k_2. D_1 and E_1 be the flux density and electric field intensity in the medium 1 and the corresponding values in the medium 2 are D_2 and E_2. The electric field is still radial but of different strengths in the two media. We know that

$$D_1 = \varepsilon_0 k_1 E_1 \text{ and } D_2 = \varepsilon_0 k_2 E_2$$

The boundary conditions at the dielectric interface is $E_{t_1} = E_{t_2}$. Since the electric field intensities are radial.

$$\frac{D_1}{\varepsilon_0 k_1} = \frac{D_2}{\varepsilon_0 k_2} \quad \text{...(4.112)}$$

Consider a Gaussian surface 'S' of hemisphere of radius r on each side of the interface. Let Q_1 and Q_2 be the charge distributions in the medium 1 and 2 respectively.

Applying Gauss's law to the hemisphere

$$\oint D_1 n_1 ds = Q_1 = D_1 \times 2\pi r^2 \quad \text{...(4.113)}$$

$$\oint D_2 n_2 ds = Q_2 = D_2 \times 2\pi r^2 \quad \text{...(4.114)}$$

From equations (4.113) and (4.114) it can be written as

$$\frac{Q_1}{D_1} = \frac{Q_2}{D_2} = 2\pi r^2$$

$$\frac{Q_1}{D_1} = \frac{Q_2}{D_2} = \frac{Q_1 + Q_2}{D_1 + D_2}$$

$$\frac{Q_1}{Q_2} = \frac{D_1}{D_2} \quad \text{...(4.115)}$$

But from the equations (4.112) and (4.115), we can write as

$$\frac{Q_1}{Q_2} = \frac{D_1}{D_2} = \frac{k_1}{k_2} \quad \text{...(4.116)}$$

$$\frac{Q_1}{k_1} = \frac{Q_2}{k_2} = \frac{Q_1 + Q_2}{k_1 + k_2} = \frac{Q}{k_1 + k_2} \quad \text{...(4.117)}$$

$$\therefore \quad Q_1 = \frac{k_1 Q}{k_1 + k_2} \quad \text{and} \quad Q_2 = \frac{k_2 Q}{k_1 + k_2}$$

The above expressions suggest that out of a total charge Q a portion $\frac{k_1 Q}{k_1 + k_2}$ comes out in the medium1 and $\frac{k_2 Q}{k_1 + k_2}$ comes out in the medium 2. It means that the flux favours medium of higher permitivity.

4.21 CAPACITORS AND CAPACITANCE

Capacitor is a device which can store electrical charge. A practical capacitor essentially consists of two conductors separated by a dielectric. If one of the conductor carries a positive charge and other conductor carries an equal negative charge. The capacitance of a device is defined as the ratio of change on either of conductor to potential difference between the conductors.

Mathematically $$C = \frac{Q}{V}$$

where, Q = charge in coulombs and
V = potential difference.

The capacitance of a capacitor depends only on the geometry of conductors and dielectric medium. If the charge on a capacitor is increased the potential difference also increases such that the ratio of charge to potential difference is capacitance remain unchanged.

4.21.1 Capacitance of Parallel Plate Capacitor

A parallel plate capacitor comprises of two metallic plates separated by small distance. The space in between the plates is filled with suitable dielectric material. Consider the parallel plate capacitor as shown in Fig. 4.21.

Let the charge on each plate be Q coulombs (+ Q on plate 1 and – Q on plate 2). Let ε = permitivity of dielectric medium in between the plates, d = distance between the plates. If the surface charge density is ρ_s and is given by

$$\rho_s = \frac{Q}{A} \text{ C/m}^2$$

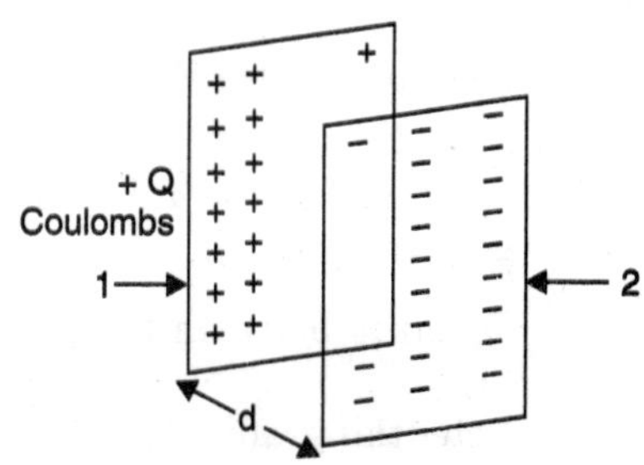

Fig. 4.21

Electric field density normal to surface of the plates is equal to the surface charge density.

Therefore $D_N = \rho_s$

Electric field intensity, $E = \frac{V}{d}$ (in magnitude)

We have $$D_N = \varepsilon E = \varepsilon \left(\frac{V}{d}\right) \text{ (in magnitude)}$$

Since, $$D_N = \rho_s = \frac{Q}{A}$$

$\therefore$ $$\frac{Q}{A} = \varepsilon\left(\frac{V}{d}\right)$$

(or) $$\frac{Q}{V} = \frac{A\varepsilon}{d}$$

By definition of capacitor,

$$C = \frac{Q}{V}$$

Putting $\frac{Q}{V} = \frac{A\varepsilon}{d}$, we get

$$C = \frac{A\varepsilon}{d} \text{ farads}$$

4.21.2 Energy in a Capacitor

The capacitor stores an electrostatic energy equal to work required to build the charge on the conductors. Thus the amount of electrostatic energy stored can be found by calculating the work in charging the condenser.

Since the potential is defined as the work done per unit charge, potential difference V is work done = VdQ.

But for a capacitor we have $V = \frac{Q}{C}$

$\therefore$ Work done $= \frac{Q}{C} dQ$

$\therefore$ The total work done in charging the capacitor to Q coulombs is given by

$$W = \int_0^Q \frac{Q}{C} dQ = \frac{1}{C}\int_0^Q Q\,dQ = \frac{Q^2}{2C} = \frac{1}{2}\frac{Q^2}{C}$$

$\therefore$ Electrostatic energy in a charged capacitor is

$$= \frac{1}{2}\frac{Q^2}{C} = \frac{1}{2}VQ = \frac{1}{2}V^2C$$

Example 4.15: *Find the capacitance of parallel plate capacitor, if the plates are of area 3.5 m², the distance between the plates is 5 mm, potential gradient is 10⁵ V/m and charge density on the plates is 4.5 μC/m².*

Solution: Given area of each plate, $A = 3.5 \text{ m}^2$

Distance between the plates, $d = 5$ mm

Potential gradient, $\frac{dV}{ds} = 10^5$ V/m

Surface charge density $\rho_s = 4.5\ \mu\text{C/m}^2$

The capacitance of parallel plate capacitor is given as

$$C = \frac{\varepsilon A}{d}$$

where ε = permitivity of medium

Electric flux density, $D = \rho_s = 4.5\ \mu\text{C/m}^2$

Also $E = \frac{dV}{ds}$ (in magnitude)

Therefore, $D = \varepsilon \frac{dV}{ds}$ (or) $\varepsilon = \frac{D}{\left(\frac{dV}{ds}\right)}$

$$= \frac{4.5 \times 10^{-6}}{10^5} = 45 \text{ pF/m}$$

Therefore capacitance,

$$C = 3.5 \times \frac{4.5 \times 10^{-11}}{5 \times 10^{-3}} = 31.5 \text{ nF}.$$

Example 4.16: *A parallel plate capacitor of 30 cm × 30 cm and d = 3 cm is charged to a potential of 2 kV with air as dielectric.*

(a) Find the energy stored.

(b) The capacitor is now disconnected from source and a dielectric slab is inserted into capacitor ($\varepsilon_R = 5$), calculate the energy stored.

Solution:

(*a*) Given $A = 30 \text{ cm} \times 30 \text{ cm} = 900 \text{ cm}^2 = 0.09 \text{ m}^2$

$d = 3 \text{ cm} = 3 \times 10^{-2} \text{ m} = 0.03 \text{ m}$ and $V = 2 \text{ kV}$

Capacitance, $$C = \frac{\varepsilon_0 A}{d} = 0.05 \times \frac{8.854 \times 10^{-12}}{0.03} \text{ F}$$

$$= 14.756 \text{ pF}.$$

Energy stored, $$W = \frac{1}{2} C V^2 \text{ joules}$$

$$= \frac{1}{2} \times 14.756 \times 10^{-12} \times (2000)^2$$

$$= 29.512 \ \mu\text{J}$$

(*b*) Given $\varepsilon_r = 5$

We have capacitance $C = \dfrac{A\varepsilon_0\varepsilon_r}{d} = 73.78 \text{ pF}.$

Energy stored, $$W = \frac{1}{2} CV^2$$

$$= \frac{1}{2} \times 73.78 \times 10^{-12} \times (2000)^2$$

$$= 147.56 \ \mu\text{J}.$$

Example 4.17: *Find the relative permitivity of dielectric material used in parallel plate capacitor if (a) C = 45 nF, d = 0.4 mm and S = 0.35 m² (b) d = 0.6 mm, E = 700 kV/m and ρ_s = 35 μC/m² (c) D = 75 μC/m² and energy density is 35 J/m³.*

Solution:

(*a*) $$C = \frac{\varepsilon_0 \varepsilon_r S}{d}$$

$\therefore$ $$\varepsilon_r = \frac{Cd}{\varepsilon_0 S} = \frac{45 \times 10^{-9} \times 0.4 \times 10^{-3}}{\varepsilon_0 \times 0.35} = 5.81$$

(*b*) $$|\overline{D}_N| = \rho_s = 35 \ \mu \text{ C/m}^2$$

(or) $$\varepsilon_0 \, \varepsilon_r \, E_N = 35 \ \mu \text{ C/m}^2$$

$$\therefore \qquad \varepsilon_r = \frac{35 \times 10^{-6}}{\varepsilon_0 E_N} = \frac{35 \times 10^{-6}}{\varepsilon_0 \times 700 \times 10^3} = 5.64$$

(c) Energy density $= \dfrac{\varepsilon_0 \varepsilon_r}{2} |\bar{E}|^2 = \dfrac{D^2}{2\varepsilon_0 \varepsilon_r}$

$$\varepsilon_r = \frac{(75 \times 10^{-6})^2}{2\varepsilon_0 (35)} = 9.079.$$

4.21.3 Parallel Plate Capacitors with Two Dielectric Media

Consider a parallel plate capacitor with space between the plates filled with two different insulating material of relative permitivities ε_{r_1} and ε_{r_2} as shown in Fig. 4.22. Let A = area of each plate and d_1, d_2 are the thickness of slabs of dielectrics. Let voltage V be applied across the capacitor. Let V_1, V_2 denote the voltages across the dielectric slabs and Q = charge on each plate.

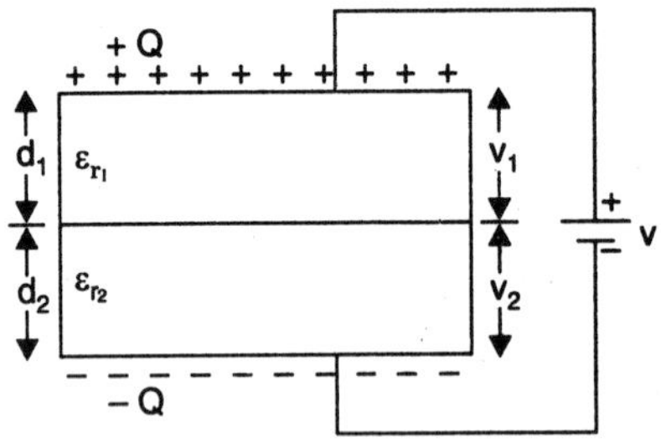

Fig. 4.22

We have $V_1 = E_1 d_1$ and $V_2 = E_2 d_2$ where E_1 and E_2 are the field intensities in region of dielectrics.

Also $\qquad V = V_1 + V_2$

Let $\qquad D$ = Normal component of field intensity

$$D = \varepsilon_1 E_1 = \varepsilon_2 E_2$$

(or) $\qquad E_1 = \dfrac{D}{\varepsilon_1} = \dfrac{D}{\varepsilon_0 \varepsilon_{r_1}}$ and $E_2 = \dfrac{D}{\varepsilon_2} = \dfrac{D}{\varepsilon_0 \varepsilon_{r_2}}$

Therefore, $\qquad V_1 = E_1 d_1 = \dfrac{Dd_1}{\varepsilon_0 \varepsilon_{r_1}}$

and $\qquad V_2 = E_2 d_2 = \dfrac{Dd_2}{\varepsilon_0 \varepsilon_{r_2}}$ (only magnitudes are considered)

$$V_1 + V_2 = \frac{Dd_1}{\varepsilon_0 \varepsilon_{r_1}} + \frac{Dd_2}{\varepsilon_0 \varepsilon_{r_2}}$$

$$= \frac{D}{\varepsilon_0}\left[\frac{d_1}{\varepsilon_{r_1}} + \frac{d_2}{\varepsilon_{r_2}}\right] = V \qquad \text{...(4.118)}$$

We know $\qquad D = \dfrac{Q}{A}$ (or) $Q = AD$

Capacitance, $\qquad C = \dfrac{Q}{V} = \dfrac{AD}{\dfrac{D}{\varepsilon_0}\left[\dfrac{d_1}{\varepsilon_{r_1}} + \dfrac{d_2}{\varepsilon_{r_2}}\right]}$

(or) $$C = \frac{A\varepsilon_0}{\left(\dfrac{d_1}{\varepsilon_{r_1}} + \dfrac{d_2}{\varepsilon_{r_2}}\right)} \qquad \text{...(4.119)}$$

Example 4.18: *A capacitor is composed of two plates separated by a sheet of insulating material 4 mm thick and $\varepsilon_r = 5$. The distance between the plates is increased to allow the insertion of second sheet 5 mm thick and of relative permeability ε_r. If the capacitance of the capacitor so formed is one-half of the original capacitance. Determine the value of ε_r.*

Solution:

(*a*) Before insertion of second sheet:

Distance between the plates, d = 4 mm = 0.004 m and Relative permeability of the medium = $\varepsilon_{r_1} = 5$

Capacitance of the capacitor,

$$C_1 = \frac{\varepsilon_0 \varepsilon_{r_1} A}{d} = \frac{5\, \varepsilon_0 A}{0.004} = \frac{5000}{4} \varepsilon_0 A$$

(*b*) After insertion of second sheet:

Thickness of first sheet, d_1 = 0.004 m; Thickness of second sheet, d_2 = 0.005 m

Relative permeability of first sheet, $\varepsilon_{r_1} = 5$; Relative permeability of second sheet = ε_{r_2} (say)

Capacitance of the capacitor,

$$C_2 = \frac{\varepsilon_0 A}{\dfrac{d_1}{\varepsilon_{r_1}} + \dfrac{d_2}{\varepsilon_{r_2}}}$$

$$= \frac{\varepsilon_0 A}{\dfrac{0.004}{5} + \dfrac{0.005}{\varepsilon_{r_2}}}$$

Given $$C_2 = \frac{1}{2} C_1$$

$$\frac{\varepsilon_0 A}{\dfrac{0.004}{5} + \dfrac{0.005}{\varepsilon_{r_2}}} = \frac{1}{2} \times \frac{5000}{4} \varepsilon_0 A$$

$$\varepsilon_{r_2} = \frac{3.125}{0.5} = 6.25.$$

Example 4.19: *Calculate the capacitance of a parallel plate capacitor with the following details: Plate area = 150 cm^2 ; Dielectric $\varepsilon_{r_1} = 5$; d_1 = 3 mm; Dielectric $\varepsilon_{r_2} = 4$, d_2 = 4 mm. If 220 V is applied across plates, what will be the voltage gradient across each dielectric?*

Solution:

Given $$A = 150 \text{ cm}^2;\ d_1 = 3 \text{ mm};\ d_2 = 4 \text{ mm}$$

$$C = \frac{\varepsilon_0 A}{\dfrac{d_1}{\varepsilon_{r_1}} + \dfrac{d_2}{\varepsilon_{r_2}}} = \frac{\varepsilon_0 \times 150 \times 10^{-4}}{\dfrac{3 \times 10^{-3}}{5} + \dfrac{4 \times 10^{-3}}{4}}$$

$$= \frac{8.854 \times 10^{-12} \times 150 \times 10^{-4}}{1.6 \times 10^{-3}} = 83.006 \text{ pF}$$

$$Q = CV = 83.006 \times 10^{-12} \times 220 = 1.82 \times 10^{-8}\text{ C}$$

$$C_1 = \frac{\varepsilon_0 \varepsilon_{r_1} A}{d_1} = \frac{8.854 \times 10^{-12} \times 150 \times 10^{-4} \times 5}{3 \times 10^{-3}} = 221.3\text{ pF}$$

$$C_2 = \frac{\varepsilon_0 \varepsilon_{r_2} A}{d_2} = \frac{8.854 \times 10^{-12} \times 4 \times 10^{-4} \times 150}{3 \times 10^{-3}} = 132.8\text{ pF}$$

$$V_1 = \frac{Q_1}{C_1} = \frac{1.82 \times 10^{-8}}{221.3 \times 10^{-12}} = 82.24\text{ V}$$

$$V_2 = 220 - 82.24 = 137.75\text{ V}$$

$$\therefore \quad E_1 = \frac{V_1}{d_1} = \frac{82.24}{3 \times 10^{-3}} = 27.41\text{ kV/m}$$

$$E_2 = \frac{V_2}{d_2} = \frac{137.75}{4 \times 10^{-3}} = 34.43\text{ kV/m.}$$

Example 4.20: *A capacitor for the type specified below has the following data:* $A_1 = 2$; $d = 0.02$; $d_1 = 0.009$ $\varepsilon_1 = 5\varepsilon_0$; $\varepsilon_2 = \varepsilon_0$, *where* ε_0 *is the permitivity of air.*

Calculate the capacitance of the system. If voltage of 4 kV is applied across the capacitor, determine the potential gradient in the two dielectrics.

Solution:

$$C = \frac{A}{\dfrac{d_1}{\varepsilon_1} + \dfrac{(d - d_1)}{\varepsilon_2}}$$

Substituting for A, d, d_1, ε_1 and ε_2 in above equation also for $\varepsilon_0 = 8.854 \times 10^{-12}$

$$C = 2660 \times 10^{-12}\ F = 8.886\text{ pF.}$$

If $\pm Q$ is the charge on plates of the capacitor, and A the area of plate surface,

$$E_1 = \frac{D}{\varepsilon_1} = \frac{Q}{A\varepsilon_1}$$

$$E_2 = \frac{D}{\varepsilon_2} = \frac{Q}{A\varepsilon_2}$$

$$V_1 = E_1 d_1 \ ; \ V_2 = E_2 (d - d_1)$$

So that,

$$\frac{V_1}{V_2} = \frac{\dfrac{d_1}{E_1}}{\dfrac{(d - d_1)}{E_2}} = \frac{9}{11} \quad \text{but as } V_1 + V_2 = 4000 \text{ volts,}$$

$$V_1 = 1799.77\ V \ ; \ V_2 = 2200 \text{ volts.}$$

Potential gradients which are equal to the field intensities E_1 and E_2

$$\therefore \quad E_1 = \frac{V_1}{d_1} = \frac{1799.77}{0.009}\text{ V/m} = 199.97\text{ kV/cm}$$

$$E_2 = \frac{V_2}{(d - d_1)} = \frac{2200}{0.011}\text{ V/m} = 200\text{ kV/cm.}$$

4.21.4 Capacitance of Isolated Sphere

Suppose a sphere of radius 'x meters' having charge Q coulombs, then by assuming a second conductor of infinite radius and zero potential, the formation of capacitor may be imagined.

Electric flux density at a distance 'x' mts from center of sphere is

$$D = \frac{Q}{4\pi x^2} \quad ...(4.120)$$

Electric field Intensity, $E = \dfrac{D}{\varepsilon_0 \varepsilon_r} = \dfrac{Q}{4\pi\varepsilon_1 x^2}$ ...(4.121)

We know that work done $= -Q \int \overline{E} \cdot \overline{dL}$ ...(4.122)

Work done to move a unit charge by a distance 'dx', the negative sign shows that work done is against a force of repulsion

$$V = -\int_a^b E\, dx$$

$$V = -\int_a^b \frac{Q}{4\pi\varepsilon_1 x^2}\, dx$$

If charge is brought from infinity to a distance x from center of sphere, potential at that point is

$$V = -Q \int_\infty^x \frac{dx}{4\pi\varepsilon_0\varepsilon_r x^2}$$

$$= \frac{-Q}{4\pi\varepsilon_0\varepsilon_r}\left[-\frac{1}{x}\right]_\infty^x = \frac{Q}{4\pi\varepsilon_0\varepsilon_r x}$$

Since capacitance, $C = \dfrac{Q}{V}$

Capacitance of sphere, $C = \dfrac{Q}{\left(\dfrac{Q}{4\pi\varepsilon_0\varepsilon_r x}\right)}$

$\therefore$ $C = 4\pi\, \varepsilon_0\, \varepsilon_r\, x$ farads. ...(4.123)

4.21.5 Capacitance Due to Two Concentric Spherical Conductors

Consider the arrangement as shown in Fig. 4.23.

It consists of two spherical conductor of radius 'a' and 'b' placed concentrically and the space in between them is filled with a dielectric material of relative permitivity ε_r.

The outer sphere is grounded and the inner sphere causes a charge of Q coulombs on its surface. Let V be the potential between the spheres. The electric flux density

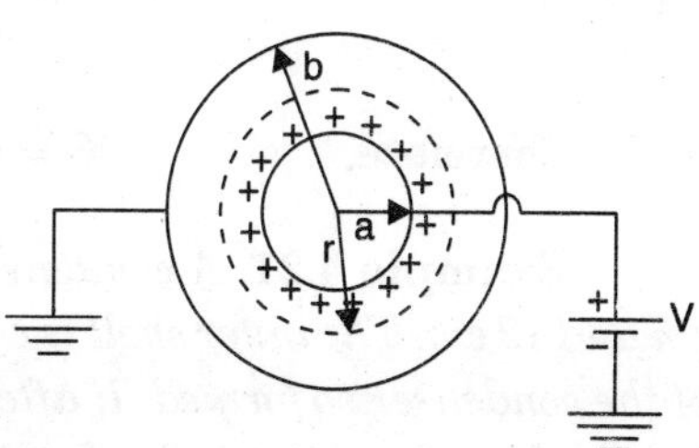

Fig. 4.23

also acts radially outwards. Consider a Gaussian surface in form of sphere of radius r, concentric with other spheres. By Gauss law, the total flux diverging from sphere surface = total charge enclosed with in the sphere.

Let charge enclosed $= Q$

Total electric flux, $\psi = \int_S D\,ds$ ...(4.124)

where, D = flux density

$$\int_S ds = 4\pi r^2 \text{ (total surface area)}$$

$\therefore$ Total electric flux, $\Psi = 4\pi r^2$ D coulombs

Therefore, $Q = 4\pi r^2 D$ (or) $D = \dfrac{Q}{4\pi r^2}$ (in magnitude)

But $D = \varepsilon E$,

where, E = Field intensity

$$E = \frac{D}{\varepsilon} = \frac{Q}{4\pi r^2 \varepsilon}$$

We have $E = -\dfrac{dV}{dr}$ (or) $dV = -E\,dr$

(or) $dV = -\dfrac{Q}{4\pi r^2 \varepsilon}\,dr$

The voltage ranges from 0 at 'b' to V_0 at 'a' also 'r' ranges from a to b.

Which implies $\displaystyle\int_0^{V_0} dV = \int_b^a -\frac{Q}{4\pi r^2 \varepsilon}\,dr$

$$= -\frac{Q}{4\pi\varepsilon}\int_b^a \frac{dr}{r^2}$$

$$V = \frac{Q}{4\pi\varepsilon}\left(\frac{b-a}{ab}\right) \quad \text{...(4.125)}$$

Let C = capacitors of spherical capacitors

We have $C = \dfrac{Q}{V}$ ...(4.126)

Substituting the value of Q from equation (4.125) in equation (4.126)

$\therefore$ $C = \dfrac{4\pi\,\varepsilon V\,ab}{V(b-a)}$ (or) $C = \dfrac{4\pi\,\varepsilon\,ab}{b-a}$

Therefore, $C = 4\pi\varepsilon\left(\dfrac{ab}{b-a}\right)$ farads ...(4.127)

Example 4.21: *A condenser is formed by two concentric thin spherical shells of radii 9 cm and 12 cm. The inner shell is charged and the outer shell is earthed. Calculate the capacitance of the condenser so formed. If after the inner shell acquire a potential of 220 V, the outer shell is removed. Determine the final potential of the inner charged shell.*

Solution: Let 'a' be the radius of inner shell = 0.09 m and b is the radius of outer shell = 0.12 m.

(*a*) Capacitor $C = 4\pi\varepsilon_0 \varepsilon_r \left(\frac{ab}{b-a}\right)$ farads

$$= 4\pi\varepsilon_0 \left(\frac{ab}{b-a}\right) \text{ farad in dielectric is air}$$

$$= 4\pi \times 8.854 \times 10^{-12} \times \frac{0.09 \times 0.12}{0.12 - 0.09}$$

$$= 40.03 \times 10^{-12} \text{ F}$$

$$= 40.03 \text{ pF}$$

(*b*) Charge acquired by the inner shell when it requires a potential of 220 volts is given by

$$Q = CV = 40.03 \times 10^{-12} \times 220$$

$$= 88.06 \times 10^{-10} \text{ coulombs}$$

New capacitor $C' = 4\pi\varepsilon_0 a$

$$= 4\pi \times 8.854 \times 10^{-12} \times 0.09 \text{ farads}$$

$$= 10.004 \times 10^{-12} \text{ F}$$

New potential acquired by the inner shell, $V' = \frac{Q}{C'}$

$$= \frac{88.06 \times 10^{-10}}{10.004 \times 10^{-12}} = 880.247 \text{ volts.}$$

Example 4.22: *Derive an expression for the capacity of a spherical capacitor consisting of two concentric spheres of radii 'a' and 'b', the dielectric medium between the two spheres being air. Hence forth show that the same expression can be written as* $C = \frac{\varepsilon_0}{d}\sqrt{A_a A_b}$ *, where* A_a *and* A_b *are the surface areas of the two spheres with radii 'a' and 'b' respectively.*

Solution: The capacitance of the spherical shell capacitor is

$$C = \frac{4\pi\varepsilon_0\, ab}{b-a} \quad \ldots(a)$$

Let the areas of the spherical surfaces be A_a and A_b and the separation = $b - a$

Then as, $A_a = 4\pi a^2$ and $A_b = 4\pi b^2$

Therefore, $a = \sqrt{\frac{A_a}{4\pi}}, b = \sqrt{\frac{A_b}{4\pi}}$ ...(*b*)

Equation (*a*) can be modified as follows:

$$C = \frac{4\pi\varepsilon_0}{d}\sqrt{\left(\frac{A_a}{4\pi} \times \frac{A_b}{4\pi}\right)} = \frac{\varepsilon_0}{d}\sqrt{A_a A_b}$$

4.22 ENERGY DENSITY IN ELECTROSTATIC FIELD

Consider a parallel plate capacitor of capacitance C as shown in Fig. 4.24. If it is charged to a potential difference V between the plates, the energy stored is

$$W = \frac{1}{2} CV^2 = \frac{1}{2} QV$$

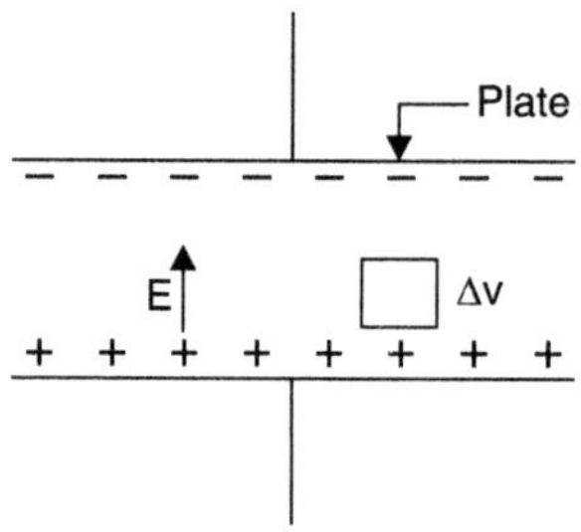

Fig. 4.24 (*a*)

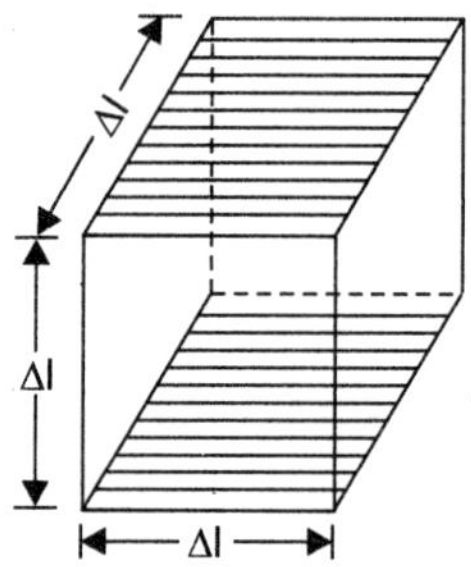

Fig. 4.24 (*b*)

The energy is stored in the form of electric field between the plates. Imagine a small cubical volume $\Delta v (= (\Delta l)^3)$ between the plates as shown in Fig. 4.24 (*a*). This elementary volume is shown in Fig. 4.24 (*b*). The length of each side is Δl. If thin sheets of metal foils are placed coincident with the top and bottom faces of the volume, the field will be undisturbed, provided the sheets are sufficiently thin. The elementary volume Δv constitutes a small capacitor of capacitance ΔC given by

$$\Delta C = \varepsilon \frac{(\Delta l)^2}{\Delta l} = \varepsilon (\Delta l) \qquad \text{...(4.128)}$$

The potential difference ΔV of the thin sheets is given by

$$\Delta V = E(\Delta l) \qquad \text{...(4.129)}$$

$\therefore$ The energy ΔW stored in the volume Δv is given by

$$\Delta W = \frac{1}{2} \Delta C \, (\Delta v)^2$$

Substituting the values of ΔC from equation (4.128)and ΔV from equation (4.129) in the above equation we get

$$\Delta W = \frac{1}{2} \varepsilon E^2 \, \Delta v$$

or

$$\frac{\Delta W}{\Delta v} = \frac{1}{2} \varepsilon E^2$$

Taking the limit of this ratio as Δv approaches zero will give energy per volume or energy density, W at the point around which Δv shrinks to zero.

$$\therefore \quad W = \underset{\Delta v \to 0}{Lt} \frac{\Delta W}{\Delta v} = \frac{1}{2} \varepsilon \bar{E} \text{ J/m}^3$$

This can be written in terms of flux density $\bar{D}$

$$\therefore \quad W = \frac{1}{2} \bar{D} \, \bar{E} \qquad (\because \; \bar{D} = \varepsilon \bar{E})$$

The total energy stored by the capacitor is

$$= \int_V W\, dv = \frac{1}{2}\int_V \varepsilon E^2\, dv$$

To distinguish it from energy density in magnetic field, it is referred as electric charge density.

Example 4.23: *Find the magnitude of electric field intensity in a sample of silver having $J = 7.17 \times 10^7$ amp/m² and $\mu_e = 0.0046$ m²/V-s if (a) The drift velocity is 3 mm/J (b) The current density is 10^7 A/m² (c) The sample is a cube 5 mm on a side, carrying a total current of 80 A (d) the sample is a cube, 5 mm on a side having a potential difference of 0.4 mV between opposite face.*

Solution:

(*a*) We know that $V_d = -\mu_e E$

$$\therefore \quad E = -\frac{V_d}{\mu_e}$$

$$|\bar{E}| = \frac{3\times 10^{-3}}{0.0046} = 0.652 \text{ V/m}$$

(*b*) $$\bar{J} = \sigma E$$

$$\therefore \quad E = \frac{\bar{J}}{\sigma}$$

$$|\bar{E}| = \frac{10^7}{7.17\times 10^7} = 0.1394 \text{ V/m}$$

(*c*) $$J = \frac{80}{\text{Area of one cube of force}} = \frac{80}{(0.005)^2} = \frac{80}{25\times 10^{-6}} \text{ A/m}^2$$

$$\therefore \quad |\bar{E}| = \frac{80}{25\times 10^{-6}\times 7.17\times 10^7} = 0.0446 \text{ V/m}$$

(*d*) $$|E| = \frac{0.4\times 10^{-3}}{5\times 10^{-3}} = 0.08 \text{ V/m.}$$

Example 4.24: *An aluminium conductor is 1000 ft long and has a circular cross-section with a diameter of 0.8 inches. If there is a d.c. voltage of 2.0 V between the ends find (a) The current density (b) The current (c) The power dissipated?*

Solution:

(*a*) For aluminium $\sigma = 3.82 \times 10^7$ mho/m

Length of conductor $= 1000$ ft $= 1000 \times 0.3048$ m $= 304.8$ m

$$E = \frac{2.0}{304.8} = 6.56\times 10^{-3} \text{ V/m}$$

$$J = \sigma E = 3.82\times 10^7 \times 6.56\times 10^{-3}$$

$$= 25.05 \times 10^4 \text{ A/m}^2.$$

(*b*) Resistance, $R = \dfrac{L}{\sigma S}$

where, S = cross sectional area of conductor.

r = radius of conductor = 0.4 × 0.0254 m = 0.010 m

Therefore, $$R = \frac{304.8}{3.82 \times 10^7 \times \pi \times (0.010)^2} = 0.0254 \text{ ohms}$$

Current, $$I = \frac{V}{R} = \frac{2.0}{0.0254} = 78.74 \text{ Amps}$$

(*c*) Power dissipated $= I^2R = (78.74)^2 \times 0.0254$

$= 157.47$ W.

Example 4.25: *Two concentric conductors sphere of radius r = 6 cm and b = 18 cm have equal and opposite charges 10^{-4} C on the inner and -10^{-4} C on the outer. Assuming $\varepsilon = \varepsilon_0$ for region between the spheres. Find (a) the maximum value of $|\overline{E}|$ between spheres (b) the potential difference between them (c) total energy stored.*

Solution:

(*a*) Consider a sphere of radius 'r' such that $0.06 < r < 0.18$. Charge enclosed by this sphere is equal to 10^{-4} C. Hence $Q = 10^{-4}$ C.

$$\therefore \qquad \overline{D} = \frac{10^{-4}}{4\pi r^2} U_r \text{ C/m}^2$$

(or) $$\overline{E} = \frac{10^{-4}}{4\pi\varepsilon_0 r^2} U_r$$

(or) $$|\overline{E}| = \frac{10^{-4}}{4\pi\varepsilon_0 r^2}$$

When 'r' is minimum $|\overline{E}|$ will be maximum. Now minimum value of r is 0.06

$$|\overline{E}| = \frac{10^{-4}}{4\pi\varepsilon_0 (0.06)^2} = 2500 \times 10^5 \text{ V/m}$$

(*b*) We have the equation $E = -\nabla V = -\dfrac{dV}{dr} U_r = \dfrac{10^{-4}}{4\pi\varepsilon_0 r^2} U_r$

(or) $$\frac{dV}{dr} = \frac{10^{-4}}{4\pi\varepsilon_0 r^2}$$

$$\therefore \qquad V = \frac{10^{-4}}{4\pi\varepsilon_0 r} + C$$

At $r = 0.06$, $$V = V_1 = \frac{10^{-4}}{4\pi\varepsilon_0 (0.06)} + C = 150 \times 10^5 + C$$

At $r = 0.18$, $$V = V_2 = \frac{10^{-4}}{4\pi\varepsilon_0(0.18)} + C = 50 \times 10^5 + C$$

Potential difference between spheres

$$= V_1 - V_2$$
$$= (150 \times 10^5 + C) - (50 \times 10^5 + C)$$
$$= 100 \times 10^5 \text{ V}$$

(c) Total energy stored, $$W_E = \frac{\varepsilon_0}{2} \int |\overline{E}|^2 \, dv$$

$$= \frac{\varepsilon_0}{2} \int_{0.06}^{0.18} \int_0^{\pi} \int_0^{2\pi} \left(\frac{10^{-4}}{4\pi\varepsilon_0 r^2} \right)^2 r^2 \sin\theta \, dr \, d\theta \, d\phi$$

$$W_E = \frac{\varepsilon_0}{2} \times \frac{10^{-8}}{16\pi^2 \varepsilon_0^2} \left[-\frac{1}{r} \right]_{0.06}^{0.18} [-\cos\theta]_0^{\pi} [\phi]_0^{2\pi} = 500.01 \text{ J.}$$

Example 4.26: *A spherical surface is free space, $r = 8$ cm, contains a uniform surface charge density of 30 $\mu C/m^2$. Determine r_A if region $0.08 < r < r_A$ contains precisely 1 MJ of energy.*

Solution: The region $0.08 < r < r_A$ lies outside to the sphere of radius 0.08 m. Consider a sphere of radius r such that $r > 0.08$ m.

The charge enclosed by this sphere $= 30 \times 10^{-6} \times 4\pi (0.08)^2$

$$= 2.413 \times 10^{-6} \text{ C}$$

$\therefore$ $$\Psi = 2.413 \times 10^{-6} \text{ C}$$

So, $$\overline{D} = \frac{2.413 \times 10^{-6}}{4\pi r^2} U_r = \frac{0.192 \times 10^{-6}}{r^2} U_r \text{ C/m}^2$$

$$W_E = \frac{1}{2\varepsilon_0} \int |\overline{D}|^2 \, dv$$

$$= \frac{1}{2\varepsilon_0} \int_{0.08}^{r_A} \int_0^{\pi} \int_0^{2\pi} \left(\frac{0.192 \times 10^{-6}}{r^2} \right)^2 r^2 \sin\theta \, dr \, d\theta \, d\phi$$

$$1 \, MJ = \frac{0.03687 \times 10^{-12}}{2\varepsilon_0} \left[-\frac{1}{r} \right]_{0.08}^{r_A} [-\cos\theta]_0^{\pi} [\phi]_0^{2\pi}$$

Hence, $$\left(-\frac{1}{r_A} + \frac{1}{0.08} \right) = 0.0382$$

(or) $$r_A = 0.0802 \text{ m.}$$

Example 4.27: *Find the capacitance between a circular conducting cylinder in air, 3 mm in radius and (a) A conducting plane 2 cm distant from cylinder axis; (b) A similar cylinder axes separated by 0.5 cm.*

Solution:

(*a*) $h = 2$ cm $= 2 \times 10^{-2}$ m and $b = 3$ mm $= 3 \times 10^{-3}$ m

$$C = \frac{2\pi\varepsilon L}{\cos h^{-1}(h/b)}$$

$$\therefore \qquad \frac{C}{L} = \frac{2\pi\varepsilon_0}{\cos h^{-1}\left(\dfrac{2\times 10^{-2}}{3\times 10^{-3}}\right)} = 21.51 \text{ pF/m}$$

(*b*) $h = 0.5$ cm $= 0.5 \times 10^{-2}$ m and $b = 3 \times 10^{-3}$ m.

The capacitance between two circular conductors separated by a distance is one half of the capacitance given by

$$\frac{C_1}{L} = \frac{2\pi\varepsilon_0}{\cos h^{-1}\left(\dfrac{0.5\times 10^{-2}}{3\times 10^{-3}}\right)} = 50.61 \text{ pF/m}$$

Hence required capacitance,

$$\frac{C}{L} = \frac{1}{2}\left(\frac{C_1}{L}\right) = \frac{50.61}{2} = 25.30 \text{ pF/m}$$

Example 4.28: *Current density in a certain region is given as* $\bar{J} = \left(\frac{6}{\rho}\right)U_\rho + \left[\frac{12}{\rho^2+1}\right]U_z$ *A/m² (a) Find the total current crossing the surface z = 4, ρ < 6 in the* U_z *direction. (b) Find* $\frac{\partial \rho_v}{\partial t}$ *(c) Calculate the total current crossing the closed surface bounded by z = 0, z = 3, ρ = 1 and ρ = 6 (d) Is* $\int_{Vol} \Delta \cdot J\, dv = \oint_S J \cdot ds$?

Solution:

(*a*) To the surface $Z = 4$, U_z is perpendicular. Hence $ds = \rho\, d\rho\, d\phi\, U_z$

$$\therefore \qquad I = \int \bar{J}\cdot\overline{ds} = \int_0^6\int_0^{2\pi} \frac{12\rho\, d\rho\, d\phi}{\rho^2+1}$$

$$= \frac{12}{2}\left[\ln(\rho^2+1)\right]_0^6 [\phi]_0^{2\pi} = 136.128 \text{ A}$$

(*b*)
$$\nabla\cdot\bar{J} = \frac{-\partial\rho_v}{\partial t}$$

$$\nabla\cdot J = \frac{1}{\rho}\frac{d}{d\rho}\left(\rho\times\frac{6}{\rho}\right) + \frac{d}{dt}\left(\frac{10}{\rho^2+1}\right) = 0$$

Hence,
$$\frac{\partial\rho_v}{\partial t} = 0$$

(c) $$\oint \bar{J}\cdot\overline{ds} = \int_{(z=0)} \bar{J}\cdot\overline{ds} + \int_{(z=3)} \bar{J}\cdot\overline{ds} + \int_{(\rho=1)} \bar{J}\cdot\overline{ds} + \int_{(\rho=6)} \bar{J}\cdot\overline{ds}$$

$$= \int \bar{J}\cdot(-\rho\, d\rho\, d\phi\, \bar{U}_z) + \int J\cdot(-\rho\, d\phi\, dz\, \bar{U}_\rho)_{(\rho=1)} \int \bar{J}\cdot\rho\, d\phi\, dz\, U_z\,_{(\rho=6)}$$

$$= -\int_0^6\int_0^{2\pi} \frac{10\rho\, d\rho\, d\phi}{\rho^2+1}\Bigg|_{(Z=0)} + \int_0^6\int_0^{2\pi} \frac{10\rho\, d\rho\, d\phi}{\rho 2+1}\Bigg|_{(Z=3)} - \int_0^{2\pi}\int_0^3 \frac{J}{\rho}\rho\, d\phi\, dz\,_{(\rho=1)}$$

$$+ \int_0^{2\pi}\int_0^3 \frac{5}{\rho}\rho\, d\phi\, dz\,_{(\rho=6)}$$

$$= -113.440 + 113.440 - 188.495 + 188.495 = 0$$

(d) $$\int (\bar{\nabla}\cdot\bar{J})\, dv = \int 0\, dv = 0$$

$$\therefore \quad \int_{vol} (\bar{\nabla}\cdot\bar{J})\, dv = \oint_S \bar{J}\cdot\overline{ds}.$$

Example 4.29: *Two perfectly conducting spherical surface are located at r = 3 and r = 11 cm. The total current passing radically outward through the medium between the spheres is 3.5 A (a) Find the V_{ts} and resistance between the spheres and $\bar{E}$ in the region between the spheres if a conducting material having σ = 0.03 mho/M is present for 3 < r < 11cm (b) Repeat if $\sigma = \frac{0.002}{r}$ for 3 < r < 11 cm (c) Show that integrating the power dissipated per unit volume in part (b) over the volume given the total dissipated power.*

Solution:

(a) As the current is flowing radically outward its direction is given by $\bar{U}_r$. Hence

$$\bar{J} = \frac{I}{Area}\cdot\bar{U}_r = \frac{3.5}{4\pi r^2}\bar{U}_r \text{ A/m}^2 = \frac{278.66}{r^2}\bar{U}_r \text{ mA/m}^2$$

$$\therefore \quad \bar{E} = \frac{\bar{J}}{\sigma} = \frac{278.66U_r\times 10^{-3}}{r^2\times 0.03} = \frac{9.288}{r^2}\bar{U}_r \text{ V/m.}$$

$$V = -\int_{0.11}^{0.03} \bar{E}\cdot\overline{dL} = -\int_{0.11}^{0.03}\left(\frac{9.288}{r^2}\bar{U}_r\right)\cdot(dr\,\bar{U}_r)$$

$$= -\int_{0.11}^{0.03}\frac{9.288}{r^2}dr = 9.288\left[\frac{1}{r}\right]_{0.11}^{0.03} = 225.17 \text{ V}$$

$$R = \frac{V}{I} = \frac{225.17}{3.5} = 64.33\ \Omega.$$

(b) $$\bar{E} = \frac{\bar{J}}{\sigma} = \frac{278.66\times 10^{-3}}{r^2}U_r\times\frac{r}{0.002} = \frac{139.33}{r}\cdot U_r \text{ V/m}$$

$$V = -\int_{0.11}^{0.03} E\cdot dL = -\int_{0.11}^{0.03}\frac{139.33}{r}dr = -139.33\,[\ln r]_{0.11}^{0.03} = 180.95 \text{ V}$$

$$R = \frac{V}{I} = \frac{180.95}{3.5} = 51.700\ \Omega.$$

(c) $$P = I^2 R = (3.5)^2 \times 51.700 = 633.325 \text{ watts}$$

$$\frac{dp}{dV} = \text{Power per unit volume} = JE$$

$\therefore$ $$dp = JE\, dv \text{ (or) } P = \int JE\, dv$$

$$P = \int_{0.03}^{0.11} \int_0^{\pi} \int_0^{2\pi} \frac{278.66 \times 10^{-3}}{r^2} \times \frac{139.33}{r} \cdot r^2 \sin\theta\, dr\, d\theta\, d\phi$$

$$= (38825.69) \cdot [\ln r]_{0.03}^{0.11} [-\cos\theta]_0^{\pi} [\phi]_0^{2\pi} \times 10^{-3}$$

$$= 633.45 \text{ watts, hence proved.}$$

Example 4.30: *A coaxial capacitor of the compressed gas type is to be designed to have 50×10^{-12} F capacitance and is to work at 220 kV DC. The maximum voltage gradient is not to exceed 300 kV/cm. If the outside diameter of the inner conductor is 6 cm. Calculate the inner diameter of the outer conductor and the length of the capacitor (take $\varepsilon_r = 1.0$ for gas).*

Solution: Given, $b - a = 0.06$ m and $C = 50 \times 10^{-12}\, F = \dfrac{Q}{V_{ba}} = \dfrac{4\pi\varepsilon_0\, ba}{(b-a)}$

$\Rightarrow$ $$ab = \frac{C(b-a)}{4\pi\varepsilon_0} = \frac{50 \times 10^{-12} \times 0.06}{4\pi \times 8.85 \times 10^{-12}} = 0.026$$

Now, we have $ab = 0.026$ and $b - a = 0.06$

$\Rightarrow$ $$b - \frac{0.026}{b} = 0.06 \quad \Rightarrow b = 0.08 \text{ m}$$

Hence, $$a = \frac{0.026}{b} = \frac{0.026}{0.08} = 0.325 \text{ m}$$

$\therefore$ The radii are; Inner radius, 'a' = 0.325 m; Outer radius, 'b' = 0.08 m.

Example 4.31: *The radii of two spheres differ by 0.06 m and the capacitance of spherical condenser is 55×10^{-12} F, if the outer sphere is earthed. Determine the radii considering air as a dielectric. Take $E_{max} = 310$ kV/cm; inner radius = a = 0.025 m and V = 200 kV.*

Solution: Inner radius = a = 2.5 cm and Outer radius = b (say)

$$C = 55 \times 10^{-12} \text{ F}; V = 200 \text{ kV}; E_{\max} = 310 \text{ kV/cm}$$

$$\varepsilon = \varepsilon_0 \varepsilon_r = \varepsilon_0 \times 1$$

$$= 8.854 \times 10^{-12} \text{ farad/m}$$

We know that; $$E_{\max} = \frac{V}{a \ln\left(\frac{b}{a}\right)}$$ (Max. value of E is at $r = a$)

$\therefore$ $$V = E_{\max}\, a \left[\ln \frac{b}{a}\right]$$

$\Rightarrow$ $$200 = 310 \times 2.5 \left(\ln \frac{b}{a}\right)$$

$$\Rightarrow \qquad \ln \frac{b}{a} = 0.2580 \text{ (or) } b = 3.23 \text{ cm}$$

Capacitance/metre length is given by

$$C = \frac{2\pi\varepsilon}{\ln\left(\frac{b}{a}\right)} = \frac{2\pi \times 8.854 \times 10^{-12}}{0.2580} = \frac{55 \times 10^{-12}}{l}$$

$$\Rightarrow \qquad l = 25.52 \text{ cm.}$$

Example 4.32: *A wire 5 mm in diameter and 6 km in length is surrounded at a constant height 15 m above the sea water. Calculate the capacitance between conductor and earth and also derive necessary equation.*

Solution: If the charge carried by conductor A is $+\lambda$ C/m; thus its image A' carries a charge of $-\lambda$ C/m. Hence potential differences between conductors A and A' is given by

$$V = \text{potential rise from } A' \text{ to } A$$

$$= -\int_{2h}^{r} \frac{+\lambda}{2\pi\varepsilon_0 x} dx - \int_{r}^{2h} \frac{-\lambda}{2\pi\varepsilon_0 x} dx$$

$$= \frac{\lambda}{2\pi\varepsilon_0} \ln \frac{2h}{r} + \frac{\lambda}{2\pi\varepsilon_0} \ln \frac{2h}{r} = \frac{\lambda}{\pi\varepsilon_0} \ln\left(\frac{2h}{r}\right)$$

Since, the earth's surface is exactly mid-way between A and A'. The potential of the actual conductor A is one half of the potential difference between A and $A' = \frac{V}{2}$.

Hence, capacitance of the wire to earth per metre length is

$$C_r = \frac{\lambda}{(V/2)} = \frac{\lambda}{\left[\frac{\lambda}{2\pi\varepsilon_0} \ln\left(\frac{2h}{r}\right)\right]}$$

$$= \frac{2\pi\varepsilon_0}{\ln(2h/r)} \text{ farads}$$

or

$$C_r = \frac{0.0241}{\log_{10}(2h/r)} \mu\text{F/km}$$

Here $\qquad r = 2.5 \times 10^{-3}$ m ; $h = 15$ m ; $2h = 30$ m

Which implies $\qquad \frac{2h}{r} = \frac{30}{2.5} \times 10^3 = 12000$

By substituting the values in the equation of C_r, we get

$$C_r = \frac{0.0241}{\log_{10}\left(\frac{30}{2.5} \times 10^{-3}\right)} = 0.0059 \ \mu\text{F/km}$$

Total conductance between conductor to earth = 6 × 0.0059 = 0.0354 μF.

Example 4.33: *A parallel plate capacitor has the region between its plates filled with a dielectric slab of relative permitivity ε_r. The plate dimensions are width 'b' and length 'l' and the plate separation is d. the capacitor is charged to a potential 'V', after which it is disconnected.*

The dielectric is now partially withdrawn in the l-dimension until only the length x remains between the plates what is now the potential across the capacitors? State the assumption made.

Solution: Referring to Fig. 4.25, we see that with the dielectric slab inserted over the entire length l, capacitance is given by

$$C = \varepsilon_0 \, \varepsilon_r \frac{l\omega}{d} \qquad ...(a)$$

Since, $A = l\omega$

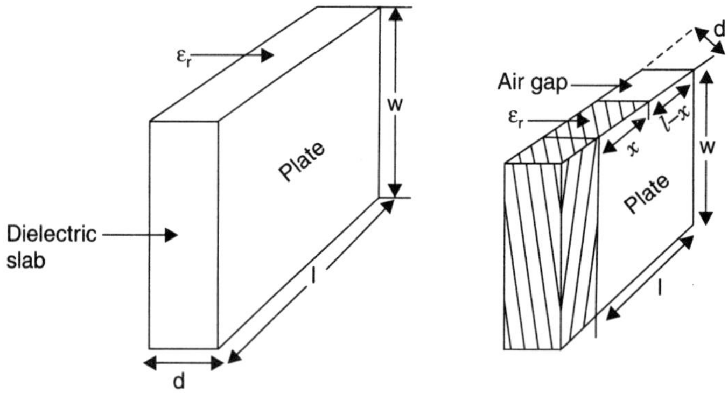

Fig. 4.25

As the capacitor is charged to a potential difference V_0, the corresponding charge Q is given by CV_0. Thus

$$Q = \varepsilon_0 \, \varepsilon_r \frac{l\omega}{d} V_0 \qquad ...(b)$$

The dielectric is now partially withdrawn as in Fig. 4.25, so that the capacitance is that of a parallel combination of two capacitors, one with relative permitivity ε_r and length x and the other with $\varepsilon_r = 1$ (air) and length. The new capacitance C' is given by

$$C' = \varepsilon_0 \, \varepsilon_r \frac{x\omega}{d} + \varepsilon_0 \frac{(l-x)\omega}{d} = \frac{\varepsilon_0 \omega}{d} [\varepsilon_r \, x + (l - x)]$$

$$= \frac{\varepsilon_0 \omega}{d} [l + (\varepsilon_r - 1)\, x] \qquad ...(c)$$

Assuming an ideal condition, viz. that there is no loss of charge, the potential difference across the capacitor under the new condition will be

$$V = \frac{Q}{C'} \qquad ...(d)$$

Substituting equations (*b*) and (*c*) in equation (*d*), we obtain

$$V = \frac{\varepsilon_r l V_0}{[l + (\varepsilon_r - 1)x]}$$

Example 4.34: *A capacitor is formed of two parallel plates distant 't' apart, one of which is at ground potential. The space between the plates is filled with a dielectric whose permitivity increases uniformly from one plate (ungrounded) to the other (grounded). Calculate the capacitance per unit area, in terms of the permittivity at the plates* ε_1 *(ungrounded) and* ε_2 *(grounded) and t. Neglect edge effects.*

Solution. The permittivity ε at distance x from the ungrounded plate $= \varepsilon_1 + \dfrac{\varepsilon_2 - \varepsilon_1}{t} x$

Electric field E at this point $= \dfrac{\rho_s}{\varepsilon}$,where ρ_s = surface charge density

$$E = \frac{\rho_s}{\left[\varepsilon_1 + \dfrac{\varepsilon_2 - \varepsilon_1}{t} x\right]}$$

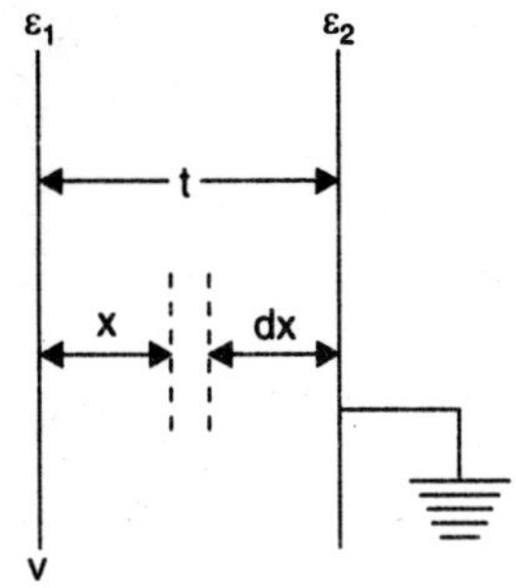

Fig. 4.26

Since, $d\phi = - E\, dx$

$$= - \frac{\rho_s}{\left[\varepsilon_1 + \dfrac{\varepsilon_2 - \varepsilon_1}{t} x\right]} dx$$

$$\therefore \quad \phi = V = - \int_{x=t}^{x=0} \frac{\rho_s}{\left[\varepsilon_1 + \dfrac{\varepsilon_2 - \varepsilon_1}{t} x\right]} dx$$

$$= \rho_s \int_0^t \frac{dx}{\left[\varepsilon_1 + \dfrac{\varepsilon_2 - \varepsilon_1}{t} x\right]}$$

$$= \frac{\rho_s t}{(\varepsilon_2 - \varepsilon_1)} \left[\log_e \left(\varepsilon_1 + \frac{\varepsilon_2 - \varepsilon_1}{t} x\right)\right]_0^t$$

$$= \frac{\rho_s t}{(\varepsilon_2 - \varepsilon_1)} \log_e \left(\frac{\varepsilon_2}{\varepsilon_1}\right)$$

Therefore, capacitance per unit area

$$= \frac{\rho_s}{V} = \frac{\varepsilon_2 - \varepsilon_1}{t \log_e (\varepsilon_2/\varepsilon_1)}, \text{ F/m}^2.$$

Example 4.35: *Derive an expression for the potential due to a quadruple with charges +Q, – 2Q, +Q at a great distance 'r' with spacing l.*

Solution: As seen in case of dipole, the potential at any point 'P' at a great distance 'r' may be calculated by applying the principle of superposition. The total potential due to $+Q$, $-2Q$ and $+Q$ is given by

$$V = \left[\frac{Q}{4\pi\varepsilon_0 r_1} - \frac{2Q}{4\pi\varepsilon_0 r} + \frac{Q}{4\pi\varepsilon_0 r_2}\right]$$

$$= \frac{Q}{4\pi\varepsilon_0}\left[\frac{1}{r_1} - \frac{2}{r} + \frac{4}{r_2}\right] = \frac{Qr}{4\pi\,\varepsilon_0}\left[\frac{r}{r_1} - 2 + \frac{r}{r_2}\right] \quad ...(a)$$

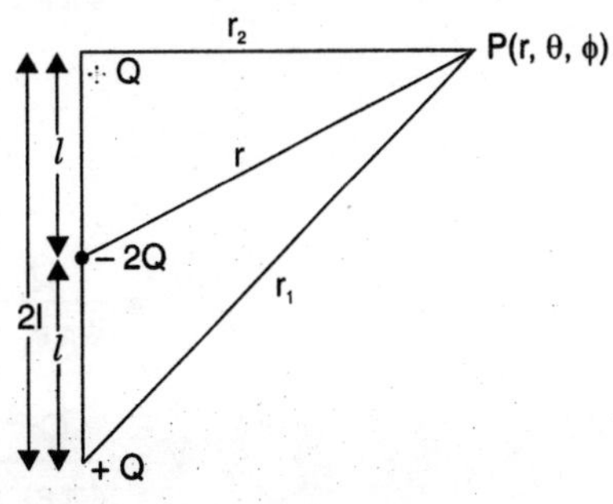

Fig. 4.27

Relation between l, r and r_1 is

$$r_1^2 = r^2 + l^2 + 2rl \cos\theta$$

or
$$r_1^2 = r^2\left[1+\left(\frac{l}{r}\right)^2+\left(\frac{2l}{r}\right)\cos\theta\right]$$

or
$$r_1 = r\sqrt{1+(l/r)^2+(2l/r)\cos\theta}$$

or
$$\frac{r_1}{r} = \sqrt{1+(l/r)^2+(2l/r)\cos\theta}$$

$$\frac{r}{r_1} = \left[1+\left(\frac{l}{r}\right)^2+2\left(\frac{l}{r}\right)\cos\theta\right]^{-1/2}$$

Expanding by Binomial Theorem

$$=\left[1-\frac{1}{2}\left(\left(\frac{l}{r}\right)^2+\frac{2l}{r}\cos\theta\right)+\frac{-\frac{1}{2}\left(-\frac{1}{2}-1\right)}{2}\left[\left(\frac{l}{r}\right)^2+\frac{2l\cos\theta}{r}\right]\right]^2 + \ldots..$$

$$\therefore \qquad \frac{r}{r_1} = 1-\frac{l}{r}\cos\theta-\frac{1}{2}\left(\frac{l}{r}\right)^2+\frac{3}{8}\times\frac{4l^2\cos^2\theta}{r^2}$$

Neglecting the higher power of l/r *i.e.*, $(l/r)^2$ and thereafter, We have

$$\left(\frac{r}{r_1}\right) = \left[1-\frac{l}{r}\cos\theta+\left(\frac{l}{r}\right)^2\left\{\frac{3}{2}\cos^2\theta-\frac{1}{2}\right\}\right]$$

Similarly from r, l, r_2 we get

$$\frac{r}{r_2} = \left[1-\frac{l}{r}\cos\theta+\left(\frac{l}{r}\right)^2\left\{\frac{3}{2}\cos^2\theta-\frac{1}{2}\right\}\right]$$

Substituting these values in equation (*a*) ; we get

$$V = \frac{Q}{4\pi\varepsilon_0 r}\left[1-\frac{l}{r}\cos\theta+\left(\frac{l}{r}\right)^2\frac{(3\cos^2\theta-1)}{2}\right.$$

$$\left.+\ 1+\frac{l}{r}\cos\theta+\left(\frac{d}{r}\right)^2\frac{(3\cos^2\theta-1)}{2}-2\right]$$

$$= \frac{Q}{4\pi\varepsilon_0 r}\left[\left(\frac{l}{r}\right)^2(3\cos^2\theta-1)\right]$$

$$V = \frac{Ql^2}{4\pi\varepsilon_0 r^3}(3\cos^2\theta-1)$$

Thus
$$V \propto \frac{1}{r^3}$$

Example 4.36: *A square 1 m of a side in air has charge $Q_1 = 2 \times 10^{-12}$ C at the upper right corner, a point charge $Q_2 = -2 \times 10^{-11}$ C at the lower right corner and a line distribution of charge density $\rho_L = 3 \times 10^{-11}$ C/m along the left edge. Find the potential at the centre of the square.*

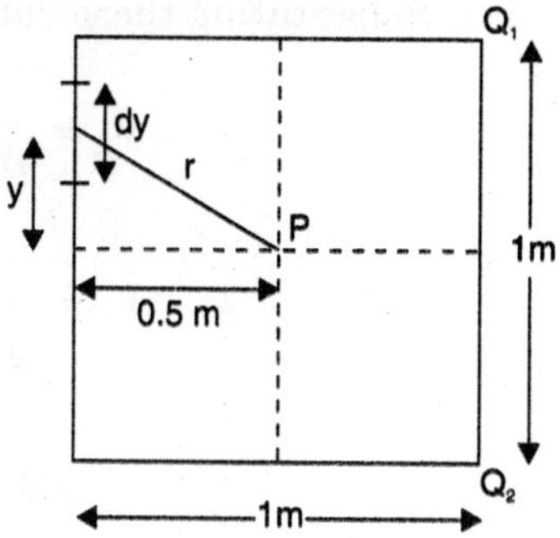

Fig. 4.28

Solution:

$$V_{(P)} = V_{Q1\,(P)} + V_{Q2\,(P)} + V_{Ql\,(P)}$$

where, $$V_{Ql\,(P)} = 9 \times 10^9 \times \frac{Q_l}{Q_l P} \text{ and } \frac{1}{4\pi\,\varepsilon_0} = 9 \times 10^9$$

distance, $$Q_l P = \sqrt{Q_1 P'^2 + PP'^2} = \sqrt{0.5^2 + 0.5^2} = \sqrt{0.5}$$
$$= 0.7$$

$$\therefore \quad V_{Q1\,(P)} = \frac{9 \times 10^9 \times 2 \times 10^{-12}}{0.7} = \frac{18}{0.7} \times 10^{-3} = 25.71 \times 10^{-3} \text{ V}$$

$$V_{Q2\,(P)} = -\frac{9 \times 10^9 \times 2 \times 10^{-11}}{PQ_2} = -\frac{18 \times 10^{-2}}{0.7} = -25.71 \times 10^{-2} \text{ V}$$

$$V_{Ql\,(P)} = 9 \times 10^9 \times Q_l \int \frac{dy}{r} = 9 \times 10^9 \times 3 \times 10^{-11} \int_{-0.5}^{0.5} \frac{dy}{\sqrt{y^2 + 0.5^2}}$$

$$= 27 \times 10^{-2} \times 1.77$$
$$= 47.79 \times 10^{-2} \text{ V}$$

$$\therefore \quad V_{(P)} = (25.71 \times 10^{-3} - 25.71 \times 10^{-2} + 47.79 \times 10^{-2}) \text{ V}$$
$$= 0.2465 \text{ V.}$$

Example 4.37: *A square of 1 m side has a point charge Q_1 (1×10^{-12} C) at the upper right corner, a point charge Q_2 (-10^{-11} C) at the lower right corner and a line distribution of charge of uniform density $\rho_L = 10 \times 10^{-11}$ C/m along the line edge. Find the potential at the centre of the square if it is placed in air.*

Solution: The potential at due to the point charges Q_1 and Q_2 is

$$V_P = \frac{1}{4\pi\varepsilon_0}\left(\frac{10^{-12}}{0.707} - \frac{10^{-11}}{0.707}\right) = -0.115 \text{ V} \qquad \text{...(a)}$$

The potential at P caused by the line distribution of charge is

$$V_L = \frac{1}{4\pi\,\varepsilon_0} \int \frac{\rho_L}{r}\, dl \qquad \text{...(b)}$$

In order to evaluate equation (*b*), let us consider an elemental length $dy = dl$, at a distance y. Then

$$\rho_L\, dl = \rho_L\, dy$$

$$r = \sqrt{y^2 + (0.5)^2}$$

Substituting these values in equation (*b*), we have

$$V_L = \frac{1}{4\pi\varepsilon_0}\int_{y=-0.5}^{y=+0.5} \frac{10^{-11}}{\sqrt{(0.5)^2 + y^2}}\,dy$$

$$V_L = \frac{1}{4\pi\times\dfrac{10^{-9}}{36\pi}}\int_{-0.5}^{0.5} \frac{10^{-11}}{\sqrt{(0.5)^2 + y^2}}\,dy$$

$$= 9 \times 10^{-2} \times 2.303 \times \log_{10} 6 = 0.158$$

The total potential at P is then,

$$V = V_P + V_L = -\,0.115 + 0.158$$

$$= 0.043 \text{ V} = 43 \text{ mV}.$$

Example 4.38: *Point charges $Q_1 = 2 \times 10^{-9}$ C and $Q_2 = 5 \times 10^{-9}$ C are located at (0, 0, 0) and (0.8, 0, 0) respectively. Determine the electric field at all points in x-y plane and also at P (0.4, 0.3) by first finding the potential due to the charges and then taking the gradient of the potential.*

Solution:

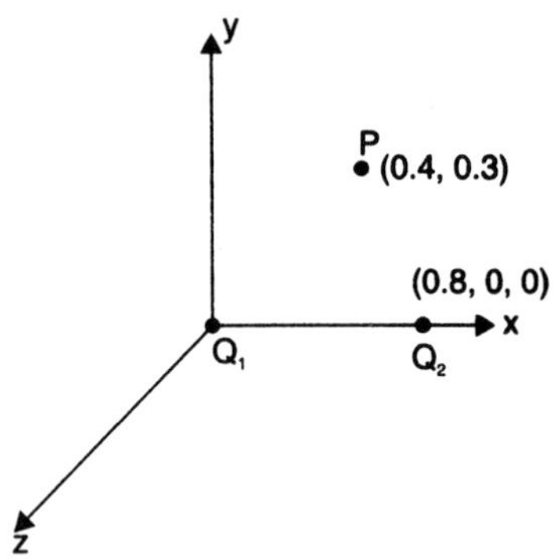

Fig. 4.29

$$V_{(P)} \text{ at } P(x, y) = V_{1(P)} + V_{2(P)}$$

$$= 9 \times 10^9 \left[\frac{Q_1}{Q_1P} + \frac{Q_2}{Q_2P}\right]$$

$$V_{(P)} \text{ at } P\,(x, y) = 9\left[\frac{2}{\sqrt{x^2 + y^2}} + \frac{5}{\sqrt{(x - 0.8) + y^2}}\right]$$

$$\therefore\quad V_{(P)} \text{ at } (0.4, 0.3) = 9\left[\frac{2}{0.5} + \frac{5}{0.5}\right]$$

$$= 9 \times 14 = 126 \text{ V}$$

$$E = -\,\nabla V = -\left(\frac{\partial}{\partial x}U_x + \frac{\partial}{\partial y}U_y + \frac{\partial}{\partial z}U_z\right) V \text{ for } x\text{-}y \text{ plane}$$

$$= -\frac{\partial\phi}{\partial x}U_x - \frac{\partial\phi}{\partial y}U_y$$

Now,

$$\frac{\partial V}{\partial x} = 9\left[2\left(-\frac{1}{2}\right)(x^2 + y^2)^{-3/2}\, 2x + 5\left(-\frac{1}{2}\right)[(x - 8)^2 + y^2]^{-3/2}\, 2\,(x - 0.8)\right]$$

$$= 9\left[\frac{2}{\sqrt{x^2 + y^2}} + \frac{5}{\sqrt{(x - 0.8) + y^2}}\right]$$

$$\frac{\partial V}{\partial y} = 9\left[2\left(-\frac{1}{2}\right)(x^2 + y^2)^{-3/2}\, 2y + 5\left(-\frac{1}{2}\right)[(x - 0.8)^2 + y^2]^{-3/2}\, 2y\right]$$

$$= -\,9\,[2\,(x^2 + y^2)^{-3/2}\, 2y + 5\,[(x - 0.8)^2 + y^2]^{-3/2}\, 2y]$$

$$\therefore \quad \vec{E} = -\left[\frac{\partial V}{\partial x} U_x + \frac{\partial V}{\partial y} U_y\right]$$

As there is no variation of V with z.

$$= 9\{2(x^2+y^2)^{-3/2} x + 5[(x-0.8)^2 + y^2]^{-3/2}(x-0.8)\} U_x$$
$$+ 9\{2(x^2+y^2)^{-3/2} y + 5[(x-0.8)^2 + y^2]^{-3/2} y\} U_y$$

Replacing the values of $x = 4$ and $y = 0.3$ in above equation

$$\vec{E} = -86.5\, U_x + 151\, U_y$$

Example 4.39: *Four positive point charges 2×10^{-9} C each are situated in x-y plane at points (0, 0) (0, 1) (1, 1) and (1, 0)m. Find the electric field and potential at $\left(\frac{1}{2}, \frac{1}{2}\right)$.*

Solution: Let A be the centre of the square at the corners of which charges Q_1, Q_2, Q_3 and Q_4 are located as shown in Fig. 4.30

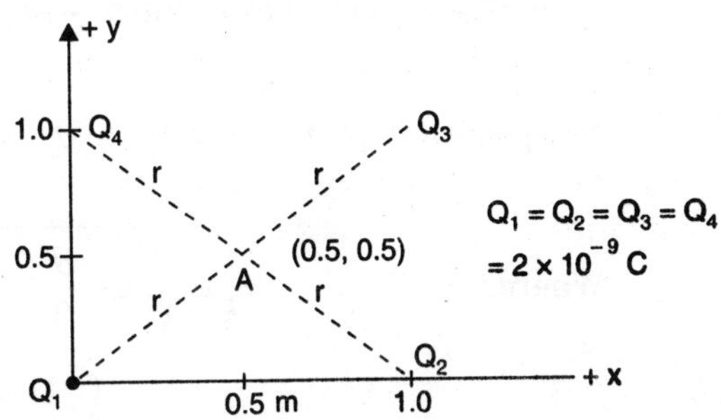

Fig. 4.30

(*a*) To find the electric field intensity E at A. The resultant field at A is the vector sum of the fields due to the individual charges considered separately.

Field due to Q_1 at A is given by

$$E_1 = \frac{Q_1}{4\pi\varepsilon_0 (OA)^2}, \text{ along } AQ_3$$

Field due to Q_3 at A is similarly equal to

$$E_3 = \frac{Q_3}{4\pi\varepsilon_0 (OA_3)^2}, \text{ along } AQ_1$$

As the charges are equal and $OA = AQ_3$

$|E_1| = |E_3|$ (in magnitude) and they are oppositely directed

Therefore, $E_1 + E_4 = 0$

So that, $E_1 + E_2 + E_3 + E_4 = 0$

The resultant field at A is, therefore, zero.

(*b*) To find the potential at A

All the charges are equal and are at the same distance of $r = (\sqrt{2}/2)$ m from A so that the total potential at A is four times that due to each. Each of the charges produces a potential at

$$A = \frac{Q}{4\pi\varepsilon_0} r = (9 \times 10^9) \frac{2 \times 10^{-9}}{\sqrt{2}} \times 2$$

$$= \frac{18}{\sqrt{2}} \text{ volts.}$$

$\therefore$ Total potential = 101.8 volts.

Example 4.40: *A + Q charge is uniformly distributed on the surface of a conducting sphere centered at* $\left(0, 0, \frac{d}{2}\right)$ *of a rectangular system. A – Q charge is uniformly distributed on the surface of a second conducting sphere centered at* $\left(0, 0, \frac{-d}{2}\right)$. *The radius of the each sphere is equal but much smaller than the spacing d, find the field intensity at P (0, y, z) as in Fig. 4.31.*

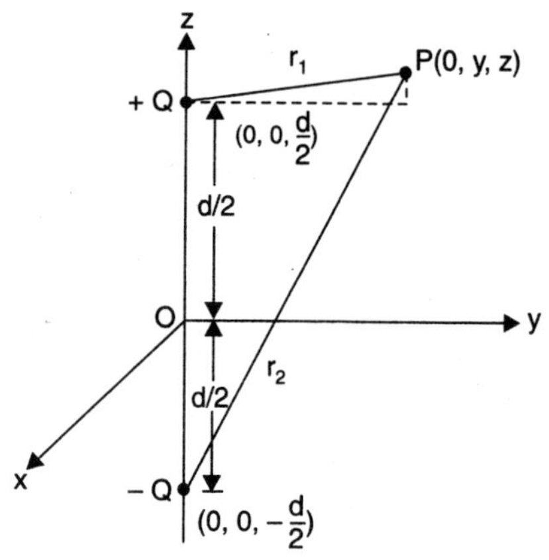

Fig. 4.31

Solution: Potential at P due to $+Q$, $V^+ = \dfrac{Q}{4\pi\varepsilon r_1}$

Similarly, potential at P due to $-Q$, $V^- = \dfrac{-Q}{4\pi\varepsilon r_2}$

Net potential at P, $V = V^+ + V^- = \dfrac{Q}{4\pi\varepsilon}\left[\dfrac{1}{r_1} - \dfrac{1}{r_2}\right]$

Where, $r_1 = \sqrt{\left(z - \frac{d}{2}\right)^2 + y^2}$ and $r_2 = \sqrt{\left(z + \frac{d}{2}\right)^2 + y^2}$

$$\therefore \qquad V = \frac{Q}{4\pi\varepsilon}\left[\frac{1}{\sqrt{\left(z - \frac{d}{2}\right)^2 + y^2}} - \frac{1}{\sqrt{\left(z + \frac{d}{2}\right)^2 + y^2}}\right]$$

Now, $\vec{E} = -\nabla V = -\left[\dfrac{\partial V}{\partial x}U_x + \dfrac{\partial V}{\partial y}U_y + \dfrac{\partial V}{\partial z}U_z\right]$

Where, $\dfrac{\partial V}{\partial x} = 0$ as ϕ is independent of x.

$$-\frac{\partial V}{\partial y} = \frac{Q}{4\pi\varepsilon}\left[\left[-\frac{1}{2}\left(z - \frac{d}{2}\right)^2 + y^2\right]^{-1/2-1} 2y - 1\left(-\frac{1}{2}\right)\left\{\left(z + \frac{d}{2}\right)^2 + y^2\right\}^{-1/2-1} 2y\right]$$

$$= \frac{Q}{4\pi\varepsilon}\left[\frac{-y}{\left(\left(z - \frac{d}{2}\right)^2 + y^2\right)} + \frac{y}{\left(\left(z - \frac{d}{2}\right)^2 + y^2\right)^{+3/2}}\right]$$

$$\frac{\partial V}{\partial z} = \frac{Q}{4\pi\varepsilon}\left\{\left(z - \frac{d}{2}\right)^2 + y^2\right\}^{-3/2}\left(-\frac{1}{2}\right)2\left(z - \frac{d}{2}\right) - \left\{\left(z + \frac{d}{2}\right)^2 + y^2\right\}^{-3/2}\frac{1}{2}\left(z + \frac{d}{2}\right)\Bigg]$$

$$= \frac{Q}{4\pi\varepsilon}\left[\frac{-\left(z-\frac{d}{2}\right)}{\left[\left(z-\frac{d}{2}\right)^2+y^2\right]^{3/2}}+\frac{\frac{1}{2}\left(z+\frac{d}{2}\right)}{\left[\left(z+\frac{d}{2}\right)^2+y^2\right]^{3/2}}\right]$$

Example 4.41: *Two parallel conducting plates 5 cm apart and situated in air are connected to a source of constant potential difference of 80 kV. Find the electric field intensity between the plates if it is within permissible value? If a mica sheet (ε_r = 5) of thickness 2 cm is introduced between the plates, determine the field intensity in air and mica.*

Solution: The field intensity between the plates in the air gap is

$$E = \frac{V}{t} = \frac{80}{5} = 16 \text{ kV/cm}$$

As the dielectric strength of air = 30 kV/cm, the above value is safe and the dielectric will not breakdown. Next, with the introduction of mica, there are two dielectrics

Air film: t_1 = 3 cm ; Mica film : t_2 = 2 cm.

Let the corresponding field intensities be E_1 and E_2 in kV/cm. Then

$$E_1 t_1 + E_2 t_2 = V \qquad ...(a)$$

Where $\qquad E_1 = \frac{D}{\varepsilon_0}$ and $E_2 = \frac{D}{\varepsilon_0\varepsilon_r} = \frac{D}{5\varepsilon_0}$

Therefore, $\qquad E_1 = 5\,E_2 \qquad ...(b)$

Substituting equation (*b*) in equation (*a*), we obtain

$$5\,E_2 \times 3 + 2\,E_2 = 80$$

$\therefore \qquad E_2 = 4.70$ kV/cm

Field intensity in air gap, E_1 = 23.52 kV/cm.

Example 4.42: *Two concentric conductor spheres of radius r = 7 cm and b = 17 cm have equal and opposite charges 10^{-9} C on the inner and -10^{-9} C on the outer. Assuming $\varepsilon = \varepsilon_0$ for region between the spheres. Find (a) The maximum value of E between the spheres and (b) Potential difference between them.*

Solution:

(*a*) Consider a sphere of radius *r* such that 0.07 < *r* < 0.17. Charge enclosed by this sphere is equal to 10^{-9} C. Hence, $\psi = 10^{-9}$ C.

$$\therefore \qquad \bar{D} = \frac{10^{-9}}{4\pi r^2} U_r \text{ C/m}^2 \quad \text{(or)} \quad \bar{E} = \frac{10^{-9}}{4\pi\varepsilon_0 r^2} U_r$$

(or) $$|\bar{E}| = \frac{10^{-9}}{4\pi\varepsilon_0 r^2}$$

When *r* is minimum $|\bar{E}|$ will be maximum and now minimum value of *r* is 0.07.

$$|\bar{E}| = \frac{10^{-9}}{4\pi\varepsilon_0 (0.07)^2} = 1.836 \text{ kV/m}$$

(*b*) We have the equation

$$E = -\nabla V = -\frac{\partial V}{\partial r} U_r = \frac{10^{-9}}{4\pi\varepsilon_0 r^2} U_r$$

(or) $$\frac{\partial V}{\partial r} = \frac{10^{-9}}{4\pi\varepsilon_0 r^2}$$

$\therefore$ $$V = \frac{10^{-9}}{4\pi\varepsilon_0 r} + C$$

At $r = 0.07$, $$V = V_1 = \frac{10^{-9}}{4\pi\varepsilon_0 (0.07)} + C = 128.57 + C$$

At $r = 0.17$, $$V = V_2 = \frac{10^{-9}}{4\pi\varepsilon_0 (0.17)} + C = 52.941 + C$$

Potential difference between spheres $= V_1 - V_2$

$$= (128.57 + C) - (52.941 + C)$$

$$= 75.628 \text{ V.}$$

Example 4.43: *Potential is given by $V = 2(x + 1)^2 (y + 1)^2 (z + 3)^2$ V in free space. At a point P (2, –1, 4) calculate: (i) the potential at point P, (ii) electric field intensity E at point P, (iii) flux density D at point P, (iv) volume charge distribution ρ_v at P.*

Solution:

(*i*) $$V = 2(x + 1)^2 (y + 1)^2 (z + 3)^2 V; \text{ at } P (2, -1, 4)$$

$$= 2 \times 3^2 \times 1^2 \times 7^2 = 882V$$

(*ii*) $$E = -\nabla V = -\left[\frac{\partial V}{\partial x} U_x + \frac{\partial V}{\partial y} U_y + \frac{\partial V}{\partial z} U_z\right]$$

$$= -4(x + 1)(y + 2)^2 (z + 3)^2 U_x - 4(x + 1)^2 (y + 2)(z + 3)^2 U_y - 4(x + 1)^2 (y + 2)^2 (z + 3) U_z$$

At $P(2, -1, 4)$ $$E = -588U_x - 1764U_y - 252U_z$$

(*iii*) $$D = \varepsilon_0 E = 8.854 \times 10^{-12} (-588U_x - 1764U_y - 252U_z)$$

$$= -5.21U_x - 15.62U_y - 2.23U_z \text{ nC/m}^2.$$

(*iv*) $\rho_v = \nabla \cdot D = \varepsilon_0 \left(\frac{\partial E_x}{\partial x} + \frac{\partial E_y}{\partial y} + \frac{\partial E_z}{\partial z}\right)$; and from (*ii*) it can be written as

$$\rho_v = \nabla \cdot D = \varepsilon_0 [-4(y + 2)^2 (z + 3)^2 - 4(x + 1)^2 (z + 3)^2 - 4(x + 1)^2 (y + 2)^2]$$

$$\rho_v = 8.854 \times 10^{-12} [-196 - 1764 - 36] = -17.67 \text{ nC/m}^3.$$

Example 4.44: *In a field $E = -50y\, u_x - 50x\, u_y + 30u_z$ V/m, calculate the differential amount of work done in moving 2 μC charge a distance 5 μm from A(1, 2, 3) to B(2, 4, 1).*

Solution: Work done

$$\Delta W = -qE \cdot dl$$

where $$dl = 5u_{AB} \text{ μm} = 5 \times \frac{(2-1)u_x + (4-2)u_y + (1-3)u_z}{\sqrt{1^2 + 2^2 - 2^2}} \text{ μm}$$

$$= \frac{5}{3}(u_x + 2u_y - 2u_z)\ \mu\text{m}$$

$$\Delta W = -2 \times 10^{-6} (-50y\, u_x - 50x\, u_y + 30u_z) \cdot \frac{5}{3} \times 10^{-6} (u_x + 2u_y - 2u_z)$$

At initial point $A(1, 2, 3)$

$$\Delta W = -2 \times 10^{-6} (-50 \times 2u_x - 50 \times 1u_y + 30u_z) \cdot \frac{5}{3} \times 10^{-6} (u_x + 2u_y - 2u_z)$$

$$= 867 \text{ p J.}$$

Example 4.45: *The potential field at any point in a space containing a dielectric material of relative permitivity of 2.1 is given by $V = 5x^2y + 3yz^2 + 6xz$ volt, where x, y, z are in meters. Calculate the volume charge density at point P(2, 5, 3)m.*

Solution: Poisson's equation is given by

$$\nabla^2 V = -\frac{\rho_v}{\varepsilon}$$

$$\rho_v = -\varepsilon \nabla^2 V$$

$$\nabla^2 V = \frac{\partial^2 V}{\partial x^2} + \frac{\partial^2 V}{\partial y^2} + \frac{\partial^2 V}{\partial z^2}$$

$$\frac{\partial^2 V}{\partial x^2} = \frac{\partial^2}{\partial x^2}(5x^2 y + 3yz^2 + 6xz)$$

$$= \frac{\partial}{\partial x}(10xy + 6z)$$

$$= 10y$$

$$\frac{\partial^2 V}{\partial y^2} = \frac{\partial^2}{\partial y^2}(5x^2 y + 3yz^2 + 6xz)$$

$$= \frac{\partial}{\partial y}(5x^2 + 3z^2)$$

$$= 0$$

$$\frac{\partial^2 V}{\partial z^2} = \frac{\partial^2}{\partial z^2}(5x^2 y + 3yz^2 + 6xz)$$

$$= \frac{\partial}{\partial z}(6yz + 6x)$$

$$= 6y$$

$$\frac{\partial^2 V}{\partial x^2} + \frac{\partial^2 V}{\partial y^2} + \frac{\partial^2 V}{\partial z^2} = 10y + 0 + 6y = 16y$$

$\therefore$ At point P(2, 5, 3), $\dfrac{\partial^2 V}{\partial x^2} + \dfrac{\partial^2 V}{\partial y^2} + \dfrac{\partial^2 V}{\partial z^2} = 80$

$$\rho_v = -\varepsilon_0 \varepsilon_r \Delta^2 V$$
$$= -8.854 \times 10^{-12} \times 2.1 \times 80$$
$$= 1.487 \times 10^{-9} \text{ C/m}^3.$$

Example 4.46: *A boundary exists at z = 0 between two dielectrics $\varepsilon_{r_1} = 2.5$, region z < 0, and $\varepsilon_{r_2} = 4$, region z > 0. The field in region of ε_{r_1} is $E = -30u_x + 50u_y + 70u_z$ V/m. Find (a) normal component of E (b) tangential component of E (c) the angle between E and normal to the surface (d) normal component of D_2 (e) tangential component of D_2 (f) value of D_2.*

Solution:

Region $\varepsilon_{r_1} = 2.5 \quad z < 0$ and $E_1 = -30u_x + 50u_y + 70u_z$

(*a*) Normal component of $E_1 = E_{n_1} = E_1 \cdot U_z = 70$ V/m.

(*b*) Tangential component of $E_1 = E_{t_1} = x$ and y components of E_1

$$= -30U_x + 50U_y$$

$$|E_{t_1}| = \sqrt{30^2 + 50^2} = 58.3 \text{ V/m.}$$

(*c*) Angle α_1 between E_1 and normal to surface

$$|E_1| \cos \alpha_1 = E_z$$

$$(\sqrt{30^2 + 50^2 + 70^2}) \cos \alpha_1 = 70$$

$$\cos \alpha_1 = \frac{70}{91.10}$$

$$\alpha_1 = 39.8°$$

(*d*) Normal component of D_2

$$D_{n_1} = D_{n_2} = 70 \times 2.5 \times 8.854 \times 10^{-12} \text{ C/m}^2$$
$$= 1.549 \times 10^{-9} \text{ C/m}^2.$$

(*e*) $E_{t_1} = E_{t_2} = -30U_x + 50U_y$

$$D_{t_2} = 4\varepsilon_0 E_{t_2} = 4 \times 8.854 \times 10^{-12} (-30U_x + 50U_y)$$
$$= -1.602 \times 10^{-9} U_x + 10771 \times 10^{-9} U_y \text{ C/m}^2$$

$$|D_{t_2}| = 2.065 \times 10^{-9} \text{ C/m}^2.$$

(*f*) $D_2 = D_{t_2} + D_{n_2} = (-1.602U_x + 1.771U_y + 1.549U_z) \times 10^{-9}$ C/m^2.

SHORT QUESTIONS AND ANSWERS

1. What is dielectric constant?

Dielectric constant is same as that of relative permitivity (or) of a dielectric and is defined as the ratio of permitivity of the dielectric to that of free space.

2. What is permitivity of the medium?

It is the ability of the medium by virtue of which it can allow the electric flux lines to pass through it.

i.e., $E = \varepsilon_0 \varepsilon_r$ where $\varepsilon_0 = 8.854 \times 10^{-12}$

ε_r = relative permitivity

3. Identify the equation $\bar{\mathbf{J}} + \frac{\partial \mathbf{P}}{\partial \mathbf{t}} = 0$.

This equation is the continuity equation.

4. State electric displacement.

Electric flux density $(\vec{D})$ is also called as electric displacement and it is defined as the electric flux per unit area.

i.e., $$D = \frac{\Psi}{area}$$

5. Write the capacitance of an spherical capacitor with inner radius '*a*' and outer radius '*b*'.

$$C = \frac{4\pi\varepsilon}{\frac{1}{a} - \frac{1}{b}}$$

6. Write the capacitance of co-axial capacitor with inner and outer radii '*a*' and '*b*' respectively

$$C = \frac{2\pi\varepsilon L}{ln\left(\frac{b}{a}\right)}$$; where L is the length of co-axial capacitor (cable)

7. State Poisson's equation in Cartesian coordinates.

$$\nabla^2 V = \frac{\partial^2 V}{\partial x^2} + \frac{\partial^2 V}{\partial y^2} + \frac{\partial^2 V}{\partial z^2} = -\frac{\rho}{E}$$

8. What is the significance of Laplace's equation?

It is used for a very important of class of problems in which the charges are restricted to the surfaces of conductor's *i.e.*, either the distribution on the surface of the conductors. Or a combination of the two are specified (boundary value field in charge free region bounded by the conductors).

9. State the difference between poisson's eqn. and Laplace's equation.

Poisson's equation is $\nabla^2 V = -\frac{\rho}{E}$ and Laplace's equation is $\nabla^2 V = 0$

Poisson's equation is applicable for charged media with defined boundaries whereas Laplace's equation is for charge free media with defined boundaries.

10. What is the unit of polarization?

The unit for polarization is coulom b/m^2 (C/m^2).

11. Define magnetic dipole.

A small bar magnet with pole strength Q m and length l may be treated as magnetic dipole whose magnetic is Q ml.

MULTIPLE CHOICES

1. I = ...

(a) $\int \bar{J} \cdot d\bar{l}$ (b) $\int_S \bar{J} \cdot d\bar{s}$

(c) $\bar{J}$ (d) $\int \bar{J}^2 \cdot d\bar{s}$

2. $\bar{J}$ = ...

(a) $\dfrac{\sigma}{\bar{E}}$ (b) $\sigma^2 \bar{E}$

(c) $\sigma \bar{E}$ (d) none of these

3. ρ_v = ...

(a) $\rho V_0\, e^{t/Tr}$ (b) $\rho v_0\, e^{-t/Tr}$

(c) $\rho V_0\, e^{-t}$ (d) none of these

4. Identify the poisson's equation from the following:

(a) $\nabla^2 V = 0$ (b) $\nabla^2 V = -\rho_v$

(c) $\nabla^2 V = -\dfrac{\rho_v}{\varepsilon}$ (d) $\nabla^2 V = 0$

5. Identify the Laplace's equation ....

(a) $\nabla^2 V = 0$ (b) $\nabla^2 V = -\rho_v$

(c) $\nabla^2 V = Q$ (d) none of these

6. For a co-axial capacitor capacitance C = ...

(a) $\dfrac{2\pi\varepsilon}{\dfrac{1}{a} - \dfrac{1}{b}}$ (b) $\dfrac{2\pi\varepsilon}{\ln \dfrac{b}{a}}$

(c) $\dfrac{2\pi\varepsilon L}{\ln \dfrac{b}{a}}$ (d) $\dfrac{\varepsilon s}{d}$

7. Identify the continuity equation ....

(a) $\nabla \cdot \bar{J} = \rho_v$ (b) $\nabla \cdot \bar{E} = -\dfrac{\partial \rho_v}{\partial t}$

(c) $\nabla \cdot \bar{H} \cdot \bar{H} = \rho_v$ (d) $\nabla \cdot \bar{J} = -\dfrac{\partial \rho_v}{\partial t}$

8. For steady currents the continuity equation is ....

(a) $\nabla \cdot \bar{J} = 0$ (b) $\nabla \cdot \bar{E} = 0$

(c) $\nabla V = -\bar{E}$ (d) none of these

9. Capacitance offered by an isolated sphere with radians r is ....

(a) $\frac{1}{4\pi\varepsilon r}$ (b) $4\pi\varepsilon r$

(c) $\frac{\varepsilon A}{d}$ (d) none of these

10. For spherical capacitor $C =$ ....

(a) $\frac{2\pi\varepsilon}{\ln\frac{b}{a}}$ (b) $\frac{2\pi\varepsilon}{ab}$

(c) $\frac{2\pi\varepsilon}{\frac{1}{a}-\frac{1}{b}}$ (d) none of these

11. Identify the point form of Ohms law

(a) $\bar{D} = \varepsilon\bar{E}$ (b) $D = \frac{\varepsilon_0}{E}$

(c) $\bar{J} = \sigma\bar{E}$ (d) $\bar{J}\frac{\sigma}{\vec{E}}$

12. For dielectric-dielectric boundary ....

(a) $\frac{E_{1t}}{\varepsilon_1} = \frac{E_{2t}}{\varepsilon_2}$ (b) $D_{1t} = D_{2t}$

(c) $\varepsilon_1 D_{1t} = \varepsilon_2 D_{2t}$ (d) $\varepsilon_2 D_{1t} = \varepsilon_1 D_{2t}$

13. For dielectric-dielectric boundary ....

(a) $D_{1n} - D_{2n}$ (b) $\varepsilon_1 D_{1n} = \varepsilon_2 D_{2n}$

(c) $\varepsilon_1 E_{1n} - \varepsilon_2 E_{2n}$ (d) none of these

14. For dielectric-dielectric boundary ....

(a) $\varepsilon_1 D_{1n} = \varepsilon_2 D_{2n}$ (b) $\varepsilon_1 E_{1n} - \varepsilon_2 E_{2n}$

(c) $\varepsilon_{1n} = E_{2n}$ (d) none of these

15. For conductor-dielectric boundary ....

(a) $E_t = 0$ (b) $E_n = 0$

(c) $D_t = 0$ (d) both (a) and (c)

16. For conductor-dielectric boundary ....

(a) $E_t = 0; D_n = \rho_s$ (b) $E_t = \rho_s; D_n = 0$

(c) $E_t \neq 0; D_n \neq \rho_s$ (d) $E_t = 0; D_n = 0$

17. Identify the Gauss's law from the following:

(*a*) $\oint \overline{D} \cdot d\overline{s} = Q_{enc}$ (*b*) $\oint \overline{D} \cdot d\overline{s} = \rho_v$

(*c*) $J = \sigma E$ (*d*) none of these

18. Which is the example of convention current:

(*a*) a moving charged belt (*b*) electronic movement in a vaccum tube

(*c*) an electron beam in a television tube (*d*) electric current flowing in a copper wire

19. When a steady potential difference is applied across the ends of a conditioning wire.

(*a*) all electrons move with constant velocity

(*b*) all electrons move with a constant acceleration

(*c*) the random electronic motion will, on the average be equivalent to a constant velocity of each electron

(*d*) the random electronic motion will, on the average be equivalent to a non-zero constant acceleration of each electron.

20. The formula $R = l\,(\sigma s)$ is for thin wires.

(*a*) True (*b*) False (*c*) not necessarily

21. Sea water has $\varepsilon_r = 80$. Its permitivity is

(*a*) 81 (*b*) 79

(*c*) 5.162×10^{-10} t/M (*d*) 7.074×10^{-4} F/m

22. Both ε_0 and x_0 are dimensionless.

(*a*) True (*b*) False

23. If $\nabla \cdot \vec{D} = \varepsilon \nabla \cdot \vec{E}$ and $\nabla \cdot \vec{J} = \sigma \nabla \cdot \vec{E}$ in a given material, the material is said to be

(*a*) linear (*b*) homogenous

(*c*) isotropic (*d*) both (*a*) & (*b*)

24. The relaxation time of mica ($\sigma = 10^{-15}$ mhos/m, $\varepsilon_r = 6$) is

(*a*) 5×10^{-10} s (*b*) 10^{-6} s

(*c*) 5 hours (*d*) 15 hours

25. Which of the following statements are incorrect:

(*a*) The conductivities of conductors and insulators vary with temperature and frequency

(*b*) A conductor is an equipotential body and *E* is always tangential to the conductor

(*c*) Non polar molecules have no permanent dipoles

(*d*) In a linear dielectric, *P* varies linearly with *E*

26. The electric conditions (charge and potential) inside and outside an electric screening are completely independent of one another.

(*a*) True (*b*) False

27. The equation of continuity for time varying fields is ....

(*a*) $\nabla \cdot \vec{J} = 0$ (*b*) $\nabla \cdot \vec{D} = 0$

(*c*) $\nabla \cdot \vec{J} = \dfrac{\partial \rho_v}{\partial t}$ (*d*) $\nabla \cdot \vec{J} = -\dfrac{\partial \rho_v}{\partial t}$

28. The volume charge density $\rho_v - 20t^2$. Find $\nabla \cdot \overline{J}$ at 10 sec?

(*a*) 400 A/M^2 (*b*) 200 A/M^3

(*c*) 0 A/m^3 (*d*) – 400 A/m^3

29. $\vec{E} = \frac{1}{\varepsilon} 20t \times 10^{-5}\ \overline{u}_y$ V/M. Find displacement current density ...., in free space.

(*a*) $20t^2 \times 10^{-5} \varepsilon_0 \overline{u}_y$ (*b*) $20 \times 10^{-5} \varepsilon_0 \overline{u}_y$

(*c*) $\frac{20}{\varepsilon_0} \times 10^{-5} \varepsilon_0 \overline{u}_y$ (*d*) $-\frac{20}{\varepsilon_0} \times 10^{-5} \varepsilon_0 \overline{u}_y$

30. $\rho_v = -10t^2$ C/m^3 ; $\nabla \cdot$ J = ... at t = 0.1 sec.

(*a*) 0.01 A/m^3 (*b*) 0.1 A/m^2

(*c*) both (*d*) none of these

31. The displacement currents, I_d = ...

(*a*) $\frac{dJ_d}{ds}$ (*b*) $\int_S \overline{J}_d \cdot d\overline{s}$

(*c*) $\frac{J_d}{s}$ (d) none of these

32. $\overline{D} = 10x^2 y\overline{u}_x + 20xy\overline{u}_z$; ρ_v = ... at (1, 1, 0).

(*a*) 20 (*b*) – 20

(*c*) $-20\overline{u}_x$ (*d*) $20\overline{u}_x$

33. The point from of Ohms law is ....

(*a*) $\overline{J} = \varepsilon\overline{E}$ (*b*) $\overline{J} = \sigma\overline{D}$

(*c*) $\overline{J} = \sigma\overline{E}$ (*d*) $\overline{J} = \overline{E}$

34. $\overline{E} = 100\overline{u}_x + 2\overline{u}_z$; Find $\overline{J}$ = ... if $\sigma = 0.1$.

(*a*) $0.1\overline{u}_x + 2\overline{u}_z$ (*b*) $10\overline{u}_x + 0.02\overline{u}_z$

(*c*) $0.1\overline{u}_x + 0.02\overline{u}_z$ (*d*) $10\overline{u}_x + 2\overline{u}_z$

35. The following is not a Maxwell's equation for a time varying fields:

(*a*) $\nabla \cdot \overline{D} = \rho_v$ (*b*) $\nabla \cdot \overline{D} = 0$

(*c*) $\nabla \times \overline{E} = 0$ (*d*) $\nabla \times \overline{H} = \overline{J} + \overline{D}$

36. The continuity equation for time varying fields in integral form is ....

(*a*) $\oint \overline{D} \cdot d\overline{s} = 0$ (*b*) $\oint \overline{H} \cdot d\overline{l} = I$

(*c*) $\oint \overline{D} \cdot d\overline{s} = -\int \frac{\partial P}{\partial t} d_v$ (*d*) $\oint \overline{D} \cdot d\overline{s} = \int \rho\, dv$

37. For dielectric-dielectric boundary ....

(*a*) $E_{1t} = E_{2t}$ (*b*) $\frac{E_{1t}}{\varepsilon_1} = \frac{E_{2t}}{\varepsilon_2}$

(*c*) $D_{1t} = D_{2t}$ (*d*) none of these

38. Identify the continuity equation for time varying fields ....

(*a*) $\nabla \cdot \vec{J} = -\frac{\partial P}{\partial t}$ (*b*) $\nabla \cdot \vec{J} = 0$

(*c*) $\oint \vec{J} \cdot d\vec{s} = -\int \frac{\partial P}{\partial t} dv$ (*d*) both (*a*) & (*c*)

39. $\vec{D} = 10t^2\bar{u}_x$; find J_D at $t = 0.1$ sec.

(*a*) $2\bar{u}_x$ (*b*) 12

(*c*) $2\bar{u}_y$ (*d*) $2\bar{u}_z$

40. $\rho_v = 20t^2$ C/m^3; $\nabla \cdot J =$ .... at t = 0.1 sec.

(*a*) – 40 (*b*) – 4

(*c*) $4\bar{u}_x$ (*d*) $-4\bar{u}_x$

41. The displacement current is ....

(*a*) the current flowing through a capacitor (*b*) the current flowing through inductor

(*c*) the current flowing in a resistor (*d*) always zero

Key

1. (*b*)	**2.** (*c*)	**3.** (*b*)	**4.** (*c*)	**5.** (*a*)	**6.** (*c*)
7. (*d*)	**8.** (*a*)	**9.** (*b*)	**10.** (*c*)	**11.** (*c*)	**12.** (*d*)
13. (*a*)	**14.** (*b*)	**15.** (*d*)	**16.** (*a*)	**17.** (*a*)	**18.** (*d*)
19. (*c*)	**20.** (*c*)	**21.** (*d*)	**22.** (*b*)	**23.** (*d*)	**24.** (*d*)
25. (*b*)	**26.** (*c*)	**27.** (*d*)	**28.** (*a*)	**29.** (*b*)	**30.** (*b*)
31. (*b*)	**32.** (*a*)	**33.** (*c*)	**34.** (*c*)	**35.** (*c*)	**36.** (*c*)
37. (*a*)	**38.** (*d*)	**39.** (*a*)	**40.** (*b*)	**41.** (*b*)	

REVIEW QUESTIONS

1. A hollow conductor in the form of a matchbox is placed in a static electric field. What is the electric field inside the conductor? Sketch the charge distribution on the inner and the outer surfaces of the conductor.
2. Derive the relation between electric field and electric potential in rectangular coordinates.
3. Explain briefly about work done.
4. Define potential difference and derive the expression for potential difference.
5. If the charge is distributed over a thin spherical shell of radius *b*, what is $\vec{E}$ inside the shell?

6. Explain what is meant by an equipotential surface?
7. Obtain the expression for the capacitances of a parallel plate condenser.
8. How much energy is required to assemble a point charge? Do you think that a point charge can really exist?
9. Derive the expression for capacitances of the spherical condenser.
10. Find the capacitance of a two concentric spherical shells.
11. Why are the equipotential surfaces perpendicular to the electric flux lines?
12. A point charge of 1 coulomb is brought from infinity to point a in free space. How much work is required to do this? If the energy expenditure in bringing a second charge of 1 coulomb from infinity to point b is 1 joule, what is the separation between the charges?
13. What do we mean by boundary conditions?
14. Show that $E = -\nabla V$.
15. Derive Poisson's and Laplace equations from fundamentals.
16. Is the solution of $\nabla^2 V = 0$ unique by itself?
17. Derive the expression for the capacitance of parallel plate capacitor with two dielectrics.
18. Obtain the general solution for the potential distribution using cylindrical coordinates of Laplace's equation.
19. Derive the expression for capacitance of an isolated sphere.
20. Derive the expression for capacitance of a coaxial cable.
21. What is the significance of boundary conditions when solving Laplace's equation?
22. What is the definition of a bound charge?
23. Find the capacitance per unit area of a condenser consisting of two infinite conducting planes placed at a distance 'h' apart.
24. Derive the relationship between field and potential in differential form.
25. Prove that for a charge of free region (*i*) $\nabla \cdot D = 0$ and (*ii*) $\nabla^2 V = 0$, where D is electric flux density and V is the electric potential.
26. Show that for any spherical distribution of charge, the field at radius r is the same as if all the charge inside the volume of radius r were concentrated at the center, and that outside of r were removed.
27. In a charge-free region, $E_x = \alpha x$, and $E_y = \beta y$. Find E_z.
28. If n capacitors are connected in series, what is the effective capacitance?
29. If n capacitors are connected in parallel, what is the effective capacitance?
30. Verify that the potential due to an electric dipole satisfies Laplace's equation.
31. Deduce the equations of Poisson and Laplace expressing the relationship of the space rate of variation of electric field component with the distributed charge field.
32. Why is the charge on each capacitor the same when capacitors are connected in series?
33. Is the charge on each capacitor the same when they are connected in parallel?
34. In what ways does the electric field intensity vary with distance for (*i*) A point charge (*ii*) An electric dipole.

EXERCISE PROBLEMS

1. Given the potential field $V = 50\ xyz\ V$ is free space. (*a*) Find the total energy stored within the cube $0 < xyz < 2$ (*b*) What value would be obtained by assuming a uniform energy density equal to value at center to cube?

2. In the region of free space where, $2 < \rho < 3, 0 < \phi < \frac{\pi}{2}$ and $5 < z < 7$ let $\overline{E} = \left(\frac{k}{\rho}\right) U_\phi$ (*a*) Find a positive value for *k* so that the total energy stored is 1 MJ (*b*) Show that surface $\phi = 0$ and $\phi = \frac{\pi}{2}$ are equipotential surface (*c*) What is potential difference between the surface at $\phi = 0$ and $\phi = \frac{\pi}{2}$?

3. A dipole having a moment $\overline{P} = 10U_x - 3U_y + 5U_z$ nC-m is located at *Q* (2, 4, 1) in free space (*a*) Find *V* at *P* (*x*, *y*, *z*) (*b*) Find *V* at (5, 1, 0).

4. A potential field is given as $V = 100\ e^{-5x} \sin 3y \cos 4z\ V$. Let point $P\ (0.1, \frac{\pi}{2}, \frac{\pi}{24})$ be located at a conductor free boundary. At point *P* find the magnitude of (*a*) *V* (*b*) $\overline{E}$ (*c*) E_N (*d*) E_z (*e*) P_s.

5. A point charge of 25 nC is located in free space at *P*(2, –3, 5) and a perfectly conducting plane at *z* = 2 find (*a*) *V* at (3, 2, 4) (*b*) $\overline{E}$ at (3, 2, 4) (*c*) *P*, at (3, 2, 2).

6. The region *y* < 0 contains a dielectric material for which $\varepsilon_{n_1} = 2.5$ while region *y* > 0 is characterized by $\varepsilon_{r_2} = 4$. Let $E_1 = -30U_x + 50U_y + 70U_z$ V/m and find (*a*) E_{N_1} (*b*) E_{t_1} (*c*) E_1 (*d*) θ_1.

7. Find the capacitor between a circular conducting cylinder in air 2 mm in radius and (*a*) Conducting plane 1 cm distant from cylinder axis (*b*) A conducting plane 1 cm distant from cylinder axis (*c*) A similar cylinder axes separated by 1 cm.

8. Point charge of 1 μC and – 1μC are located at (0, 0, 0.5) and (0, 0, – 0.5) respectively. Treating these two charges as a dipole at the origin calculate (*a*) *V* at *P* (3, 0, 4) (*b*) $|\overline{E}|$ at *P*. Now find the exact values for (*c*) *V* at P (*d*) $|\overline{E}|$ at *P*.

9. For the point (3, 60°, 2) is cylindrical coordinates and potential field $V = 10\ (P + 1)\ z^2 \cos \phi$ V is free space. Find at P (*a*) V (*b*) $\overline{E}$ (*c*) $\overline{D}$ (*d*) P_v (*e*) $\frac{dv}{dr}$ (*f*) U_N.

10. Let $V = 2\ (x + 1)^2\ (y + 2)^2\ (z + 3)^2$ *V* is free space. At *P* (2, – 1, 4). Find (*a*) *V* (*b*) $\overline{E}$ (*c*) $|\overline{E}|$ (*d*) $\overline{D}$ (*e*) $|\overline{D}|$ (*f*) ρ_v (*g*) $\overline{U}_N$.

11. Potential difference between two spherical conductors separated by a uniform dielectric with $\varepsilon_r = 4.0$ is 200 V, the outer conductor being at ground potential. Inner conductor, has radius = 5 cm and the outer 40 cm, determine the surface change distribution on the conductor.

12. For two spherical shells with radii 0.15 M and 0.25 M respectively, the potentials are 150 V and zero respectively. Assuming free space between the shells, determine E and D in free space. Also evaluate maximum value E.

13. Using Laplace's equation, for a parallel plate capacitor with plate surface, normal to x-axis, find the potential at any point between the plates $V = V_1$ and $x = x_1$ and $V = V_2$ at $x = x_2$.

14. Electric flux density in a region is given in spherical system of coordinates as follows $\overline{D} = 1.0\ R^2\ U_r$ C/m^2, verify divergence theorems for value enclosed by $R = 5$ m and $\theta = \frac{\pi}{6}$.

15. A dipole with $\rho = 0.1U_z$ μC. M is located at point A (0, 0, 1) in free space and the $Z = 0$ plane is perfectly conducting (*a*) Find V at ρ (0, 1, 2) (*b*) Find the equation of 200 V equipotential surface in Cartesian coordinates.

16. Two concentric conducting spheres of radii $a = 5$ cm, $b = 20$ cm have dielectric $\varepsilon_{R_1} = 2, 0 < \theta < \frac{\pi}{2}$ and $\varepsilon_{R_2} = J, \frac{\pi}{2} < \theta < \pi$, separating them. Assume $E_{r_1} = 200$ V/m at $r = 10$ cm, $\theta = 28°$ and $\phi = 106°$ and find (*a*) $\overline{E}_1$ (*b*) $\overline{E}_2$ (*c*) V_{ab} (*d*) the total change on the inner sphere.

17. Find the capacitance of parallel plate capacitor, if the plates are of area 4.5 m^2, the distance between the plates is 3 mm, potential gradient is 20^5 V/m and charge density on the plates is 4.5 μC/m^2.

18. In cylindrical coordinates, $V = 50$ at $\rho = 4$ mm and $V = 0$ at $\rho = 56$ mm. Find the voltage at $\rho = 160$ mm, if the potential depends only on ρ.

CHAPTER

5

Static Magnetic Fields

5.1 INTRODUCTION

Now we discuss forces produced by charges in steady motion. Because current is rate of change of charge. The field produced by this type of currents is called static magnetic field. Here the current should be *D.C*. The term magnetic field is the space (or) region around a magnet where the effect of magnet is felt .When a steady direct current flows in a conductor, a magnetic field is produced and it is a steady magnetic field.

A magnetic field resembles an electric field but there is one main difference. In an electric field, lines of flux originate from a positive charge and these spread out all around charge. If charge is negative, the flux lines converge on charge. If charge is an isolated charge, the flux lines originating from it do not return but magnetic poles exist in pairs, lines of magnetic flux originate from a north pole, journey through the surrounding medium until they reach a South pole and they terminate on the North pole after traversing through body of magnet if field is due to permanent magnet.

If magnetic field is produced by current carrying conductor, then lines of flux form concentric circles around the conductor. So each lines of force form a closed path. Hence the total flux having a surface is same as flux entering the surface.

5.2 MAGNETIC FLUX AND FLUX DENSITY

Magnetic field can be visualized as comprising of lines of force. These lines of force are referred as magnetic flux (ϕ) measured in Weber. The magnetic flux crossing unit normal surface area is called as magnetic flux density (B) measured in Weber/m^2 (or) Tesla.

i.e.,
$$B = \frac{\phi}{A} \qquad \text{...(5.1)}$$

5.2.1 Magnetic Field Intensity (H)

Force experienced by a unit north pole placed at that point and is denoted by H, measured in Newton/Weber.

$$B = \mu_0 H \text{ in free space (or) air}$$

Permeability, $\mu_0 = 4\pi \times 10^{-7}\ H/\text{Weber}$

Other medium, $B = \mu_0\, \mu_r\, H$

Where, μ_r = relative permeability of the medium.

5.3 BIOT-SAVART'S LAW

Let it be assumed that source of steady magnetic filed is a current carrying conductor. The field due to a conductor when carrying a steady direct current can be obtained on basis of **Biot-Savart's law.**

This law states that at any point P, the magnitude of magnetic field intensity (H) produced by differential element is proportional to product of current and magnitude of differential length dL and sine of angle lying between the filament and a line connecting the filament to point P where the field is desired.

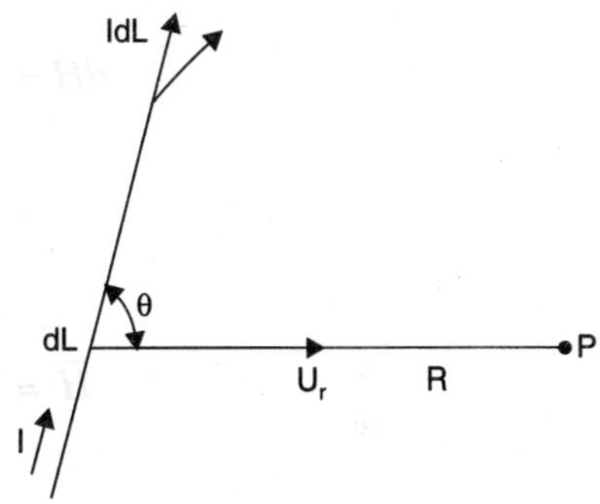

Fig. 5.1

Consider different current element IdL. This may be seen as extremely small section of current carrying filament of wire of small radius. Let P be a point at a distance R from element IdL. Also let θ be angle made by differential current element $I\Delta L$ with line joining IdL and point P.

According to definition,

$$dH \propto \frac{IdL \sin\theta}{R^2} \qquad \text{...(5.2)}$$

$$= K\frac{IdL \sin\theta}{R^2}$$

Where, $K = \dfrac{1}{4\pi}$

$$\therefore \quad dH = \frac{IdL \sin\theta}{4\pi R^2}\ \text{A/m} \qquad \text{...(5.3)}$$

In the expression θ is the angle between two vectors IdL and U_r, where U_r is unit vector directed along line joining IdL and P. So, $IdL \sin\theta$ is the cross product of vector IdL and U_r.

$$dH = \frac{IdL \times U_r}{4\pi R^2}\ \text{A/m} \qquad \text{...(5.4)}$$

Then flux density at P

$$dB = \mu\, dH$$

$$= \frac{\mu}{4\pi R^2} IdL \times U_r\ \text{wb/m}^2$$

Field intensity due to entire length of conductor is

$$H = \frac{IdL \times U_r}{4\pi R^2} \qquad \text{...(5.5)}$$

Equation (5.5) is known as integral form of Biot-Savart's law.

Example 5.1: *A filamentary current of 15 A is directed from infinity to the origin on the positive x-axis, and then back out to infinity along the position y-axis. Use the Biot-Savart's law of find $\overline{H}$ at P (0, 0, 1).*

Solution: Consider current which lies on the x-axis

$$d\overline{H} = \frac{Id\overline{L} \times \overline{U}_{12}}{4\pi R^2}$$

$$dL = dxU_x \,;\, 2\,(0, 0, 1), 1\,(x, 0, 0)$$

$$U_{12} = -xU_x + U_z$$

$$R = (x^2 + 1)^{1/2}$$

$$\therefore \quad dH = \frac{15dx\, U_x}{4\pi} \times \left[\frac{-xU_x + U_z}{(x^2+1)^{3/2}}\right]$$

$$= \frac{15dx\, U_y}{4\pi\,(x^2+1)^{3/2}}$$

$$\therefore \quad \overline{H} = \frac{15}{4\pi}\int_{\infty}^{0} \frac{dxU_x}{(x^2+1)^{3/2}} = \frac{-15}{4\pi}\left[\frac{y}{\sqrt{y^2+1}}\right]_0^{\infty} U_y$$

$$= \frac{-15}{4\pi}\,[0-1]\,U_y = 1.194\,U_y \text{ A/m}$$

Similarly for the current which lies on the y-axis we have

$$dH = \frac{15dy\, U_y}{[4\pi\sqrt{y^2+1^2}\,]^2} \times \left[\frac{-yU_y + U_z}{\sqrt{y^2+1}}\right] = \frac{15dy\, U_y}{4\pi(y^2+1)^{3/2}}$$

$$\therefore \quad \overline{H} = \frac{15}{4\pi}\int_{\infty}^{0} \frac{dyU_x}{(y^2+1)^{3/2}} = \frac{15}{4\pi}\left[\frac{y}{\sqrt{y^2+1}}\right]_0^{\infty} U_x$$

$$= \frac{15}{4\pi}\,[1-0]U_x = 1.194U_x \text{ A/m}$$

$$\overline{H}_{\text{total}} = [1.194U_x + 1.194U_y] \text{ A/m.}$$

Example 5.2: *Given points A(2, 3, 5) ,B(– 3, – 2, 4) and C(4, 2, – 3). Let a differential current element with I = 5 A and* $|\,d\overline{L}\,| = 10^{-4}$ *m be located at A. The direction of dL is from A to B. Find* $d\overline{H}$ *at C.*

Solution:

$$\overline{R}_{AB} = (-3-2)U_x + (-2-3)U_y + (4-5)U_z$$

$$= -5U_x - 5U_y - U_z$$

$$U_{AB} = \overline{U}_L = \frac{\overline{R}_{AB}}{|\,R_{AB}\,|} = \frac{-5U_x - 5U_y - U_z}{\sqrt{(-5)^2 + (-5)^2 + (-1)^2}}$$

$$= -0.700U_x - 0.700U_y - 0.140U_z$$

$$\therefore \quad Id\bar{L} = I\ |d\bar{L}|\ \bar{U}_L$$
$$= 5 \times 10^{-4}(-0.700U_x - 0.700U_y - 0.140U_z)\ \text{A/m}$$
$$d\bar{H} = \frac{Id\bar{L} \times \bar{U}_{AC}}{4\pi R^2}$$
$$\bar{R}_{AC} = (4-2)U_x + (2-3)U_y + (-3-5)U_z$$
$$= 2U_x - U_y - 8U_z$$
$$U_{AC} = \frac{\bar{R}_{AC}}{|R_{AC}|} = \frac{2U_x - U_y - 8U_z}{\sqrt{(2)^2 + (-1)^2 + (-8)^2}}$$
$$= 0.2407U_x - 0.1203U_y - 0.963U_z$$
$$Id\bar{L} \times U_{AC} = 5 \times 10^{-4} \begin{vmatrix} U_x & U_y & U_z \\ -0.700 & -0.700 & -0.140 \\ 0.2407 & -0.1203 & -0.963 \end{vmatrix}$$
$$= 5 \times 10^{-4}(0.6572U_x - 0.7077U_y + 0.2527U_z$$
$$d\bar{H} = \frac{5 \times 10^{-4}(0.6572U_x - 0.7077U_y + 0.2527U_z)}{4\pi(\sqrt{69})^2}$$
$$= (0.379U_x - 0.4083U_y + 0.1457\ \mu\ U_z)\ \text{A/m.}$$

Example 5.3: *Find the magnetic field intensity at the origin due to a current element $6\pi\ U_z$ μA/m at the point (3, – 6, 0) in free space.*

Solution: Given $Idl = 6\pi\ U_z\ \mu$ A/m, point $P = (3, -6, 0)$ and so distance of P from the origin is
$$R = \sqrt{3^2 + (-6)^2} = 6.708$$

Directed line segment from point P to the origin, $R = -3U_x + 6U_y$

Let U_R be a unit vector and so it can be written as
$$\bar{U}_R = \left(-\frac{3}{\sqrt{45}}\right)U_x + \left(\frac{6}{\sqrt{45}}\right)U_y = -0.447U_x + 0.894U_y$$
$$\therefore \quad (Idl \times U_R) = \begin{vmatrix} U_x & U_y & U_z \\ 0 & 0 & 6\pi \\ -0.447 & 0.894 & 0 \end{vmatrix}$$
$$= -16.84\ U_x + 8.421\ U_y$$

Now, the magnetic field intensity is
$$d\bar{H} = \frac{Idl \times \bar{U}_R}{4\pi R^2} = \frac{(-16.84U_x + 8.421U_y) \times 10^{-6}}{4\pi \times 45}$$
$$= (-0.0297U_x + 0.0148U_y) \times 10^{-6}\ \text{A/m}$$
$$= (-0.0297U_x + 0.0148U_y)\ \mu\ \text{A/m.}$$

Example 5.4: *Find the magnetic field at A(3, 4, – 3) due to a current element $I\bar{dl} = \pi(0.7u_x - 0.4u_y + 0.9u_z)$ μA/m situated at B(4, – 3, 5).*

Solution: According to given data $I\bar{dl} = \pi(0.7u_x - 0.4u_y + 0.9u_z)$ μA/m is situated at $B(4, -3, 5)$ and the coordinates of point A are $A(3, 4, -3)$

So, $\overline{R}_{BA} = R_A - R_B$

$= (3-4)u_x + (4+3)u_y + (-3-5)u_z$

$= -1u_x + 7u_y - 8u_z$

Magnitude of $\overline{R}_{BA} = \sqrt{(-1)^2 + 7^2 + (-8)^2} = \sqrt{114}$

$$\overline{u}_r = \frac{-u_x + 7u_y - 8u_z}{\sqrt{144}} = -0.0833u_x + 0.5833u_y - 0.666u_z$$

So, the vector product of $I\overline{dl}$ and $\overline{u}_r$ can be written as

$$(I\overline{dl} \times \overline{U}_r) = \begin{vmatrix} u_x & u_y & u_z \\ 0.7\pi & -0.4\pi & 0.9\pi \\ -0.0833 & 0.5833 & -0.666 \end{vmatrix}$$

$$= 2.4844u_x + 1.2275u_y - 1.3866u_z$$

Magnetic field $dH = \dfrac{I\overline{dl} \times \overline{u}_r}{4\pi R^2}$ A/m

$$= \frac{2.4844u_x + 1.2275u_y - 1.3866u_z}{4\pi \times 144} \times 10^{-6}$$

$$= (1.373u_x + 0.6786u_y - 0.766u_z) \text{ nA/m.}$$

5.4 APPLICATIONS OF BIOT-SAVART'S LAW

This law helps us in determining the magnetic field intensity due to several different conductor configurations as follows:

5.4.1 Field Intensity Along the Axis of Circular Wire

Let I be the current flowing in the circular loop and P be the point at a distance x on the axis of the loop at which magnetic field intensity is to be found. Consider two elements dl_1 and dl_2 each of length dL diametrically opposite to each other as shown in the Fig. 5.2. Since the flow of current in the two elements is in opposite direction to each other, the direction of magnetic induction by Biot-Savart's law can be shown as

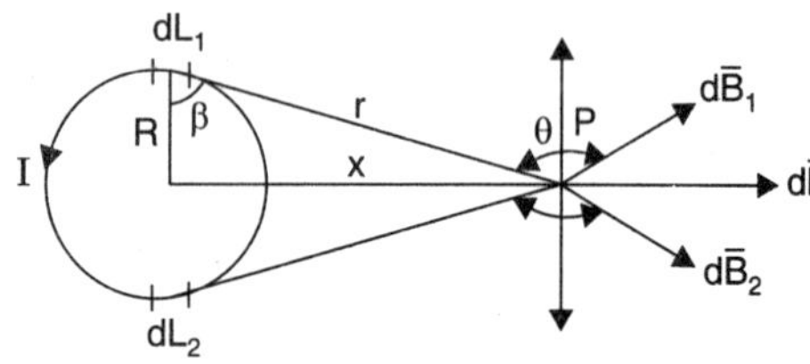

Fig. 5.2

$$dB_1 = \frac{\mu I dl \sin\theta}{4\pi r^2} = \frac{\mu I dl}{4\pi r^2} \qquad (\because\ \theta = 90°) \qquad ...(5.6)$$

$$dB_x = \frac{\mu I dl}{4\pi r^2}\cos\beta$$

$$= \frac{\mu I R dl}{4\pi r^3} \qquad \left(\because\ \cos\beta = \frac{R}{r}\right) \qquad ...(5.7)$$

Since at point P, the horizontal components due to dB_1 and dB_2 add each other and vertical components cancel each other, we have the net magnetic flux density in the horizontal direction only.

$$\therefore \quad B = \Sigma dB_x = \frac{\mu IR\,(2\pi R)}{4\pi r^3} = \frac{\mu IR^2}{2r^3}$$

$$\therefore \quad \overline{B} = \frac{\mu IR^2}{2(R^2+x^2)^{3/2}}\,\overline{u}_x \qquad (\because \quad r = \sqrt{R^2+x^2}\,)$$

$$\therefore \quad \overline{H} = \frac{\overline{B}}{\mu} = \frac{IR^2}{2(R^2+x^2)^{3/2}}\,\overline{u}_x \qquad \text{...(5.8)}$$

In order to find the magnetic field intensity at the center of the current loop, putting $x = 0$

we get
$$\overline{H} = \frac{I}{2R}\,\overline{u}_x \qquad \text{...(5.9)}$$

Example 5.5: *Find the field intensity at a point on the axis, 6 m from the centre of a circular coil of area 125 cm² and carrying a current of 45 A.*

Solution: Given current, $I = 45$ A

Distance of the field point, $x = 6$ m

Area of the coil = 125 cm² = 125×10^{-4} m² = 0.0125 m².

Radius of the coil, $R = \sqrt{\dfrac{Area}{\pi}}$

$$= \sqrt{\frac{0.0125}{\pi}} = \sqrt{3.9808 \times 10^{-3}} = 0.0630 \text{ m}$$

The field intensity at a point on the axis of a circular current loop of radius R, distant x from the centre of the loop is given by the expression.

$$H = \frac{IR^2}{2(R^2+x^2)^{3/2}} \text{ A/m (in magnitude)}$$

Putting $I = 45$ A, $R = 0.0630$ m and $x = 6$ m, we get

Field intensity,
$$H = \frac{45 \times 0.0630^2}{2(0.0630^2 + 6^2)^{3/2}} = 4.1336 \times 10^{-4} \text{ A/m.}$$

Example 5.6: *A single turn circular coil of 45 m diameter carries a current of 25 × 10⁴ A. Determine the magnetic field intensity h at a point on the axis of the coil and 125 m away from the coil. The relative permeability of free space surrounding the coil is unity.*

Solution: We have $H = \dfrac{IR^2}{2(R^2+x^2)^{3/2}}$ A/m

where R = radius of the coil,

I = current carried by the coil, and x = distance from the coil along the axis.

Given diameter = 45 m or $R = \dfrac{45}{2} = 22.5$ m

and also $I = 25 \times 10^{-4}$ A and $x = 125$ m.

Substituting the given values in the above equation, we get

$$H = \frac{(25 \times 10^4) \times 22.5^2}{2(22.5^2 + 125^2)^{3/2}} = 30.886 \text{ A/m}$$

H acts along the axis of the coil and away from it.

5.4.2 Field Intensity Due to Infinite Long Straight Conductor

Consider an infinitely long straight conductor lying along z-axis. Let it carry a steady direct current I in the direction shown in Fig. 5.3. Let it be required to find field intensity at point P, a distance ρ from conductor.

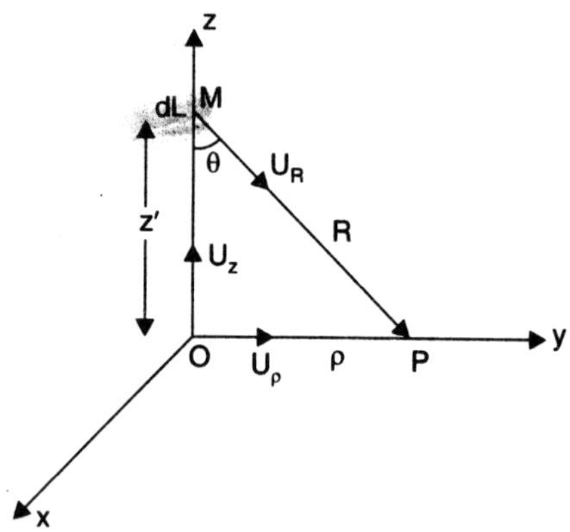

Fig. 5.3

Consider an elemental length dL $(= dz)$ at a point M on conductor

Let $MP = R$ and $OM = Z'$

We have $R = R_{MP} = R_P - R_M$

Where, $R_p = OP$ and $R_M = OM$

$$U_R = \frac{R_{MP}}{|R_{MP}|} = \frac{\rho U_p - Z'U_z}{\sqrt{\rho^2 + (Z')^2}} \qquad ...(5.10)$$

By Biot-Savart's law the field intensity at P due to differential current IdL is

$$dH = \frac{IdZ \times U_R}{4\pi R^2} \qquad ...(5.11)$$

$$= \frac{Id\bar{Z} \times (\rho U_p - Z'U_Z)}{4\pi[\sqrt{\rho^2 + (Z')^2}]\sqrt{\rho^2 + (Z')^2}} = \frac{Id\bar{Z} \times (\rho U_p - Z'U_Z)}{4\pi(\rho^2 + (Z')^2)^{3/2}}$$

Since vector $d\bar{Z}$ can be written as $dZ\, U_z$

$$= \frac{IdZU_z \times (\rho U_P - Z'U_z)}{4\pi(\rho^2 + Z'^2)^{3/2}} \qquad ...(5.12)$$

Total field intensity

$$H = \int_{-\infty}^{\infty} dH$$

And also $\rho U_p - Z'U_z = R$; Also $\sin\theta = \frac{\rho}{R} \Rightarrow R\sin\theta = \rho$

But $R \sin\theta$ can be written as $U_\rho \times R$ (cross product)

$\Rightarrow$ $U_\rho \times R = \rho\, U_\phi$ (result of cross product of U_ρ with R)

$$\int_{-\infty}^{\infty} dH = \int_{-\infty}^{\infty} \frac{I\rho dz}{4\pi(\rho^2 + (Z')^2)^{3/2}} U_\phi$$

$$H = \frac{I}{4\pi}\int_{-\infty}^{\infty} \frac{\rho dZU_\phi}{(\rho^2 + (Z')^2)^{3/2}} \qquad ...(5.13)$$

When U_ϕ is the unit vector along ϕ.

Since U_ϕ is constant

$$H = \frac{I\rho U_\phi}{4\pi}\int_{-\infty}^{\infty}\frac{dZ}{(\rho^2+(Z')^2)^{3/2}}$$

$$= \frac{I\rho U_\phi}{4\pi}\left[\frac{Z'}{\rho^2\sqrt{\rho^2+(Z')^2}}\right]_{-\infty}^{\infty}$$

$\therefore$ $$H = \frac{I}{2\pi\rho}U_\phi \text{ A/m} \qquad ...(5.14)$$

Example 5.7: *Find H at P (3, 4, 5) in Cartesian coordinates if there is an infinity long current filament passing through the origin and point C. The current of 45 A is directed from origin to C, where the location of C is (a) C (0, 0, 5) (b) C (0, 3, 0).*

Solution:

(*a*) Given current, I = 45 A

$$U_I = \frac{\overline{R}_{OC}}{|R_{OC}|} = U_z \text{ ; } P(3, 4, 5) \text{ ; } C(0, 0, 5)$$

$$R_{CP} = 3U_x + 4U_y \text{ ; } d = |R_{CP}|$$

$$= \sqrt{(3)^2+(4)^2} = 5$$

$$\overline{H} = \frac{I}{2\pi d}[U_I \times U_{CP}]$$

$$= \frac{45}{2\pi(5)}\left[U_z \times \frac{(3U_x+4U_y)}{5}\right]$$

$$= -5.732U_x + 4.229U_y \text{ A/m.}$$

(*b*) Given current I = 45 A

$$U_I = \frac{R_{OC}}{|R_{OC}|} = U_z \text{ ; } P(3, 4, 5) \text{ ; } CP(0, 3, 0)$$

$$R_{CP} = 3U_x + U_y + 5U_z \text{ ; } d = |R_{CP}|$$

$$= \sqrt{(3)^2+(1)^2+(5)^2} = 5.916$$

$$\overline{H} = \frac{45}{2\pi(5.916)}\left[U_y \times \frac{(3U_x+U_y+5U_z)}{5.916}\right]$$

$$= 6.055U_x - 3.633U_z \text{ A/m.}$$

Example 5.8: *A single phase circuit comprises of two parallel conductors A and B, each 2 m diameter and spaced 2 m apart. The conductors carry currents of +150 and –150 amps respectively. Determine the field intensity at the surface of each conductor and also in space exactly mid-way between A and B.*

Solution: Field at any point between A and B is

$$H = H_A + H_B = \text{sum of contributions due to } A \text{ and } B$$

$$= \frac{I}{2\pi r_1} + \frac{I}{2\pi r_2}$$

It is to be noted that each conductor produces field directed downwards. Therefore,

$$H = \frac{I}{2\pi}\left(\frac{1}{r_1} + \frac{1}{r_2}\right)$$

$$H \text{ at conductor surface } = \frac{150}{2\pi}\left(\frac{150}{1}\right) = \frac{11250}{\pi} \text{ A/m}$$

$$H \text{ mid-way between } A \text{ and } B = \frac{150}{2\pi}\left(\frac{1}{1} + \frac{1}{1}\right) = \frac{150}{\pi} \text{ A/m.}$$

5.4.3 Field Intensity Due to Current Filament of Finite Length

Consider a conductor of finite length L lying along Z axis. Let it carry a steady direct current I in the direction shown in Fig. 5.4. Let it be required to find field intensity at point P, distant R from conductor.

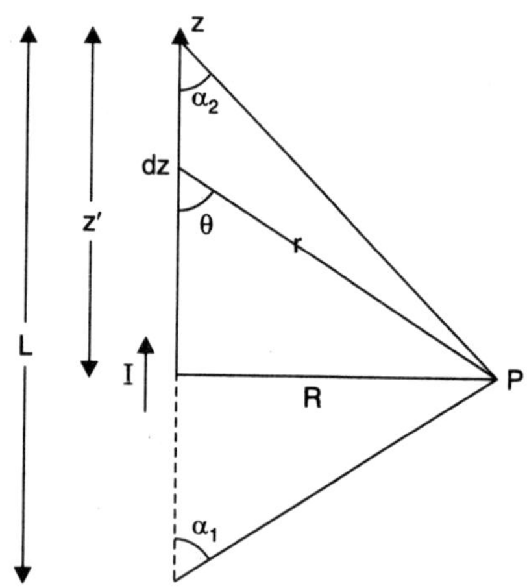

Fig. 5.4

Consider a small element of length dl at a distance Z from upper end of conductor, differential current element = Idl directed along Z-axis. Field intensity at P due to this current element is given by

$$dH = \frac{Idl \times U_r}{4\pi r^2} \text{ A/m} \qquad \text{...(5.15)}$$

$$dH = \frac{Idl \sin\theta}{4\pi r^2}$$

From Fig. 5.4, it can be seen as

$$\cot\theta = \frac{Z' - Z}{R}$$

Which implies $\quad Z' - Z = R\cot\theta$

Differentiating the above expression w.r.t. θ, we get

$$dZ = R\,\text{cosec}^2\,\theta\, d\theta$$

Also $$\sin\theta = \frac{R}{r};\ r = \frac{R}{\sin\theta} = R\,\text{cosec}\,\theta$$

Substituting all these values in equation (5.15), we get

$$dH = \frac{IR\,\text{cosec}^2\,\theta\, d\theta \sin\theta}{4\pi\,(R^2\,\text{cosec}^2\,\theta)} \qquad \text{...(5.16)}$$

$$= \frac{I\sin\theta\, d\theta}{4\pi R}$$

Total field intensity at P due to entire length is given by

$$H = \int dH = \int_{\alpha_1}^{\pi-\alpha_2} \frac{I \sin\theta \, d\theta}{4\pi R}$$

$$= \frac{I}{4\pi R}[\cos\alpha_1 + \cos\alpha_2] \quad \text{...(5.17)}$$

5.4.4 Field Intensity at Centre of Square Current Loop

It is required to find intensity at P, the centre of square loop. Consider side AD of loop. Let it be lying along y-axis. If a perpendicular line is drawn from P to AD it bisects AD at M.

Therefore M is mid-point of AD. It is obvious that

$$AD = R \quad \text{and} \quad AM = PM = DM = \frac{R}{2}.$$

Now field intensity at a point distant R from a current carrying conductor of finite length D from equation (5.17) is

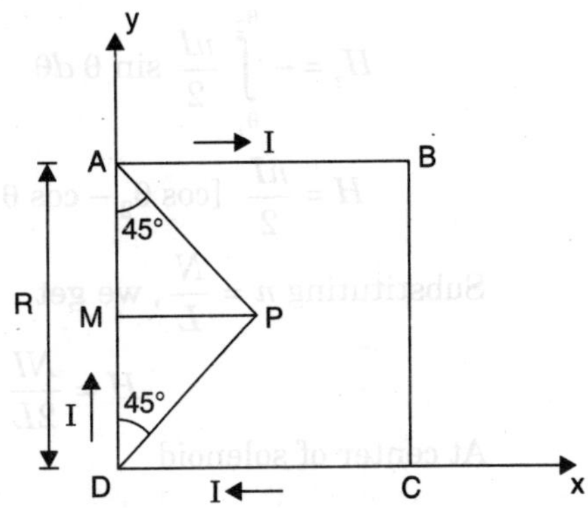

Fig 5.5

$$H = \frac{I}{4\pi R}[\cos\alpha_1 + \cos\alpha_2] \quad \text{...(5.18)}$$

$$H_1 = \frac{1}{4\pi(R/2)}[\cos 45° + \cos 45°]$$

$$= \frac{I}{\sqrt{2}\pi R} \quad \text{...(5.19)}$$

The direction of H_1 can be obtained from right handed corkscrew rule. It is seen that H_1 acts normally to plane containing conducts AD and point and hence inwards. So, field intensity due to side AB, BC and CD are equal to H_1.

So, that, total $H = H_1 + H_2 + H_3 + H_4$

$= 4H_1$ inwards

$$H = 4\left(\frac{I}{\sqrt{2}\pi R}\right)(-U_Z) \text{ A/m}$$

$$H = -\frac{2\sqrt{2}I}{\pi R} U_Z \text{ A/m} \quad \text{...(5.20)}$$

5.4.5 Magnetic Field Intensity Due to a Solenoid Current Carrying Wire

Consider cross-section of solenoid shown in the Fig. 5.6. Since solenoid consists of circular loop, we can directly write the equati on as

$$dH_z = \frac{I\,dL\,a^2}{2[a^2 + z^2]^{3/2}} \quad \text{...(5.21)}$$

$$= \frac{I\,a^2\,ndz}{2[a^2 + z^2]^{3/2}} \quad \text{...(5.22)}$$

Where $dL = ndz$ and where $n = \left(\frac{N}{l}\right)$

From the Fig. 5.6, $\tan\theta = \frac{a}{z}$

$$dz = -a\ \text{cosec}^2\ \theta\ d\theta$$

$$= -\frac{[Z^2 + a^2]^{3/2}}{a^2}\sin\theta\ d\theta \quad ...(5.23)$$

$$dH_z = -\frac{nl}{2}\sin\theta\ d\theta$$

$$H_z = -\int_{\theta_1}^{\theta_2}\frac{nI}{2}\sin\theta\ d\theta$$

$$H = \frac{nI}{2}\ [\cos\theta_2 - \cos\theta_1]\ U_z$$

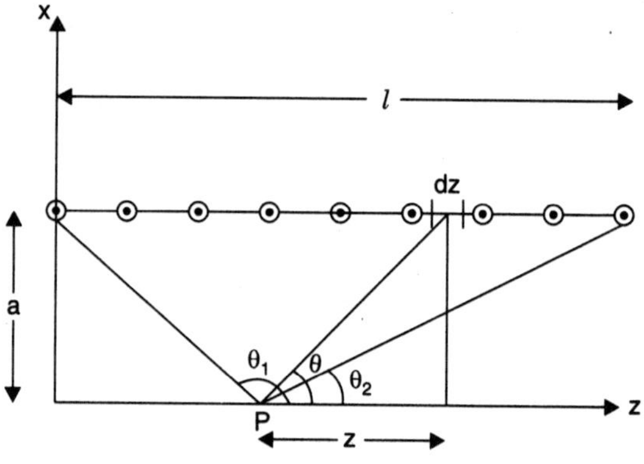

Fig. 5.6

Substituting $n = \frac{N}{L}$, we get

$$H = \frac{NI}{2L}\ [\cos\theta_1 - \cos\theta_2]\ U_z \quad ...(5.24)$$

At center of solenoid

$$\cos\theta_2 = \frac{\frac{L}{2}}{\left[a^2 + \frac{L^2}{4}\right]^{1/2}} = -\cos\theta_1$$

And

$$H = \frac{InL}{\left[a^2 + \frac{L^2}{4}\right]^{1/2}}\ U_Z \quad ...(5.25)$$

5.4.6 Magnetic Field Intensity on the Axis of an Infinite Solenoid

A solenoid of radius R with N turns per meter is shown in the Fig. 5.7. It is required to find out the magnetic field intensity 'H' on the axis of the infinite solenoid.

In the elemental distance dx there are $N\ dx$ turns. We have

$$dH = \frac{dB}{\mu} = \frac{IR^2}{2(R^2 + x^2)^{3/2}}\ Ndx$$

From the Fig. 5.7, $x = -R\cot\theta$

$$R^2 + x^2 = R^2 + R^2\cot^2\theta = R^2\ \text{cosec}^2\ \theta$$

$$dx = d(-R\cot\theta) = R\ \text{cosec}^2\ d\theta$$

$$\therefore \quad dH = \frac{NIR^2}{2R^3\ \text{cosec}^2\ \theta}\ R\ \text{cosec}^2\ \theta\ d\theta$$

$$= \frac{NI}{2}\sin\theta\ d\theta$$

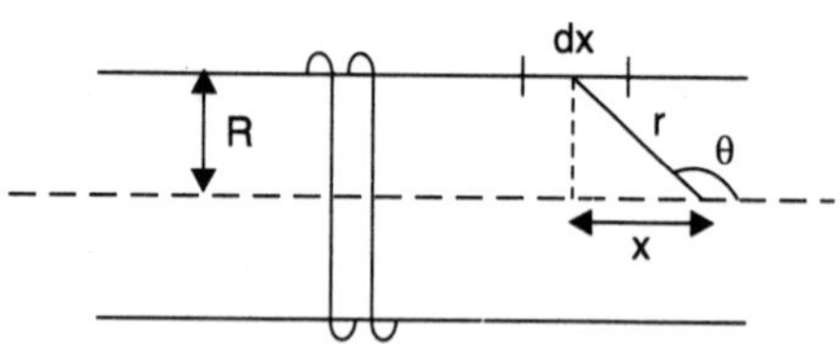

Fig. 5.7

$$\therefore \; H \text{ at the center} \quad = \frac{NI}{2}\left[\int_0^{\pi} \sin\theta \, d\theta\right] = NI$$

$$H \text{ at the extreme right end} = \frac{NI}{2}\int_{\pi/2}^{\pi} \sin\theta \, d\theta = \frac{NI}{2} \quad \text{...(5.26)}$$

$$H \text{ at the extreme left end} \quad = \frac{NI}{2}\int_{0}^{\pi/2} \sin\theta \, d\theta = \frac{NI}{2} \quad \text{...(5.27)}$$

5.5 AMPERE'S CIRCUITAL LAW

From the previous sections the field intensity at a point, distance 'r' from the long straight filament conductor carrying I has been found to be

$$H = \frac{I}{2\pi r} \cdot \bar{u}_\phi$$

Where, I = current in wire

$\bar{u}_\phi$ = unit vector indicating direction of H

r = radial distance (radius)

If integral of H is taken around a closed path with radius 'r' enclosing the conductor once, then it can be written as

$$\oint H \cdot dL = \frac{I}{2\pi r} \; 0183018 \cdot 2\pi\, r = I$$

Therefore Ampere's circuital law can be stated that line integral of H about any closed path is exactly equal to direct current enclosed by that path.

$$\oint H \cdot dL = I \quad \text{...(5.28)}$$

This law is analogous to Gauss law in electrostatics and it is very useful in determining the magnetic field around a current carrying conductor by symmetrical configuration. For determining 'H' by use of Ampere's law, there must be a considerable degree of symmetry in the case. Two conditions must be met in order to do so.

(*i*) H is either tangential or normal to the path at each point on the closed path.

(*ii*) Wherever H is tangential, H has the same value at all points of that path.

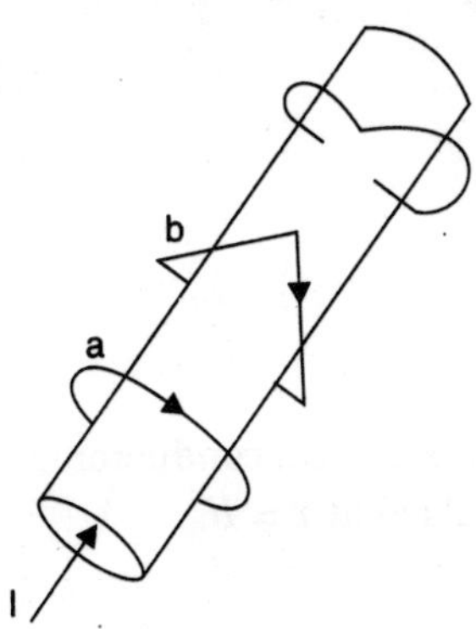

Fig. 5.8(*a*)

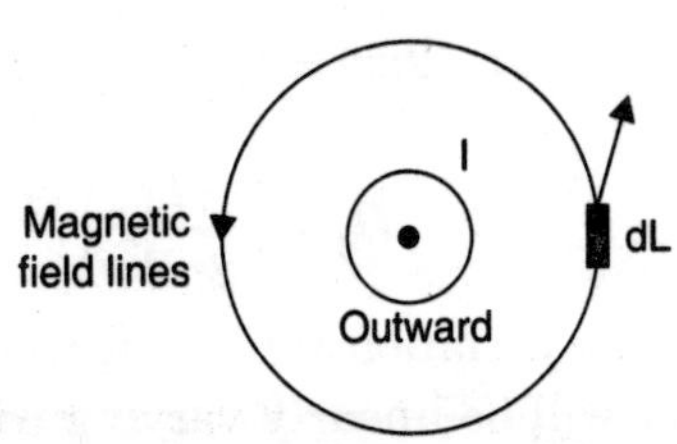

Fig. 5.8(*b*)

Figure 5.8 (*a*) shows a circular wire carrying a direct current *I*, the line integral of *H* about the closed paths *a* and *b* results in an answer of current '*I*'. Referring to Fig. 5.8 (*b*), consider a small length element $d\overline{L}$ on the curve and let $\overline{H}$ be the resultant *MFI* at the position of $d\overline{L}$. Performing the scalar product of $\overline{H} \cdot d\overline{L}$ and on integrating by varying $d\overline{L}$ on the closed curve, this integration is called line integral (or) circulation of $\overline{H}$ along the curve and is represented by $\oint H \cdot dL$

According the Ampere's circuital law

$$\oint \overline{H} \cdot d\overline{L} = I$$

$$\oint H \cdot d\overline{L} = \int_S \overline{J} \cdot d\overline{s} = I$$

Ampere's law constitutes one of Maxwell's equation for static magnetic field in integral form.

Example 5.9: *A solid conductor of radius R has uniform current density. Derive the expressions for H both inside and outside the conductor and also show the variation of H as a function of radial distance from the center of the wire.*

Solution: Ampere's circuital law is given by $\oint \overline{H} \cdot \overline{dl} = I$

As shown in Fig. 5.9 since the current density is uniformly distributed throughout the cylindrical conductor, $\overline{J}$ will be constant (where $\overline{J}$ is the current density).

Then, $$J = \frac{I_R}{\pi R^2} = \frac{I_r}{\pi r^2}$$

Where, I_R is the current corresponding to a radius R_1 and $I_R = I$

$$\therefore \quad I_r = \frac{Ir^2}{R^2}, \; r < R$$

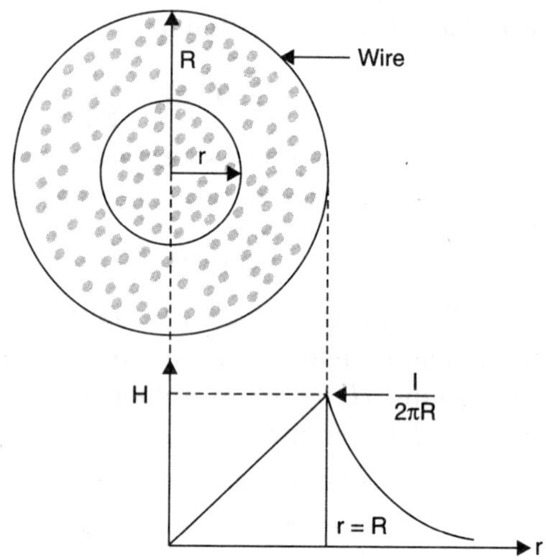

Fig. 5.9

Now, for constant *H* over a circular path of integration (where $r < R$), we have

$$H_1 (2\pi r) = I_r = \frac{r^2}{R^2} I$$

$$\therefore \quad H_1 = \left(\frac{I}{2\pi R^2}\right) r$$

From this the variation of *H* inside the solid cylindrical conductor ($r < R$) with the distance '*r*' from its center will be linearly varying with '*r*'. Also at $r = R$.

$$H = \frac{I}{2\pi R}$$

For $r > R$; $I_r = I$

$\therefore$ $$H(2\pi r) = I, \text{ which implies } H = \frac{I}{2\pi R}.$$

5.6 CURL

Just like divergence, curl is a vector operator and is important in variety of applications.

Consider an incremental closed path whose sides are Δx and Δy. It is perimeter of differential rectangular surface lying in x-y plane. Let field intensity at centre of rectangle P be denoted by H_0.

Let H_{x_0}, H_{y_0} and H_{z_0} be components of H_0 along x, y and z directions respectively.

$\therefore$ $$H_0 = H_{x_0} U_x + H_{y_0} U_y + H_{z_0} U_z \quad ...(5.29)$$

Direction be taken as 1 – 2 – 3 – 4 – 1

So, closed line integral of H about this closed path is given as sum of values of line integral of H along four sides *i.e.*, 1 – 2, 2 – 3, 3 – 4 and 4 – 1.

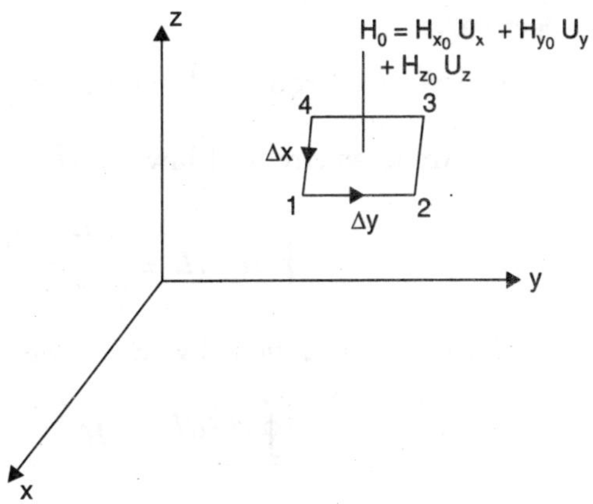

Fig. 5.10

(*a*) Path 1–2

We have $$(H \cdot \Delta L)_{1-2} = H_{y_{1-2}} \Delta y$$

But $$H_{y_{1-2}} = H_{y_0} + \frac{\partial H_y}{\partial x}\left(\frac{\Delta x}{2}\right)$$

Where, $\frac{\Delta x}{2}$ is distance between P and center of path and $\frac{\partial H_y}{\partial x}$ in space rate of change of H_y along x-axis

$$(H \cdot \Delta L)_{1-2} = \left(H_{y_0} + \frac{1}{2}\frac{\partial H_y}{\partial x}\Delta x\right)\Delta y \quad ...(5.30)$$

(*b*) Path 2–3

$$(H \cdot \Delta L)_{2-3} = H_{x_{2-3}} (-\Delta x)$$

$$= -\left[H_{x_0} + \frac{1}{2}\frac{\partial H_x}{\partial y}\Delta y\right]\Delta x \quad ...(5.31)$$

(*c*) Path 3–4

$$(H \cdot \Delta L)_{3-4} = H_{y_{3-4}} (-\Delta y)$$

$$= -\left[H_{y_0} - \frac{\partial H_y}{\partial y}\frac{\Delta x}{2}\right]\Delta y \quad ...(5.32)$$

(*d*) Path 4–1

$$(H \cdot \Delta L)_{4-1} = H_{x_{4-1}}(\Delta x)$$

$$= \left(H_{x_0} - \frac{1}{2}\frac{\partial H_x}{\partial y}\Delta y\right)\Delta x \quad ...(5.33)$$

Adding the equations (5.30), (5.31), (5.32) and (5.33), we get

$$\oint H \cdot dL = \left(\frac{\partial H_y}{\partial x} - \frac{\partial H_x}{\partial y}\right)\Delta x\, \Delta y$$

Let J denote current density and its components be J_x, J_y and J_z alone x, y and z-axes respectively.

Therefore current enclosed by path 1 – 2 – 3 – 4 – 1 is $\Delta I = J_z \Delta x\, \Delta y$

By Amperes circuital law $\oint H \cdot dL = \Delta I$

$$\oint H \cdot dL = \left(\frac{\partial H_y}{\partial x} - \frac{\partial H_x}{\partial y}\right)\Delta x\, \Delta y = J_z \Delta x\, \Delta y$$

Dividing throughout by $\Delta x\, \Delta y$, we get

$$\frac{\oint H \cdot dL}{\Delta x\, \Delta y} = \frac{\partial H_y}{\partial x} - \frac{\partial H_x}{\partial y} = J_z \quad ...(5.34)$$

The above relationship is only appropriate and it becomes exact only if we let the surface $\Delta x\, \Delta y$ shrink.

$$\underset{\Delta x,\, \Delta y \to o}{\text{Lt}} \frac{\oint H \cdot dL}{\Delta x\, \Delta y} = \frac{\partial H_y}{\partial x} - \frac{\partial H_x}{\partial y} = J_z$$

By similar procedure considering closed rectangle path in planes normal to other two axes we get

$$\underset{\Delta y,\, \Delta z \to 0}{\text{Lt}} \frac{\oint H \cdot dL}{\Delta y\, \Delta z} = \frac{\partial H_z}{\partial y} - \frac{\partial H_y}{\partial z} = J_x$$

$$\underset{\Delta x,\, \Delta z \to 0}{\text{Lt}} \frac{\oint H \cdot dL}{\Delta z\, \Delta x} = \frac{\partial H_x}{\partial z} - \frac{\partial H_z}{\partial x} = J_y$$

Since, it can be written as $J = J_x U_x + J_y U_y + J_z U_z$

$$J = \left(\frac{\partial H_z}{\partial y} - \frac{\partial H_y}{\partial z}\right)U_x + \left(\frac{\partial H_x}{\partial z} - \frac{\partial H_z}{\partial x}\right)U_y + \left(\frac{\partial H_y}{\partial x} - \frac{\partial H_x}{\partial y}\right)U_z \quad ...(5.35)$$

Expression on R.H.S. side is called as Curl H.

5.7 MAXWELL'S SECOND EQUATION DERIVED FROM GAUSS'S LAW [Div B = 0]

We know that magnetic flux through a surface is given by

$$\psi = \int_S \overline{B} \cdot \overline{ds} \qquad \text{...(5.36)}$$

Where magnetic flux ψ is in Weber's (Wb) and magnetic flux density is in Wb/m^2 (or) Tesla. In a electrostatic field, the flux density passing through a closed surface is same as charge enclosed, that is

$$\psi = \oint \overline{D} \cdot \overline{ds} = Q \qquad \text{...(5.37)}$$

But we know that in case of magnetic fields the surface integral of $\overline{B}$ over a closed surface 'S' must be zero; that is

$$\oint_C \overline{B} \cdot \overline{ds} = 0 \qquad \text{...(5.38)}$$

This equation is called to as law of conservation of magnetic flux (or) Gauss's law for magnetic fields just as Gauss's law of equation (5.37). Although the magneto static field is not conservative, magnetic flux is conserved.

By applying the divergence theorem to equation (5.38), we obtain

$$\oint_C \overline{B} \cdot \overline{ds} = \int_v \overline{B} \cdot dv = 0 \qquad \text{...(5.39)}$$

$$\therefore \qquad \nabla \cdot \overline{B} = 0$$

(or) $$\text{div}\,(B) = 0 \qquad \text{...(5.40)}$$

This equation is Maxwell's second equation. Equations (5.38) and (5.40) show that magneto static field have no sinks or sources.

Example 5.10: *(a) Evaluate the closed line integral of $\overline{H}$ from P_1 (5, 4, 1) to P_2 (5, 6, 1) to P_3 (0, 6, 1) to P_4 (0, 4, 1) to P_1, using straight line segments if $\overline{H} = [0.1\,y^3 \overline{U}_x + 0.4]\,\overline{U}_z$ A/m.(b) Determine the quotient of the closed line integral and they are enclosed by the path as an approximation to $(\overline{\nabla} \times \overline{H})_z$ (c) Determine $(\overline{\nabla} \times \overline{H})_z$ at the center of the area.*

Solution:

(*a*) $$\int_{P_1}^{P_2} \overline{H} \cdot d\overline{L} = \int_{P_1}^{P_2} (\overline{H} \cdot dy\, \overline{U}_y) = \int_4^6 0 = 0$$

$$\int_{P_2}^{P_3} \overline{H} \cdot d\overline{L} = \int_{P_2}^{P_3} (\overline{H} \cdot dx\, \overline{U}_x) = \int_5^0 (0.1\, y^3\, dx);\; y = 6$$

$$= 0.1\,(6)^3\,[x]_5^0 = -\,108$$

$$\int_{P_3}^{P_4} \overline{H} \cdot d\overline{L} = \int_{P_3}^{P_4} (\overline{H} \cdot dy\, \overline{U}_y) = \int_6^4 0 = 0$$

$$\int_{P_4}^{P_1} \overline{H} \cdot d\overline{L} = \int_{P_4}^{P_1} (\overline{H} \cdot dx\, \overline{U}_x) = \int_0^5 (0.1\, y^3\, dx);\; y = 4 = 0.1\,(4)^3\,[x]_5^0 = 32$$

$\therefore \qquad \oint \overline{H} \cdot d\overline{L} = 0 - 108 + 0 + 32 = -76$ A.

(*b*) Area enclosed by the path $P_1 P_2 P_3 P_4 P_1 = (5-0)(6-4) = 10$ m^2

$$\therefore \qquad (\overline{\nabla} \times \overline{H})_z = \frac{-76}{10} = -7.6 \text{ A/m}^2$$

(*c*)
$$\overline{\nabla} \times \overline{H} = \begin{vmatrix} u_x & u_y & u_z \\ \dfrac{\partial}{\partial x} & \dfrac{\partial}{\partial y} & \dfrac{\partial}{\partial z} \\ 0.1y^3 & 0 & 0.4x \end{vmatrix}$$

$$(\overline{\nabla} \times \overline{H})_z = \overline{u}_z\,(0.1 \times 3y^2) = 0.3y^2\,\overline{u}_z$$

The center is $\left(\dfrac{5+0}{2}, \dfrac{4+6}{2}, 1\right) = (2.5, 5, 1)$

$\therefore \qquad (\overline{\nabla} \times \overline{H})_z$ at the center $= 0.3\,(5)^2\,\overline{u}_z = 7.5\overline{u}_z$

$$= 7.5\overline{u}_z \text{ A/m}^2.$$

Example 5.11: *A surface current density $K = 30U_x$ A/m flows in the $y = 0$ plane throughout the region $-5 < z < 5$m, $-\infty < x < \infty$. Find $\overline{H}$ at P (0, 15, 0) in free space.*

Solution: Surface current density does not lie on infinite plane

So $d\overline{H} = \dfrac{Kds \times U_{12}}{4\pi d^2}$ but $\overline{K}$ lies on $y = 0$ plane and U_y is perpendicular to the plane $y = 0$.
Hence, $ds = dx\,dz\,U_y$ (or) $ds = dx\,dz$.

Point 2 is that point at which $\overline{H}$ is desired = (0, 15, 0).

Point 1 is general point on surface current density = $(x, 0, z)$.

$$\therefore \qquad \overline{R}_{12} = (0-x)U_x + (15-0)U_y + (0-z)U_z$$

$$= -xU_x + 15U_y - zU_z$$

$$d = |\overline{R}_{12}| = \sqrt{(-x)^2 + (15)^2 + (-z)^2} = \sqrt{x^2 + z^2 + 225}$$

$$U_{12} = \frac{\overline{R}_{12}}{|R_{12}|} = \frac{-xU_x + 15U_y - zU_z}{\sqrt{x^2 + z^2 + 225}}$$

$$d\overline{H} = \frac{(30U_x\,dx\,dz) \times (-xU_x + 15U_y - zU_z)}{4\pi(x^2 + z^2 + 225)^{3/2}} = \frac{7.5(10U_z + zU_y)\,dxdz}{\pi(x^2 + z^2 + 225)^{3/2}}$$

As it can be written as $\overline{H} = \int d\overline{H}$

$$\overline{H} = \frac{5}{\pi}\left[\int_{-\infty}^{\infty}\int_{-5}^{5} \frac{75U_z\,dxdz}{(x^2 + z^2 + 225)^{3/2}} + \int_{-\infty}^{\infty}\int_{-5}^{5} \frac{7.5zU_y\,dxdz}{(x^2 + z^2 + 225)^{3/2}}\right.$$

$$\overline{H} = \frac{5}{\pi}\left[\int_{-5}^{5} \frac{75U_z\,dz}{Z^2 + 225} + \int_{-5}^{5} \frac{7.5zU_z\,dz}{Z^2 + 225}\right]$$

$$= \frac{5}{\pi}\,\frac{75U_z}{15}\left[\tan^{-1}\left(\frac{Z}{15}\right)_{-5}^{5} + U_y\,[7.5\ln(Z^2 + 225)_{-5}^{5}] = 293.4\,\overline{U}_Z.\right.$$

Example 5.12: *Find the magnetic flux density at point P due to currents, I_1, I_2 and I_3 as shown in Fig. 5.11?*

Solution: The magnitude of the flux density at point P due to an infinitely long, straight, and thin conductor carrying current I_1 = 50 A is given by

$$B_1 = \frac{\mu_0 I_1}{2\pi R_1} = \frac{4\pi \times 10^{-7} \times 50}{2\pi \times \sqrt{(0.5)^2 + (0.75)^2}}$$

$$= 11.1 \times 10^{-6} \text{ Wb/m}^2$$

and the direction of B_1 will be perpendicular to the line AP and will follow the direction in accordance with the right hand rule.

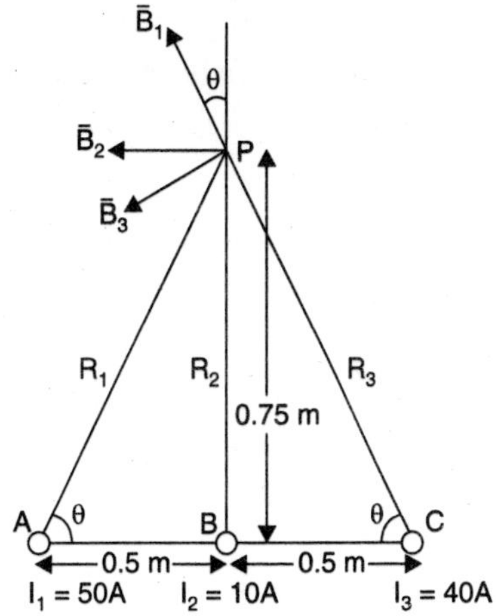

Fig. 5.11

Resolving B_1 along the unit vectors $\bar{u}_x$ and $\bar{u}_y$

$$\overline{B_1} = (B_1 \sin \theta)(-\bar{u}_x) + (B_1 \cos \theta)\bar{u}_y$$

$$= (-9.23\bar{u}_x + 6.15\bar{u}_y) \times 10^{-6} \text{ Wb/m}^2$$

Similarly, $\overline{B_2} = 2.67\bar{u}_x \times 10^{-6}$ Wb/m^2

$$\text{Magnitude of } B_3 = \frac{4\pi \times 10^{-6} \times 40}{2\pi \times 0.901} = 8.88 \times 10^{-6}$$

$$\therefore \qquad \overline{B_3} = (-8.88 \sin \theta\bar{u}_x - 8.88 \cos \theta\bar{u}_y) \times 10^{-6} \text{ Wb/m}^2$$

$$= (-7.38\bar{u}_x - 4.93\bar{u}_y) \times 10^{-6} \text{ Wb/m}^2$$

$\therefore$ The resultant magnetic flux density at P is

$$\overline{B} = \overline{B_1} + \overline{B_2} + \overline{B_3} = (-19.28\bar{u}_x + 1.22\bar{u}_y) \times 10^{-6} \text{ Wb/m}^2$$

$$= 19.4 \times 10^{-6} \angle 179.5° \text{ Wb/m}^2.$$

Example 5.13: *If $\overline{F} = x^2y\overline{U}_x + (3xy + 5z)U_y + (4xy+5z)\overline{U}_z$, find $\overline{\nabla} \times [\overline{\nabla}(\overline{\nabla} \cdot \overline{F})]$*

Solution: $$\overline{\nabla} \cdot \overline{F} = \frac{\partial F_x}{\partial x} + \frac{\partial F_y}{\partial y} + \frac{\partial F_z}{\partial z}$$

$$= \frac{\partial}{\partial x}(x^2y) + \frac{\partial}{\partial y}(3xy + 5z)U_y + \frac{\partial}{\partial z}(4xy + 5z)U_z = 2xy + 3x + 6$$

$$\overline{\nabla}(\nabla \cdot F) = \frac{\partial}{\partial x}(2xy + 3x + 6)U_x + \frac{\partial}{\partial y}(2xy + 3x + 6)U_y + \frac{\partial}{\partial z}(2xy + 3x + 6)U_z$$

$$= (2y + 3)Ux + 2x\,U_y$$

$$\nabla \times [\overline{\nabla}(\overline{\nabla} \cdot \overline{F})] = \begin{vmatrix} U_x & U_y & U_z \\ \dfrac{\partial}{\partial x} & \dfrac{\partial}{\partial y} & \dfrac{\partial}{\partial z} \\ 2y+3 & 2x & 0 \end{vmatrix} = U_z(2-2) = 0$$

Example 5.14: *Determine the current density function J associated with the magnetic field defined by the relations:*

(*a*) $H = 4u_x + 6yu_y + 3xu_z$ *A/m (Cartesian system)*

(*b*) $H = (5r)u_r + (3r)u_\phi + 7u_z$ *(Cylindrical system)*

(*c*) $H = (4r)u_R + 5u_\theta + \cos\theta\, u_\phi$ *(Spherical system).*

Solution:

(*a*) Cartesian system. The current density J may be expressed in point form as Curl of the magnetic field intensity H *i.e.*, $J = \nabla \times H$ expressed in terms of its components, H is given by

$$H = H_x u_x + H_y u_y + H_z u_z$$

According to the data $H_x = 4; H_y = 6y; H_z = 3x$

$$J = \text{Curl } H = \begin{vmatrix} U_x & U_y & U_z \\ \dfrac{\partial}{\partial x} & \dfrac{\partial}{\partial y} & \dfrac{\partial}{\partial z} \\ 4 & 6y & 3x \end{vmatrix}$$

$$= \left[\frac{\partial}{\partial y}(3x) - \frac{\partial}{\partial z}(6y)\right]u_x + \left[\frac{\partial}{\partial z}(4) - \frac{\partial}{\partial x}(3x)\right]u_y + \left[\frac{\partial}{\partial x}(6y) - \frac{\partial}{\partial y}(4)\right]u_z$$

$$= (0)u_x + (-3)u_y + (0)u_z = -3u_y \text{ A/m}^2$$

(*b*) Cylindrical system

$$H = H_r u_r + H_\phi u_\phi + H_z u_z$$

$$= (5r)y_r + (3r)u_\phi + 7u_z$$

In cylindrical coordinate system Curl can be written as

$$J = \nabla \times H = \frac{1}{r}\begin{vmatrix} U_r & rU_\phi & U_z \\ \dfrac{\partial}{\partial x} & \dfrac{\partial}{\partial \phi} & \dfrac{\partial}{\partial z} \\ 5r & (r)(3r) & 7 \end{vmatrix}$$

$$= \frac{1}{r}\left[\frac{\partial}{\partial r}(3r^2) - \frac{\partial}{\partial \phi}(5r)\right]u_z + (0)ru_\phi + (0)u_z$$

$$J = 6u_z \text{ A/m}^2.$$

(*c*) Spherical system

$$H = (4r)u_r + (5)u_\theta + (\cos\theta)u_\theta$$

As $\qquad H_r = 4r : H_\theta = 5 : H_\phi = \cos\theta$

In spherical coordinate system Curl can be written as

$$J = \nabla \times H = \frac{1}{r^2 \sin\theta}\begin{bmatrix} u_r & ru_\theta & r\sin\theta u_\phi \\ \dfrac{\partial}{\partial r} & \dfrac{\partial}{\partial \theta} & \dfrac{\partial}{\partial \phi} \\ H_r & rH_\theta & r\sin\theta H_\phi \end{bmatrix}$$

$$= \frac{1}{r^2 \sin\theta} \begin{bmatrix} u_r & ru_\theta & r\sin\theta u_\phi \\ \frac{\partial}{\partial r} & \frac{\partial}{\partial \theta} & \frac{\partial}{\partial \phi} \\ 4r & 5r & \frac{r\sin 2\theta}{2} \end{bmatrix}$$

$$= \frac{1}{r^2 \sin\theta}\left[\frac{\partial}{\partial\theta}\left(\frac{r}{2}\sin 2\theta\right) - 0\right]u_R + \frac{1}{r^2\sin\theta}\left[0 - \frac{\partial}{\partial r}\left(\frac{r}{2}\sin 2\theta\right)\right]ru_\theta$$

$$+ \frac{1}{r^2\sin\theta}\left[\frac{\partial}{\partial r}(5r - 0)\right]r\sin\theta u_\phi$$

$$= \frac{1}{r^2\sin\theta}\left(\frac{r}{2}2\cos 2\theta\right)u_R + \frac{1}{r^2\sin\theta}(-\sin\theta\cos\theta)ru_\theta + \frac{1}{r^2\sin\theta}(5)r\sin\theta\, u_\phi$$

$$= \frac{1}{r}\left[\left\{\frac{1}{\sin\theta} - 2\sin\theta\right\}u_R - (\cos\theta)u_\theta + 5u_\phi\right]$$

Example 5.15: *Find the magnetic field of current in a straight circular cylindrical conductor of radius 'a'. (Express the magnetic field as a vector interms of the current density J).*

Solution: The current density in the conductor will be uniform (= J). Consider a point P at a distance r_1 from the centre of the conductor. By Ampere's law,

$$\oint_{C_1} H \cdot dI = \text{enclosed current } I = J\,\pi\,a^2$$

or $$2\pi\, r_1\, H = I$$

$\therefore$ $$H = \frac{I}{2\pi r_1} = \frac{I}{2\pi a^2}\frac{a^2}{r_1} = \frac{J}{2}\frac{a^2}{r_1}$$

The direction of H will be circumferential, *i.e.*, normal to that of I or J and r_1.

$\therefore$ $$H = \frac{J \times r_1}{2}\left(\frac{a}{r_1}\right)^2, \text{ for } r_1 \geq a, \textit{ i.e.}, \text{ outside the conductor.}$$

Inside the conductor, at a radius r_2, such $r_2 < a$,

$$\oint_{C_2} H \cdot dl = \text{enclosed current}$$

$$2\pi r_2\, H = \frac{1}{\pi a^2}\pi r_2^2 = J\pi r_2^2$$

$$H = \frac{Jr_2}{2}$$

Hence, $$H = \frac{J \times r_2}{2}, r_2 \leq a.$$

Example 5.16. *Find the magnetic field due to a current I in a co-axial cable whose inner conductor has radius 'a' and the outer conductor has the radii 'b', c (b < c). (Express the magnetic field as a vector in terms of the current density).*

Solution: Once again, the current I is uniformly distributed in both the conductors. Current density in the inner conductor, $J_i = \frac{I}{\pi a^2}$ and the current density in the outer conductor is given by

$$J_0 = \frac{I}{\pi(c^2 - b^2)}$$

So, for the region inside the inner conductor ($r < a$), we have by Ampere's law

$$\oint_{C_1} H \cdot dl = \frac{I}{\pi a^2}\pi r^2 = J_i \pi r^2$$

or $$2\pi r H = J_i \pi r^2$$

$\therefore$ $$H = \frac{J_i}{2} r \quad \text{or} \quad H = \frac{J_i \times r}{2}, \text{ for } r < a$$

For the angular region between the inner and the outer conductors, $a < r < b$, we have by Ampere's law

$$\oint_{C_2} H \cdot dl = \text{Enclosed current} = I = \frac{I}{\pi a^2}\pi a^2 = J_i \pi a^2$$

$\therefore$ $$H = \frac{J_i}{2}\frac{a^2}{r} \quad \text{or} \quad H \frac{J_i \times r}{2}\left(\frac{a}{r}\right)^2, \text{ for } a < r < b$$

In the outer conductor ($b \le r \le c$), we have by Ampere's law

$$\oint_{C_3} H \cdot dl = \text{Enclosed current} = I - \frac{I\pi(r^2 - b^2)}{\pi(c^2 - b^2)} = \frac{I}{\pi(c^2 - b^2)}\pi(c^2 - r^2)$$

$\therefore$ $$H = \frac{J_0}{2}\left(\frac{c^2}{r} - r\right) \quad \text{or} \quad H = \frac{J_0 \times r}{2}\left\{\left(\frac{c}{r}\right)^2 - 1\right\}, \text{ for } b \le r \le c$$

Outside the outer conductor ($r > c$), we have $H = 0$.

Example 5.17: *Find the magnetic field intensity at the origin due to a current element $Idl = 3\pi(2U_x + 3U_y + 4U_z)$ μA-m at the point P(2, 4, 6) in free space.*

Solution: The magnetic field intensity at the origin is

$$dH = \frac{Idl \times U_R}{4\pi R^2}$$

According to the given data, $Idl = 3\pi(2U_x + 3U_y + 4U_z)$ μA-m and the coordinates of point P are given by (2, 4, 6).

So, distance of the origin from $P = \sqrt{2^2 + 4^2 + 6^2} = \sqrt{56} = R$

$$U_R = \frac{(0-2)U_x + (0-4)U_y + (0-6)U_z}{\sqrt{56}} = -0.267U_x - 0.534U_y - 0.801U_z$$

$\therefore$ $$(Idl \times U_R) = \begin{vmatrix} U_x & U_y & U_z \\ 6\pi & 9\pi & 12\pi \\ -0.267 & -0.534 & -0.801 \end{vmatrix}$$

$$= (-2.508U_x + 5.029U_y - 2.514U_z) \text{ μA-m}$$

$$\therefore \quad dH = \frac{(-2.508U_x + 5.029U_y - 2.514U_z)}{4\pi \times 56} \times 10^{-6} \text{ A/m}$$

$$= -3.565U_x + 7.149U_y - 3.574U_z \text{ nA/m.}$$

Example 5.18: *Find the magnetic field intensity at the origin due to a current element 8π Uz μA-m at the point (4, – 5, 1) in free space.*

Solution: Given $Idl = 5\pi\ U_z$ A-m, point $P = (4, -5, 1)$ and so distance of P from the origin is given by $R = \sqrt{4^2 + (-5)^2 + 1^2} = 6.480$

Directed line segment from point P to the origin is $R = -4U_x + 5U_y - U_Z$

Let U_R be a unit vector and so it can be written as

$$U_R = \left(-\frac{4}{\sqrt{42}}\right)U_x + \left(\frac{5}{\sqrt{42}}\right)U_y + \left(\frac{-1}{\sqrt{42}}\right)U_z$$

$$= -0.6172U_x + 0.7715U_y - 0.1543U_z$$

$$\therefore \quad (Idl \times U_R) = \begin{vmatrix} U_x & U_y & U_z \\ 0 & 0 & 8\pi \\ -0.6172 & -0.7715 & -0.1543 \end{vmatrix} = -19.38U_x - 15.504U_y$$

Now, the magnetic field intensity can be found out by

$$dH = \frac{Idl \times U_R}{4\pi R^2} = \frac{(-19.38U_x - 15.504U_y) \times 10^{-6}}{4\pi \times 42}$$

$$= (-0.0367U_x - 0.0293U_y) \times 10^{-6} \text{ A/m}$$

$$= (-0.0367U_x - 0.0293U_y)\ \mu\text{A/m.}$$

Example 5.19: *Find the field intensity at a point on the axis, 3 m from the center of a circular coil of area 150 cm² and carrying a current of 65 A.*

Solution: Given current $I = 65$ A

Distance of the field point $x = 3$ m

Area of the coil = 150 cm² = 150×10^{-4} m² = 0.015 m²

$$\text{Radius of the coil, } R = \sqrt{\frac{\text{Area}}{\pi}} = \sqrt{\frac{0.015}{\pi}} = \sqrt{4.777 \times 10^{-3}} = 0.0691 \text{ m.}$$

The field intensity at a point on the axis of a circular current loop of radius R, distant x from the centre of the loop is given by the expression.

$$H = \frac{IR^2}{2(R^2 + x^2)^{3/2}} \text{ A/m (in magnitude)}$$

Putting $I = 65$ A, $R = 0.0691$ m and $x = 3$ m, we get

$$H = \frac{65 \times 0.0691^2}{2(0.0691^2 + 3^2)^{3/2}} = 5.7432 \times 10^{-3} \text{ A/m.}$$

SHORT QUESTIONS AND ANSWERS

1. **Give the differential relation between $\vec{H}$ and $\vec{J}$.**

   $\nabla \cdot \overline{H} = \overline{J}$, which is the point form of Ampere's law.

2. **State advantages of transposition of conductors.**

   Advantages of transposition of conductors:

   (*a*) Proximity and corona effects can be reduced.

   (*b*) Line parameters like inductance can be balanced.

3. **Write down the expression for energy stored in magnetic field per unit volume.**

   Energy stored in a magnetic field/unit volume = $\frac{1}{2}(\vec{B} \cdot \vec{H})$

4. **From which law the following relation is derived from : curl $\vec{B}$ = μo $\vec{J}$?**

   Ampere's law *i.e.*, $\nabla \times \vec{H} = \vec{J}$.

5. **Comment on Ampere's circuital law.**

   Ampere's circuital law tells that always a current carrying conductor is associated with magnetic field.

6. **Justify that there cannot be any isolated magnetic pole.**

   We know that $\nabla \cdot \vec{B} = 0$; which tells that magnetic field is continuous and hence the existence of mono pole is not possible.

7. **What will be the magnetic field at the center of a loop?**

   The magnetic field at the center of loop is zero.

8. **What is $\vec{H}$ field outside a coaxial cable?**

   $\vec{H} = \frac{J(a^2 - b^2)}{2r} u_F$ where '*a*' and '*b*' are the inner and outer radii of the cable.

9. **Explain, why $\nabla \cdot \vec{H} = 0$?**

   $\nabla \cdot \vec{H} = 0$, because the magnetic field is always continuous, so net out flow of field per unit volume (*i.e.*, divergence) is zero.

10. **What is meant by magnetic flux density?**

    (A) The magnetic flux crossing unit normal surface area is called as magnetic flux density.
    (B) Measured in Weber/m^2 (or) Tesla.

11. **What is meant by magnetic field intensity?**

    Force experienced by a unit North pole placed at that point and is denoted by *H*, measured in Newton/Weber.

12. **State Biot-Savart's law.**

    This law states that at any point *P*, the magnitude of magnetic field intensity (*H*) produced by differential element is proportional to product of current and magnitude of differential length *dL* and sine of angle lying between the filament and a line connecting the filament to point *P* where the field is desired.

MULTIPLE CHOICES

1. Significance of Gauss's law for electrostatics ....
   (*a*) non-existence of monopole
   (*b*) existence of source and sinks
   (*c*) both (*a*) and (*b*)
   (*d*) none of these
2. Significance of Gauss's law for magnetostatics ....
   (*a*) non-existence of mono pole
   (*b*) existence of source and links
   (*c*) both (*a*) and (*b*)
   (*d*) none of these
3. Point form of Ampere's law is ....
   (*a*) Curl $\overline{E} = Q$
   (*b*) Curl $\overline{H} = \overline{J}$
   (*c*) Div $\overline{E} = 1$
   (*d*) none of these
4. Curl $\overline{H}$ = ...
   (*a*) E
   (*b*) D
   (*c*) J
   (*d*) none of these
5. Curl $\overline{H}$ = ... for time varying
   (*a*) J
   (*b*) conduction current density + div-placement current density
   (*c*) both (*a*) and (*b*)
   (*d*) none of these
6. Div $\overline{D}$ = ... for static fields.
   (*a*) charge density
   (*b*) charge
   (*c*) current density
   (*d*) none of these
7. Div $\overline{D}$ = ...
   (*a*) charge density
   (*b*) charge
   (*c*) zero
   (*d*) none of these
8. Div $\overline{B}$ = ... for static fields.
   (*a*) one
   (*b*) zero
   (*c*) both (*a*) and (*b*)
   (*d*) none of these
9. Div $\overline{B}$ = ... for free space.
   (*a*) one
   (*b*) zero
   (*c*) both (*a*) and (*b*)
   (*d*) none of these
10. Div (curl $\overline{H}$) = ...
    (*a*) I
    (*b*) $\overline{J}$
    (*c*) zero
    (*d*) none of these
11. Curl (grad V) = ...
    (*a*) I
    (*b*) $\overline{J}$
    (*c*) zero
    (*d*) none of these

12. Gradient of a scalar is ....

(*a*) scalar (*b*) vector

(*c*) both (*a*) and (*b*) (*d*) none of these

13. Curl of scalar is ....

(*a*) scalar (*b*) vector

(*c*) both (*a*) and (*b*) (*d*) none of these

14. Divergence of a vector is ....

(*a*) scalar (*b*) vector

(*c*) both (*a*) and (*b*) (*d*) none of these

15. The magnetic flux through a surface '*s*' is given by ....

(*a*) $\psi = \int_S \overline{B} \cdot \overline{ds}$ (*b*) $\psi = \int_S \overline{D} \cdot \overline{ds}$

(*c*) $\psi = \int_L \overline{H} \cdot \overline{dl}$ (*d*) none of these

16. $\nabla \cdot \overline{B} = 0$ implies ....

(*a*) isolated magnetic pole (*b*) isolated electric dipole doesn't exit

(*c*) isolated magnetic dipole doesn't exit (*d*) none of these

17. Identify the wrong statement ....

(*a*) $\nabla \cdot \overline{B} = 0$ (*b*) $\oint \overline{B} \cdot \overline{ds} = 0$

(*c*) $\overline{B} = \overline{\nabla} \cdot \overline{A}$ (*d*) none of these

18. Identify the correct statement for fields

(*a*) $\nabla \cdot \overline{D} = Q_{\text{enc}}$ (*b*) $\psi = \int_C \overline{H} \cdot \overline{dl}$

(*c*) $\overline{B} = \varepsilon\, \overline{H}$ (*d*) none of these

19. The magnetic field intensity at a point *P* due to a straight current carrying conductor of infinite length is ....

(*a*) $H = \frac{1}{2\pi r} U_n$ (*b*) $\frac{1}{2\pi r^3} U_n$

(*c*) $\frac{1}{2\pi r^2} U_n$ (*d*) 0

20. The current carrying conductor is kept along *x*-axis. The point where *H* is interested is on the *z*-axis. Then the *H* will be along ....

(*a*) u_x (*b*) $-u_x$

(*c*) $-u_y$ (*d*) u_z

21. Identify the Biot-Savart's law from the following:

(*a*) $d\overline{H} = \frac{I \times u_R}{4\pi R^2}$ (*b*) $d\overline{H} = \frac{I d\bar{l} \cdot u_R}{4\pi R^2}$

(*c*) $d\overline{H} = \frac{I\, \overline{dl} \times u_R}{4\pi R^2}$ (*d*) none of these

22. Which of the following statement is the correct representation of H ....

(a) $H = \int \frac{I\bar{dl} \times u_R}{4\pi R}$ (b) $H = \int \frac{K\bar{d}s \times u_R}{4\pi R^2}$

(c) $H = \int \frac{\bar{J}\, dv \times u_R}{4\pi R^2}$ (d) all of these

23. Identify the correct statement ....

(a) The circulation of $\overline{H}$ equals to the charge enclosed

(a) The circulation of $\overline{H}$ equals to the current enclosed

(c) $\oint \overline{H} \cdot \bar{d}l = \bar{j}$ (d) Both (a) and (b)

24. Identify the Ampere's law

(a) $\nabla \times \overline{H} = 0$ (b) $\nabla \times \overline{H} = \bar{J}$

(c) $\nabla \times \overline{B} = 0$ (d) $\oint \overline{B} \cdot \bar{d}s = 0$

25. For an infinite current sheet $\overline{H}$ = ...

(a) $\frac{1}{2}\overline{K} \times u_n$ (b) $-\overline{K} \times u_n$

(c) $\overline{K} \times u_n$ (d) 0

26. $\oint \overline{B} \cdot \bar{d}s = 0$, implies = ...

(a) law of conservation of magnetic flux
(b) non-existence of magnetic monopole
(c) both (a) and (b)
(d) none of these

27. Give the integral form of Ampere's law = ...

(a) $\oint \overline{B} \cdot \bar{d}s = 0$ (b) $\oint \overline{B} \cdot d\bar{s} = \bar{J}$

(c) $\oint \overline{H} \cdot \bar{d}l = \int \bar{J} \cdot \bar{d}s$ (d) none of these

28. B can be defined as ....

(a) force per unit current element (b) force per unit voltage

(c) force for unit charge (d) both (a) and (b)

29. Unit of magnetic charge ....

(a) A-m^2 (b) Coulomb

(c) Ampere (d) Ampere-meter

30. The magnetic flux coming out of a closed surface is ....

(*a*) zero (*b*) equal to charge

(*c*) not possible to find (*d*) depends on the problem

31. Biot-Savart's law is magneto static's in analogous to .... in electro static fields.

(*a*) Gauss's law (*b*) Coulomb's law

(*c*) Ohm's law (*d*) none of these

32. As per ampere's law the line integral of the magnetic intensity around a closed path is equal to ....

(*a*) the charge enclosed (*b*) the current enclosed

(*c*) the uncharged enclosed (*d*) the flux enclosed

33. The unit of magnetic flux is ....

(*a*) Debye (*b*) Weber

(*c*) Weber/m (*d*) Weber-m

KEY

1. (*b*)	**2.** (*a*)	**3.** (*b*)	**4.** (*c*)	**5.** (*b*)	**6.** (*a*)
7. (*c*)	**8.** (*b*)	**9.** (*b*)	**10.** (*c*)	**11.** (*c*)	**12.** (*b*)
13. (*b*)	**14.** (*a*)	**15.** (*a*)	**16.** (*d*)	**17.** (*d*)	**18.** (*d*)
19. (*b*)	**20.** (*c*)	**21.** (*c*)	**22.** (*b*)	**23.** (*b*)	**24.** (*b*)
25. (*a*)	**26.** (*c*)	**27.** (*c*)	**28.** (*a*)	**29.** (*d*)	**30.** (*a*)
31. (*b*)	**32.** (*b*)	**33.** (*b*)			

REVIEW QUESTIONS

1. A wire is carrying a steady current I. How is the current density $\bar{J}$ affected if (*a*) its length is doubled, (*b*) its area is doubled, (*c*) its length is doubled, but its area is halved?
2. A charge q is moving with a velocity V in free space. Write an expression for the magnetic field produced at any point by this charge.
3. Does a charge at rest establish a magnetic field?
4. State the Biot-Savart's law in your own words.
5. Two charges are moving in the same direction. What is the force experienced by either one of them?
6. A charge is moving in the direction of the $\bar{B}$ field. What is the force experienced by the charge?
7. A charged particle passes through a magnetic field without experiencing any force. What can you conclude about the magnetic field?
8. An electron moving in the x direction enters a region in which the magnetic field is in the y direction. What is the direction of motion of the electron in the region?
9. When can we use Ampere's circuital law to determine the magnetic field?
10. Is the magnetic field intensity, in general, a conservative field?

11. A wire 10 cm in diameter carries a current of 100 A. Find the magnetic field intensity just outside the wire if (*a*) the current is uniformly distributed within the wire, and (*b*) the current flows just over the surface of the wire.

12. Explain Ampere's force law.

13. What must be the condition for $\bar{J}$ to exist at the boundary between free space and a magnetic material?

EXERCISE PROBLEMS

1. The value of $\bar{A}$ within a solid nonmagnetic conducted do radius carrying a total current I in the U_z direction may be found easily. Using the known value of $\bar{H}$ (or)$\bar{B}$ for $\rho < a$, then may be solved for $\bar{A}$. Select $\bar{A}$ $(\mu_0 I \ln J)/2\pi$ at $\rho = a$ and find A at $\rho =$: (*a*) 0 (*b*) 0.3 a (*c*) 0.6 a (*d*) a.
2. Two current filaments are parallel to the z-axis. One passes through (0, 0.5, 0) and has 100 A in the U_Z direction, the other passes through (0, – 0.5, 0) and has 100 A is the $= U_z$ direction.
3. A current filament I is formed into a circle $\rho = a$ in the $Z = Z'$ plane. Find (*a*) H_Z at P (0, 0, z) is I flows in the U_ϕ direction. (*b*) Find H_z at P caused by a uniform surface current density, $\bar{k} = K_0\, U_\phi$ flowing on the cylindrical surface $\rho = a$, $0 < z < h$.
4. A filamentary current I flows is the U_ϕ direction is the circular path $\rho = a$, $z = 0$ in free space. (*a*) Find $\bar{H}$ at $\rho\,(0, 0, z)$. (*b*) Now let identified filamentary currents of 5 A flow is the U_ϕ direction at $\rho = 2$ m, $z = 1$ M.
5. A surface current $\bar{k} = 10\, U_z$ A/m flows on that portion of the plane $y = 4$ in free space bounded by $x = \pm\infty$, $Z = \pm$ K. Find $\bar{H}$ at the origin if $k =$: (*a*) 3 (*b*) 30 (*c*) ∞.
6. Let 20 A flow is a filament is the U_z direction on the z-axis, let 3 U_z A/m flow on the infinite cylinder $\rho = 2$ M and let – 2.5 U_z A/m be at $\rho = 3$. Sketch H_ϕ vs ρ, $0 < \rho < 5$.
7. The planes $Z = 0$ and $Z = 0.1$ m an perfect conductors, A voltage of 2 V is maintained between then such that $\bar{E}$ is in the U_z direction. For $0 < z < 0.1$ m there is conducting material for which $\sigma = 1000\, e^{-100\rho}$ mho/m. (*a*) Find $\bar{E}$ and $\bar{J}$ (*b*) find I_{total} is a cylindrical of radius ρ.
8. The cylindrical surface $\rho = 20$ mm causes the current $\bar{k} = 100\, U_Z$ A/m while the surface $\rho = 40$ mm has the solenoid current $k = 80\, U_\phi$ A/m calculate $|\bar{H}|$ at ρ : (*a*) 10 mm (*b*) 30 mm (*c*) 50 mm.
9. Let $\bar{H} = \left(\dfrac{2}{\pi\rho}\right)\left[1+\left(10^7\,\dfrac{\rho^3}{6}\right)\right]U_\phi + 8U_z\ A/m$ for $0 \le \rho \le 0.01$ m, $\bar{H} = \left(\dfrac{16}{3}\pi\rho\right)U_\phi + 8\,U_Z$ A/m for $\rho > 0.01$ m. (*a*) Find $\bar{J}$ everywhere (*b*) what is $\bar{J}$ at $\rho = 0$ (*c*) Is there a filamentary current at $\rho = 0$, If so what is its value (*d*) what is $|J|_{max}$ for $\rho > 0$?

10. Given points $A(3, 5, 5)$, $B(-2, -2, 5)$ and $C(5, 5, -3)$. Let a differential current element with $I = 6$ A and $|\bar{dL}| = 10^{-4}$ m be located at A. The direction of dL is from A to B. Find $\bar{dH}$ at C.

11. Find H at $P(4, 5, 6)$ in Cartesian coordinates, if there is an infinity long current filament passing through the origin and point C. The current of 40 A is directed from origin to C, where the location of C is (*a*) $C(0, 0, 5)$ (*b*) $C(1, 1, 0)$?

12. Find the magnetic field at $A(4, 5, -3)$ due to a current element $Idl = \pi(0.7u_x - 0.4u_y + 0.9u_z)$ μA-m situated at $B(5, -3, 8)$.

CHAPTER

6

Magnetic Forces

6.1 INTRODUCTION

Even though accurate quantitative results can be predicted only through the use of quantum theory, the simple atomic model which assumes that there is a central positive nucleus surrounded by electron in various circular orbits yields reasonable quantitative results and provides a satisfactory qualitative theory. An electron in an orbit can be referred to a small current loop and as a result it experiences a torque in an external magnetic field and this torque tends to align the magnetic field which is produced by orbiting electron with external magnetic field. If there were no other magnetic moments to be considered, we can thereby conclude that all the orbiting electrons in the material would shift in such a way to add their magnetic fields to applied field and thus the resultant magnetic field at any point in the material would be greater than it would be at that point if the material were not present.

Consider those atoms in which small magnetic fields produced by motion of electrons in their orbits and those produced by electron spin combine to produce a net field of zero. Here the fields are produced by the electron magnetic field; such a material is termed as diamagnetic. Let an orbiting electron whose moment '*M*' is in same direction as applied field B_0 be selected. The magnetic field produces an outward force on the orbiting electron. Since the orbital radius is quantized and cannot change, the inward coulomb force of attraction is also unchanged. Therefore the force unbalance created by the outward magnetic force must be compensated by a reduced orbital velocity. Hence the orbital moment decreases and a smaller internal field results.

Metallic bismuth shows a greater diamagnetic effect than most other diamagnetic materials among which are hydrogen, helium and other elements. Here diamagnetic effect is present in all materials, because it arises from an interaction of the external magnetic field with every orbiting electron. But if we consider an atom in which the effects of electron spin and orbital motion do not quite cancel. The atom as a whole has a small magnetic moment, but the random orientation of atoms in a larger sample produces an average magnetic moment of zero. The material shows no magnetic effects in the absence of external field. When an external field is applied, however there is a small torque on each atomic moment and these moments tend to become aligned with external field. This alignment acts to increase the value of B within the material over the external value. However the diamagnetic effect is still operating on the orbiting electron and may counteract the above increase. If net result is a decrease in B, the material is still called diamagnetic. If there is an increase is B, the material is termed as paramagnetic.

In ferromagnetic material each atom has a relatively larger dipole moment, caused primarily by uncompensated electron spin moments. Inter atomic forces causes their moments to line up in a parallel fashion over regions containing a layer number of atoms. These regions, are called domains and they may have a variety of shaper and sizes ranging from one micrometer to several centimeters, depending on the size, shape material and magnetic history of sample. Upon application of an external magnetic field, however those domains which have moments in direction of applied field increase their size at the expense of their neighbours and the internal magnetic field increases greatly over that of the external field alone. When the external field is removed, a completely random domain alignment is not usually attained and a residual (or) remnant, dipole field remains in the macroscopic structure. The fact that the magnetic moment of material is different after the field has been removed (or) that the magnetic state of material is a function of its magnetic history is called hysteresis.

The only elements that are ferromagnetic at room temperature are iron, nickel and cobalt and they lose all their Ferromagnetic characteristics above a temperature called the curie temperature. At lower temperatures some of rare earth elements such as gadolinium and dysprosium are ferromagnetic. In anti ferromagnetic materials, the force between adjacent atoms cause the atomic moments to line up in an anti parallel fashion. The net magnetic moment is zero and anti ferromagnetic material is affected only slightly by the presence of an external magnetic field. Super paramagnetic materials are composed of an assemblage of ferromagnetic particles in a non-ferromagnetic matrix. Although domain exists within the individual particle, the domain walls cannot penetrate the intervening matrix material to the adjacent particle.

6.2 STOKE'S THEOREM

From Ampere's circuital law one of Maxwell's equations *i.e.*, $\nabla \times H = J$ was derived. This latter equation should be considered as point form of Amperes circuital law and applied on a 'per unit area' basis. In this concept the Ampere circuital law will be derived from $\nabla \times H = J$. In other words the integral form will be obtained from point form (or) the point form will be obtained from integral form.

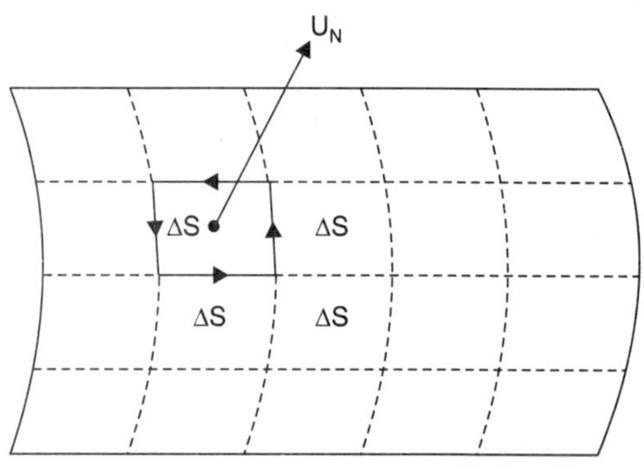

Fig. 6.1

Consider the surface S of Fig. 6.1 which is broken up into incremental surfaces of areas Δs. If the definition of curl is applied to one of the incremental surfaces then

$$\frac{\oint H \cdot dL_{\Delta s}}{\Delta s} = (\nabla \times H)_N \qquad ...(6.1)$$

Where, N subscript indicates the right hand normal to the surface. The subscript on $dL_{\Delta s}$ indicates that closed path is perimeter of an incremental area Δs. It can be written as

$$\frac{\oint H \cdot dL_{\Delta s}}{\Delta s} = (\nabla \times H) \cdot U_N \qquad ...(6.2)$$

(or)
$$\oint H \cdot dL_{\Delta s} = (\nabla \times H) \cdot U_N \, \Delta s$$
$$= (\nabla \times H) \cdot \Delta s \qquad ...(6.3)$$

Now to determine this circulation for every Δs comprising S and sum the results. As the closed line integral is evaluated for each Δs, some cancellation will occur because every interior wall is covered once is each direction. Therefore it can be written as

$$\oint H \cdot dL = \int_S (\nabla \times H) \cdot ds \qquad ...(6.4)$$

6.3 MAXWELL'S THIRD EQUATION (∇ × H = J)

From Ampere's circuital law, we can write as

$$\oint \overline{H} \cdot \overline{dL} = \int_S \overline{J} \cdot \overline{ds} \qquad ...(6.5)$$

By Stokes theorem it can be written as

$$\oint \overline{H} \cdot \overline{dL} = \int \text{Curl } H \cdot ds \qquad ...(6.6)$$

Equating R.H.S. of equations (6.5) and (6.6), we get

$$\int_S \text{Curl } H \cdot ds = \int_S \overline{J} \cdot \overline{ds} \qquad ...(6.7)$$

Canceling area on both sides it can be written as

$$\text{Curl } H = J$$

(or)
$$\nabla \times H = J \qquad ...(6.8)$$

This equation is called field form (or) differential form (or) vector form (or) point form of basic Ampere's law.

6.4 MAGNETIC FIELD INTENSITY (H) DUE TO A LONG CURRENT CARRYING FILAMENT

Consider a long, straight conductor carrying current I amperes. It is required to calculate magnetic field intensity (*MFI*) at a point P which is at a distance'd' from the wire. Also the current will be perpendicular to the plane of picture and will be coming out of it.

Let a circle be drawn passing through point 'P' and with center at 'O'. The radius of circle is $OP = d$. The *MFI* due to long, straight current at any point on the circle is along the tangent as shown in Fig. 6.2. By symmetry, all points of circle are equivalent and hence the magnitude of H should be the same at all these points the circulation of H along the circle is

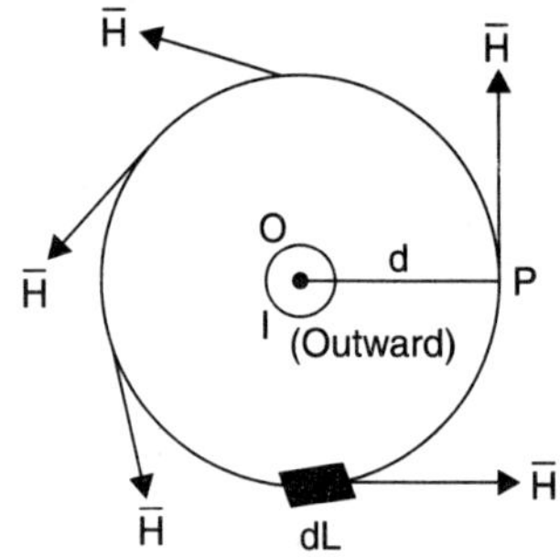

Fig. 6.2

$$\oint \overline{H} \cdot \overline{dL}$$

(∵ Angle between $\overline{H}$ and $\overline{dL}$ = 90°)

(or) $H \oint dL$

(or) $H \cdot 2\pi d$

The current crossing the area bounded by circle is I. Thus, from the Ampere's law

$$H \cdot 2\pi d = I$$

(or) $$H = \frac{I}{2\pi d} \quad ...(6.9)$$

6.5 MAGNETIC FORCE

If a charged particle is in motion in a magnetic field of flux density $\overline{B}$ (or) if a charged particle is projected in a magnetic field, in general, it experiences a magnetic force. When charged particle having charge Q travels with velocity 'v' in a magnetic field $\overline{B}$ it experiences a force $\overline{F}_m$ which is given by

$$\overline{F}_m = Q\,(\overline{v} \times \overline{B}) = Q\,vB \sin\theta\; U_n \quad ...(6.10)$$

By measuring the magnetic force $\overline{F}$ acting on the charge 'Q' which is moving at a velocity $\overline{v}$, $\overline{B}$ can be obtained. If $\overline{v}$ is parallel to $\overline{B}$, the force is zero.

By taking the magnitude in equation (6.10), we see that force is proportional to $|v \sin\theta|$

By rules of vector product, the force B perpendicular to both $\overline{B}$ and $\overline{v}$.

Example 6.1: *A current element $I_1 \Delta L_1 = 10^{-5}\, U_z$ A-m is located at P_1 (2, 0, 0), while a second element $I_2 \Delta L_2 = 10^{-5}\,(0.5U_x - 3U_y + 4U_z)$ A-m is at P_2 (– 2, 0, 0), both in free space (a) Find the vector force exerted on $I_2 \Delta L_2$ by $I_1 \Delta L_1$.*

Solution: Magnetic field intensity at point 2 due to $I_1 \Delta L_1$ is equal to

$$d\overline{H} = \frac{I_1 \Delta \overline{L}_1 \times \overline{U}_{p_1p_2}}{4\pi[d(p_1p_2)]^2}$$

$$\overline{R}_{p_1p_2} = (-2-2)U_x = -4U_x$$

$$d\,(p_1\,p_2) = |\overline{R}_{p_1p_2}| = 4$$

$$\overline{U}_{p_1p_2} = \frac{\overline{R}_{p_1p_2}}{|R_{p_1p_2}|} = \frac{-4U_x}{4} = -U_x$$

$$\therefore \qquad d\overline{H} = \frac{(10^{-5}\, U_z)\times(-U_x)}{4\pi(4)^2} = \frac{-10^{-5}}{64\pi} U_y \text{ A/m}$$

$$d\overline{B} = \mu_0 d\overline{H} = \frac{-10^{-5}\times 4\pi\times 10^{-7}}{64\pi} U_y$$

$$= -0.0625 \times 10^{-12}\, U_y \quad \text{Wb/m}^2$$

Force on $I_2\, \Delta L_2$ due to $I_1\, \Delta L_.$ is given by

$$\overline{F} = I_2\, \Delta \overline{L}_2 \times d\overline{B}$$

$$= 10^{-5}\,(0.5U_x - 3U_y + 4U_z) \times (-0.0625 \times 10^{-12} U_y)$$

$$= 10^{-17}\,(-0.03125U_z + 0.25U_x) = (0.3125U_x - 2.5U_z)\, 10^{-18} \text{ N.}$$

Example 6.2: *An electron ($Q_e = -2.40218 \times 10^{-19}$ C, $m = 8.20339 \times 10^{-31}$ kg) is moving at constant velocity $\overline{v} = 5 \times 10^7\, U_x$ m/s along the negative x-axis. At the origin, it encounters the uniform magnetic field $\overline{B} = -3 \times 10^{-3}\, U_y$ T and remains in it up to x = 3 cm. It we assume that electron remains on x-axis while it is in magnetic field, find its y and z coordinate values where x = 0.4 cm.*

Solution: We know that

$$\overline{F} = Q\, \overline{v} \times \overline{B}$$

$$= -2.40218 \times 10^{-19}\, [(5 \times 10^7 U_x) \times (-3 \times 10^{-3} U_y)$$

$$= 3.6032 \times 10^{-14}\, U_z \text{ N.}$$

Now

$$\overline{F} = m\overline{a}$$

$$\therefore \qquad \overline{a} = \frac{\overline{F}}{m} = \frac{3.6032\times 10^{-14}\, U_z}{8.20339\times 10^{-3}} = 4.392 \times 10^{16}\, U_z \text{ m/s.}$$

Example 6.3: *If a point charge of 3 coulombs moves with a velocity of ($7u_x + 4u_y - 6u_z$)* m/s, *find the force exerted (a) if the electric field intensity is ($12u_x + 7u_y - 6u_z$)*V/m *(b) if the flux density is ($6u_x + 5u_y - 7u_z$)* Wb/m².

Solution: (*a*) Given Q = 3 coulombs, velocity, $v = (7u_x + 4u_y - 6u_z)$ m/s and $E = (12u_x + 7u_y - 6u_z)$V/m.

So, the force exerted can be written as $F = QE$

$$\therefore \qquad F = 3(12u_x + 7u_y - 6u_z) \text{ N}$$

$$= (36u_x + 21u_y - 18u_z) \text{ N}$$

$$\text{Magnitude of the force} = \sqrt{36^2 + 21^2 + (-18)^2} = 45.39 \text{ N}$$

Here the direction of F is same as that of E.

(*b*) Given $B = 6u_x + 5u_y + 7u_z$ Wb/m², Q = 3 coulombs, velocity $v = (7u_x + 4u_y - 6u_z)$ m/s

So, the expression for force can be written as $F = Q(v \times B)$

$$v \times B = \begin{vmatrix} u_x & u_y & u_z \\ 7 & 4 & -6 \\ 6 & 5 & 7 \end{vmatrix}$$

$$= 58u_x - 85u_y + 11u_z$$

So,
$$F = Q(v \times B)$$
$$= 3\,(58u_x - 85u_y + 11u_z)$$
$$= 174u_x - 255u_y + 33u_z \text{ N}$$

Magnitude of
$$F = \sqrt{174^2 + (-255)^2 + 33^2} = 286.33 \text{ N}$$

6.6 MOTION OF CHARGED PARTICLE IN A UNIFORM MAGNETIC FIELD

When a charged particle having charge 'Q' travels with velocity $\bar{v}$ in magnetic field $\bar{B}$, it experiences a force $\bar{F}_m$ given by $\bar{F}_m = Q\,(\bar{v} \times \bar{B}) = Q\,v\,B \sin\theta$.

Now let us consider the following cases:

(*i*) A particle will experience no force if it is at rest in magnetic field (*i.e.*, $v = 0$).

(*ii*) The particle will not experience any force if charged particle is moving along the line of magnetic field since $\bar{v}$ and $\bar{B}$ are parallel (*i.e.*, $\theta = 0$).

(*iii*) The particle will experience a maximum force denoted by $\bar{F}_m$ if it is moving perpendicular to magnetic field.

Now to describe the motion of charged particle in a magnetic field. Consider the case of charged particle where a charge enters in a magnetic field at the right angle to the direction of field as shown in Fig. 6.3. In the Fig. 6.3 the magnetic field $\bar{B}$ is perpendicular to the paper and going into it. Let a charge be projected with a velocity $\bar{v}$ in the plane of paper. The velocity is perpendicular to the magnetic field.

The force is $F_m = Q\,v\,B$ in the direction perpendicular to both $\bar{v}$ and $\bar{B}$. This force will deflect the charged particle without changing the speed and the particle will move along the circle perpendicular to the field. The magnetic force in turn provides the centripetal force.

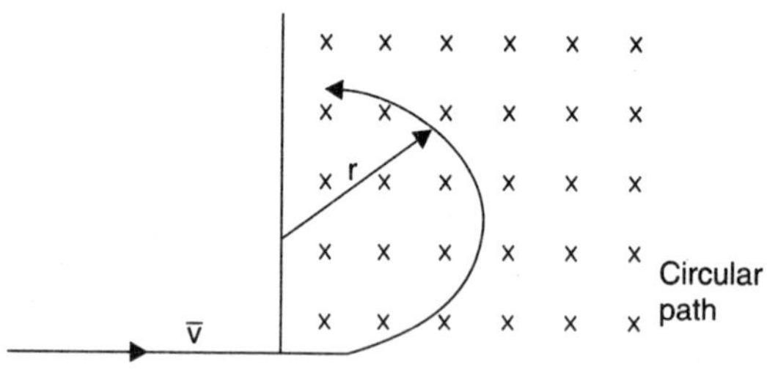

Fig. 6.3

If 'r' be the radius of circle and 'a' be the acceleration produced by force $\bar{F}_m$, then

$$F_m = ma = m\,\frac{v^2}{r}$$

$$\Rightarrow \quad Q\,v\,B = \frac{mv^2}{r}$$

(or)
$$r = \frac{mv}{QB} \qquad \text{...(6.11)}$$

The time taken to complete the cycle (time period) is

$$T = \frac{2\pi r}{v} = \frac{2\pi m}{QB} \qquad \text{...(6.12)}$$

The frequency of revolution is

$$f = \frac{1}{T} = \frac{QB}{2\pi m} \quad ...(6.13)$$

This frequency is called as cyclotron frequency and it is independent of $\bar{v}$.

6.7 FORCE ON MOVING CHARGES IN MAGNETIC FIELD (LORENTZ FORCE EQUATION)

In an electric field the force on a charged particle is given by

$$\bar{F} = Q\bar{E} \quad ...(6.14)$$

Where Q is the charge in coulombs and E is the electric field intensity. The force is in the same direction as the electric field intensity for a positive charge and is directly proportional to both E and Q. If the charge is in motion, the force at any point in its trajectory is then given by equation (6.14).

A charged particle in motion in a magnetic field of flux density $\bar{B}$ experiences a force whose magnitude is proportional to the product of the magnitudes of the charge Q, its velocity $\bar{v}$, the flux density $\bar{B}$ and to the sine of the angle between the vectors $\bar{v}$ and $\bar{B}$. The direction of the force is proportional to both $\bar{v}$ and $\bar{B}$ and is given by a unit vector in the direction of $\bar{v} \times \bar{B}$. The force may therefore be expressed as

$$\bar{F} = Q\bar{v} \times \bar{B} \quad ...(6.15)$$

From this it is clear that a force in direction at right angles to the direction in which the particle is proceeding can never change the magnitude of the velocity of the particle. In other words, the acceleration vector is always normal to the velocity vector. The kinetic energy of the particle remains unchanged and it follows that the steady magnetic field is incapable of transferring energy to the moving charge. The electric field on the other hand exerts a force on the particle which is independent of the direction in which the particle is progressing and therefore effects an energy transfer between field and particle in general.

The force on a moving particle due to combined electric and magnetic fields is obtained easily by superposition principle

$$\bar{F} = Q(\bar{E} + \bar{v} \times \bar{B}) \quad ...(6.16)$$

This equation is called as **Lorentz force equation.**

Example 6.4: *A negative point charge $Q = -40$ nC is moving with a velocity of 6×10^6 m/s in a direction specified by unit vector $U_v = -0.48U_x - 0.6U_y + 0.64U_z$. Find the magnitude of vector force exerted on the moving particle by field (a) $\bar{B} = (2U_x - 3U_y + 5U_z)$ mT (b) $\bar{E} = (3U_x - 4U_y + 6U_z)$ k V/m (c) $\bar{B}$ and $\bar{E}$ acting together.*

Solution:

(*a*)

$$\bar{v} = 6 \times 10^6 (-0.48U_x - 0.6U_y + 0.64U_z) \text{ m/sec}$$
$$= (-2.88U_x - 3.6U_y + 3.84U_z) \times 10^6$$

$$F = Q\bar{v} \times \bar{B} = -40 \times 10^{-6} \begin{vmatrix} U_x & U_y & U_z \\ -2.88 \times 10^6 & -3.6 \times 10^6 & 3.84 \times 10^6 \\ 0.002 & -0.003 & 0.005 \end{vmatrix}$$

$$= (259.2U_x + 883.2U_y - 633.6U_z) \times 10^{-6} \text{ N}$$

$$|\bar{F}| = \sqrt{(259.2)^2 + (883.2)^2 + (-633.6)^2} = 1117.442 \ \mu\text{N}$$

(*b*)

$$F = Q\bar{E} = -40 \times 10^{-9} (3U_x - 4U_y + 6U_z) \times 10^3$$

$$= (-120U_x + 160U_y - 240U_z) \ \mu\text{N}$$

$$|\bar{F}| = \sqrt{(-120)^2 + (160)^2 + (-240)^2} = 312.40 \ \mu\text{N}$$

(*c*)

$$\bar{F} = Q\bar{E} + Q\bar{v} \times \bar{B}$$

$$= [(-120U_x + 160U_y - 240U_z) + (259.2U_x + 883.2U_y - 633.6U_z)] \times 10^{-6}$$

$$= (139.2U_z + 1043.2U_y - 873.6U_z) \ \mu\text{N}$$

$$|\bar{F}| = \sqrt{(139.2)^2 + (1043.2)^2 + (-873.6)^2} = 1367.779 \ \mu\text{N}.$$

Example 6.5: *A charge $Q = 6 \times 10^{-18}$ C is moving through a uniform magnetic field $B = (-0.5U_x + 0.3U_y - 0.2U_z)$ T with a velocity $\bar{v} = (3U_x - 4U_y + 7U_z)\ 10^5$ m/s at t = 0 (a) what electric field is present at t = 0 if net force on the electron is zero (b) if electric field intensity is entirely in the U_x direction and $|F_{net}| = 2$ p N at t = 0, Find E_x.*

Solution:

(*a*)

$$\bar{F} = Q(\bar{E} + \bar{v} \times \bar{B}) = 0$$

Hence,

$$\bar{E} + \bar{v} \times \bar{B} = 0$$

or

$$\bar{E} = -\bar{v} \times \bar{B}$$

$\therefore$

$$\bar{E} = -10^5 \times \begin{vmatrix} U_x & U_y & U_z \\ 3 & -4 & 7 \\ -0.5 & 0.3 & -0.2 \end{vmatrix}$$

$$= (130\ U_x + 290\ U_y + 110\ U_z) \text{ kV/m}$$

(*b*)

$$\bar{E} = E_x U_x$$

$$\bar{F} = Q(\bar{E} + \bar{v} \times \bar{B})$$

$$= 6 \times 10^{-18} [E_x U_x - 10^3 (130U_x + 290U_y + 110U_z)]$$

$$= 6 \times 10^{-18} [(E_x - 130000)U_x - 290000U_y - 110000U_z]$$

$$|\bar{F}| = 2 \times 10^{-12}$$

$$\therefore \quad (E_x - 130000)^2 + (290000)^2 + (110000)^2 = (333333.33)^2$$

$$E_x - 130000 = \pm 122111.05$$

$$E_x = -7888.95 \text{ V/m (or) } 252.111 \text{ kV/m}.$$

6.8 FORCE ON A DIFFERENTIAL CURRENT ELEMENT

The force on a charged particle moving through a steady magnetic field can be written as differential force exerted on a differential element of charge.

$$dF = dQ\, v \times B \qquad \text{...(6.17)}$$

Physically the differential element of charge consists of large number of very small discrete charges occupying a volume which although small, is much larger than the average separation between the charges. The differential force is thus merely the sum of force on the individual charges.

Now, it can be written as

J = conduction current density.

ρ_v = volume charge density.

We have $dQ = \rho_v\, dv$ ...(6.18)

and also $J = \rho_v\, v$ ...(6.19)

(in the case of convention current and here the expression of v means velocity)

Substituting $dQ = \rho_v\, dv$ in equation (6.17), we get

$$dF = \rho_v\, dv\, (v \times B) \qquad \text{...(6.20)}$$

Again substituting equation (6.19) in equation (6.20)

$$dF = J \times B\ dv \qquad \text{...(6.21)}$$

$I\, dL$ is the line current and similarly it can be defined as K = surface current density and J = volume current density. So, it can be given as $I\, dL \simeq K\, ds \simeq J\, dv$

So, equation (6.21) gets modified as

$$dF = I\, dL \times B, \text{ for differential current element} \qquad \text{...(6.22)}$$

Also, $dF = K \times B\ ds$ for differential current surface ...(6.23)

Integrating equation (6.21) over the volume

$$F = \int_{vol} J \times B\, dv \qquad \text{...(6.24)}$$

Integrating equation (6.22) over a closed path

$$F = \oint I\, dL \times B \text{ and it can be written as}$$

$$F = -I \oint B \times dL \qquad \text{...(6.25)}$$

Integrating equation (6.23) over the surface

$$F = \int_S K \times B\, ds \qquad \text{...(6.26)}$$

Looking into I we can say that if a conductor of length L carrying a current I is situated in uniform magnetic field of flux density B, it would experience a force given by

$$F = B\, I\, L \sin \theta \qquad \text{...(6.27)}$$

Where θ is the angle between the vectors representing the direction of current flow and direction of magnetic flux density.

6.9 FORCE BETWEEN TWO CURRENT CARRYING WIRES

Consider two straight, long, parallel conductors situated in space and spaced D meters apart. Let us determine the force between them per meter length.

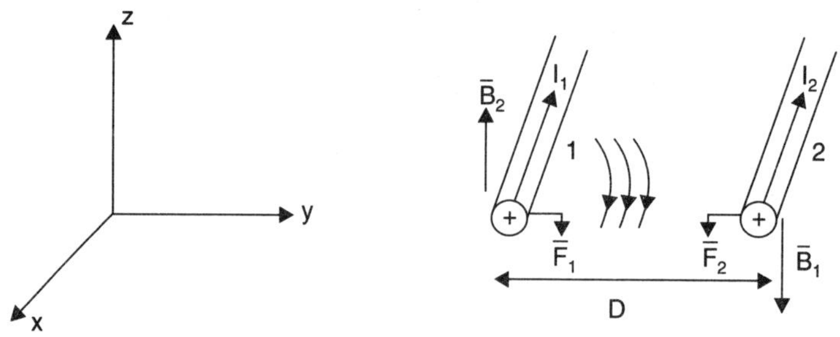

Fig. 6.4

Consider the conductors (1) and (2) as shown in Fig. 6.4. Conductor (1) produces a field around and its value at the location of conductor (2) is given by,

$$B_1 = \mu_0 \frac{I_1}{2\pi D}$$

Conductor (2) carries a current I_2 and is situated in a field with flux density B_1. From the force expression,

$$d\overline{F} = Id\overline{l} \times \overline{B}$$

For a length of 1 metre, taking I as a vector, the force on conductor (2) can be written as

$$\overline{F}_2 = \overline{I}_2 \times \overline{B}_1$$

As the conductor (2) and B_1 are mutually perpendicular,

$$\overline{F}_2 = (-\overline{u_x} I_2 \times -\overline{u_z} B_1)$$

$$= -\overline{u_y} I_2 B_1$$

Force is along the x-axis in the negative direction. This means there is a force of attraction between the conductors (1) and (2) carrying currents in the same direction.

The force of attraction is given by, $F_2 = I_2 B_1$

$$= I_2 \left(\frac{\mu_0 I_1}{2\pi D}\right) = \frac{\mu_0 I_1 I_2}{2\pi D}$$

In a similar way the field produced at (1) due to the conductor (2) is, $B_2 = \dfrac{\mu_0 I_2}{2\pi D}$

and force on (1) is $\qquad F_1 = \dfrac{\mu_0 I_1 I_2}{2\pi D} = F_2$

$\therefore \qquad \overline{F}_1 = -\overline{F}_2$

6.10 FORCE AND TORQUE ON A CLOSED CIRCUIT

The general expression for force exerted on a filamentary closed circuit (on current circuit) is

$$\overline{F} = -I\oint \overline{B} \times d\overline{l} \qquad ...(6.28)$$

If $\overline{B}$ is assumed to be uniform magnetic flux density, then $\overline{B}$ may be removed from the integral, then

$$\overline{F} = -I\,\overline{B}\oint d\overline{l}$$

However, the closed line integral in any electrostatic potential field is zero, *i.e.*, $\oint d\overline{l} = 0$, and therefore the force on a closed filamentary circuit in uniform magnetic field is zero. If the field is not uniform, the total force need not be zero. This result for uniform fields is not restricted to filamentary circuits only is also applicable to circuits containing surface currents or volume current density. If the total current is divided into filaments, the force on each is zero, as seen above, and the total force is again zero. Therefore any real closed circuit carrying direct current experiences a total vector force of zero magnitude in a uniform magnetic field. Although the force is zero, the torque is generally not equal to zero.

The torque of a vector force is given by

$$\overline{T} = \overline{R} \times \overline{F} \qquad ...(6.29)$$

Where, $\overline{F}$ is the force and $\overline{R}$ is the lever arm.

Consider a differential current loop placed in a magnetic field $\overline{B}$ as shown in the Fig. 6.5.

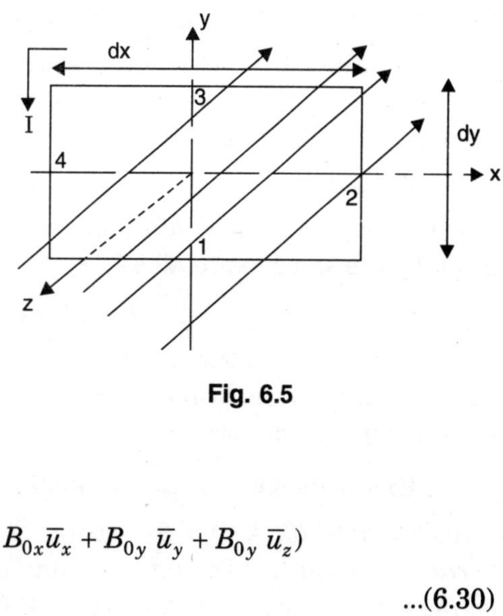

Fig. 6.5

The loop lies in the *x*-*y* plane, the sides of the loop are parallel to the *x* and *y*-axes are of length *dx* and *dy*. The value of the magnetic field at the centre of the loop is taken as B_0. Since the loop is of differential size value of $\overline{B}$, at all points on the loop may be taken as $\overline{B}_0$. The total force on the loop is therefore zero, and we can choose the origin for the torque at the centre of the loop.

The vector force on side 1 is

$$d\overline{F}_1 = Idx\overline{u}_x \times \overline{B}_0 \quad (\because \; \overline{B}_0 = B_{0x}\overline{u}_x + B_{0y}\,\overline{u}_y + B_{0y}\,\overline{u}_z)$$

or

$$d\overline{F}_1 = Idx\,(B_{0y}\overline{u}_z - B_{0y}\overline{u}_y) \qquad ...(6.30)$$

For this side of the loop the lever arm $\overline{R}$ extends from the origin to the mid-point of the side, $\overline{R}_1 = -\frac{1}{2}dy\overline{u}_y$, and the contribution to the total torque is

$$d\overline{T}_1 = \overline{R}_1 \times d\overline{F}_1$$

$$= -\frac{1}{2}dy\overline{u}_y \times Idx\,(B_{0y}\overline{u}_z - B_{0z}\overline{u}_y)$$

$$= -\frac{1}{2} dx\, dy\, IB_{0y}\bar{u}_x$$

The torque contribution on side 3 is found to be the same,

$$d\bar{T}_3 = R_3 \times d\bar{F}_3 = \frac{1}{2} dy\bar{u}_y \times (-Id \times \bar{u}_x \times \bar{B}_0)$$

$$= \frac{1}{2} dx\, dy\, IB_{0y}\, \bar{u}_x = dT_1$$

$\therefore \quad d\bar{T}_1 + d\bar{T}_3 = -dx\, dy\, I\, B_{0y}\, \bar{u}_x$

Similarly, $\quad d\bar{T}_2 + d\bar{T}_4 = dx\, dy\, I\, B_{0x}\, \bar{u}_y$

Then, the total torque is

$$dT = I\, dx\, dy\, (B_{0x}\, \bar{u}_y - B_{0y}\bar{u}_x)$$

or

$$d\bar{T} = I\, dx\, dy\, (\bar{u}_z \times \bar{B}_0)$$

$$\therefore \quad d\bar{T} = I\, d\bar{s} \times \bar{B} \qquad \text{...(6.31)}$$

Where, $d\bar{s}$ is the vector area of the differential current loop and the subscript on $\bar{B}_0$ has been dropped. We now define the product of the loop current and the vector area of the loop as the differential magnetic dipole $d\bar{m}$, with units of A-m^2, Thus

$$d\bar{m} = Id\bar{s} \quad \text{and} \quad d\bar{T} = d\bar{m} \times \bar{B} \qquad \text{...(6.32)}$$

Equation (6.32) is general result which result for differential loops of any shape, not just rectangular ones. The torque on a circular or triangular loop is also given in terms of vector surface or the moment of equation (6.32).

Since we have selected a differential current loop so that $\bar{B}$ is constant throughout it, it follows that the torque on a planar loop of any size or shape in a uniform magnetic field is given by the same expression

$$\bar{T} = I\bar{S} \times \bar{B} = \bar{m} \times \bar{B} \qquad \text{...(6.33)}$$

The torque on the current loop always tends to turn the loop so as to align the magnetic field produced by the loop with the applied magnetic field that is causing the torque. This gives the direction of the torque.

Example 6.6: *A straight solid wire segment carrying a current 2 $\bar{U}_y$ Amps extends from A (0, 1, 2) to B (0, 4, 2) is free space. This wire is subjected to the magnetic field of an infinite current filament lying along z-axis and carrying 35 A is the U_z direction. Find the vector torque on the wire segment about an origin at (a) P_A (0, 0, 2), (b) P_B (0, 0, 0).*

Solution: The wire segment extends from A (0, 1, 2) to B (0, 4, 2). In points A and B, x and z-coordinate are constant hence the general point on wire segment is given by (0, y, 2). At this point we will consider a differential current element $Id\bar{L} = 2\ dy\ U_y$ A.m. At this point (0, y, 2) the magnetic field intensity due to infinite current filament lying along z-axis is given by

$$\bar{H} = \frac{-35}{2\pi y}\, U_x \text{ A/m}$$

(or) $$\bar{B} = \frac{-35\mu_0}{2\pi y} U_x \text{ Wb/m}^2$$

$$d\bar{F} = 2dy\, U_y \times \left(\frac{-35\mu_0}{2\times y} U_x\right) = \frac{35\mu_0\, dy U_z}{\pi y} \text{ N}$$

(*a*) Now, $$d\bar{T} = \bar{r} \times d\bar{F}$$

$\bar{r}$ = a vector directed from (0, 0, 2) towards (0, *y*, 2)

$= y\, U_y$

Now, $$d\bar{T} = y\, U_y \times \left[\frac{-35\mu_0\, dy}{\pi y} U_z\right] = \frac{35\mu_0\, dy}{\pi} \bar{U}_x$$

$$\bar{T} = \int d\bar{T} = \int_1^4 \frac{35\mu_0\, dy}{\pi y} U_x = \frac{35}{\pi} \mu_0\, U_x\, (4-1)$$

$= 33.43\, \bar{U}_x\, \mu_0$ N–m (newton metre).

(*b*) $\bar{r}$ = a vector directed from (0, 0, 0) towards (0, *y*, 2).

$= y\, U_y + 2U_z$

$$d\bar{T} = (y\, U_y + 2U_z) \times \left[\frac{35\mu_0\, dy}{\pi y} U_z\right]$$

$$= \frac{35\mu_0\, d\bar{y}}{\pi y} U_x$$

$$\therefore \quad \bar{T} = \int dT = \int_1^4 \frac{35\mu_0}{\pi}\left[\frac{1-2}{y}\right] dy\, \bar{U}_x$$

$$= \frac{35\mu_0}{\pi} [y - 2\ln y]_1^4\, \bar{U}_x = 2.530\, \bar{U}_x\, \mu_0 \text{ N-m (newton metre).}$$

6.11 FORCE ON A STRAIGHT CURRENT CARRYING CONDUCTOR IN A MAGNETIC FIELD

Consider the case of current carrying conductor placed in a magnetic field (perpendicular to paper) it will experience a force as shown in Fig. 6.6.

If the charge is small, then the differential force can be written as

$$d\bar{F} = dQ\, (\bar{v} \times \bar{B}) \qquad ...(6.34)$$

There $\bar{v}$ is displacement of charge per unit time

$$\therefore \quad d\bar{F} = dQ\left(\frac{dl}{dt} \times \bar{B}\right)$$

$$= \frac{dQ}{dt} (\overline{dl} \times \bar{B})$$

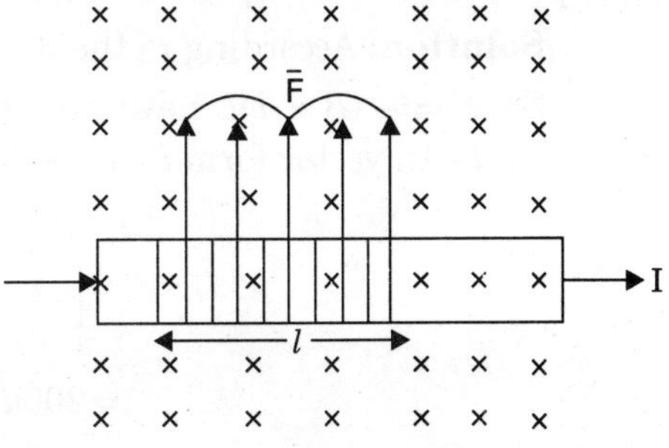

Fig. 6.6

(or) $$d\overline{F} = I\,(d\overline{L} \times \overline{B}) \qquad ...(6.35)$$

Where $\overline{B}$ is uniform magnetic field over length of conductor. Integrating equation (6.35) it can be seen as

$$\overline{F} = I(\overline{l} \times \overline{B}) \qquad ...(6.36)$$

The force $\overline{F}$ is perpendicular to both $\overline{l}$ and $\overline{B}$

$$\overline{F} = I\overline{l} \times \overline{B} \qquad ...(6.37)$$

The magnitude of force is given by familiar expression

$$|F| = BIL \sin \theta \qquad ...(6.38)$$

Where θ is angle between the vectors representing the direction of current flow and direction of magnetic flux density.

Example 6.7: *A conductor of 250 cm long carries a current of 25 amps at right angles to a uniform field produced by the pole core of an electrical machine. Now if core has a circular cross-section of 140 mm diameter and total flux in core is 15 m Wb, then find (a) Force developed on the conductor (b) Power required to move conductor at a speed of 20 m / sec in a plane at right angles to the field.*

Solution: Given, the length of conductor = 250 cm = 2.5 m and current through the conductor = 25 A

Diameter of pole core = 140 mm = 0.14 m

So, cross-sectional area of the core = $\frac{\pi}{4} \times (0.14)^2 = \frac{3.13}{4} \times 0.0196 = 0.0153\ \text{m}^2$

∴ Total magnetic flux = 15 m Wb

So, flux density $B = \frac{\text{Total flux}}{\text{polecore area}} = \frac{0.015}{0.0153} = 0.980\ \text{Wb/m}^2$

(*a*) Force on conductor = BIL = 0.980 × 25 × 2.5 = 61.25 newton

(*b*) Speed of the conductor = 20 m/sec

Power required = force × distance traveled per sec

= 61.25 × 20 = 1225 watts.

Example 6.8: *A long straight conductor lying along the z-axis carries a current of 25 A in the positive direction of z-axis. A if the magnetic flux density is $(5u_x + 8u_y)$ Wb/m², find the force per meter length of the conductor.*

Solution: According to the data given I = 25 A along z-axis and $B = (5u_x + 8u_y)$ Wb/m²

So, $F = -(B \times I)L$ newtons, if we consider current as a vector. But if L = 1 m, then $F = -(B \times I)$. In vector form I can be written as $I = 0u_x + 0u_y + 25u_z$

$$\therefore \qquad B \times I = \begin{vmatrix} U_x & U_y & U_z \\ 5 & 8 & 0 \\ 0 & 0 & 25 \end{vmatrix}$$

$$= 200u_x - 125u_y$$

∴ Force, $$F = -(B \times I) = -200u_x - 125u_y$$

Magnitude of $F = \sqrt{(-200)^2 + (125)^2} = 235.84$ N

The direction of F is along the unit vector u_r,

Where
$$u_r = \frac{-200}{235.84} u_x + \frac{125}{235.84} u_y$$
$$= -0.8480 u_x + 0.5300 u_y$$

6.12 FORCE BETWEEN TWO PARALLEL CURRENT CARRYING CONDUCTORS OF FINITE LENGTH

Let the conductor be of equal length L and let them carry currents, I_1 and I_2 in opposite direction. In the last concept the experience for force on a conductor of length L carrying a current I was derived as

$F = BIL \sin \theta$ and in vector form it can be written as

$$F = L\,(I \times B) \qquad \text{...(6.39)}$$

$$dF = dL\,(I \times B) \qquad \text{...(6.40)}$$

So, it can be seen that flux density vector B acts inwards and director of current is upward. So, as a result $\theta = 90°$ and $\sin 90° = 1$ which implies

$$dF = BI\,dL \text{ newton} \qquad \text{...(6.41)}$$

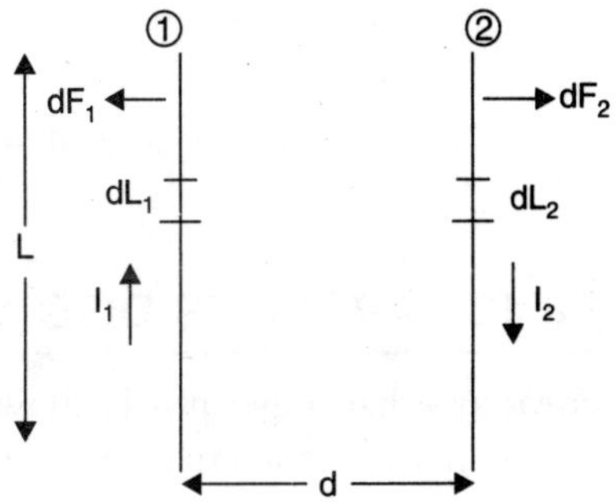

Fig. 6.7

Force acting on an elemental length dL_2 of conductor 2 due to current I_2 is given as

$$dF_2 = B_1 I_2\, dL_2, \qquad \text{...(6.42)}$$

Where B_1 is magnetic flux density at conductor 2 due to current I_1

We have $$B_1 = \left(\frac{\mu_0 I_1}{2\pi d}\right) \text{Wb/m}^2 \qquad \text{...(6.43)}$$

Substituting equation (6.43) in equation (6.42), we get

$$dF_2 = \frac{\mu_0 I_1 I_2}{2\pi d}\, dL_2 \qquad \text{...(6.44)}$$

Total force F_2 on conductor 2 exerted by current I_1 is given by

$$F_2 = \int_0^L \frac{\mu_0 I_1 I_2}{2\pi d}\, dL_2$$

Which implies $$F_2 = \frac{\mu_0 I_1 I_2}{2\pi d} L \text{ newtons} \qquad \text{...(6.45)}$$

Putting $I_1 = I_2 = I$ and $L = 1$ m the equation (6.45) becomes

$$F_2 = \frac{\mu_0 I^2}{2\pi d}$$

Similarly, it can be shown that force F_1 on conducts 1 excited by current I_2 is given by

$$F_1 = \frac{\mu_0 I_2 I_1 L}{2\pi d} \text{ newtons} \qquad \text{.....(6.46)}$$

Here the direction of F_1 is opposite to F_2. If currents, I_1 and I_2 flow in same direction, the force is attractive force (or) it is a repulsive force.

Example 6.9: *Two long parallel wires separated 4 m apart carry currents of 45 A and 95 A respectively in the same direction. Determine the magnitude and direction of the force between them per unit length.*

Solution: Given I_1 = 45 A, I_2 = 95 A and d = 4 m. Since the currents flow in the same direction, the force between the parallel wires is an attractive force.

We have force $F = \dfrac{\mu_0 I_1 I_2 L}{2\pi d}$ newtons, in free space.

Substituting the given values, we have

$$F = \frac{(4\pi \times 10^{-7})45 \times 95 \times L}{2\pi \times 4} \text{ newtons}$$

$$= 213.75\ L\ \mu \text{ newtons.}$$

$$\therefore \quad \text{Force/metre length} = \frac{213.75L}{L} = 213.75\ \mu\text{N.}$$

6.13 FARADAY'S DISC GENERATOR

Faraday's disc generator is shown in Fig. 6.8. It essentially consists of a rotating disc of radius 'a' which is attached to a shaft or axis. The disc is placed in a uniform magnetic field. There are two brushes which make contact with disc, one brush is pressed at the shaft and the other is pressed over edge of disc. The plane of disc is perpendicular to magnetic field.

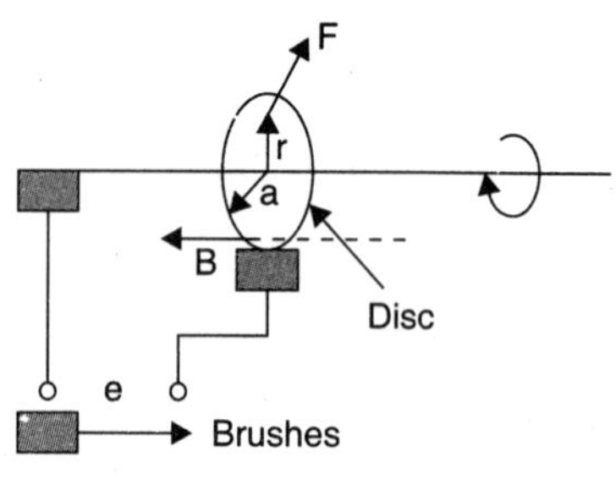

Fig. 6.8

Let a = radius of disc and B = flux density let the disc be rotated at a constant angular velocity 'ω' rad/sec.

When the disc rotates, it cuts the flux and an emf is induced in the disc. The function of brushes is to collect emf from rotating disc. Consider an electron at a point on the disc at a distance 'r' from the center. When the disc rotates at an angular velocity of 'ω' rad/sec, the electron moves at a velocity.

$$v = \omega\, r/\text{ms}$$

The force exerted on the electron is, $\bar{F} = Q(\bar{v} \times \bar{B})$

Electric field intensity, $\bar{E} = \dfrac{\bar{F}}{Q} = (\bar{v} \times \bar{B})\ N/C$

Where, Q = charge on the electron

Putting $v = \omega r$, we get, $E = \omega\, r\, B$ in magnitude where the field acts radially inward.

The e.m.f. developed between the center and rim ϕ disc is given as

$$e = \int_0^a E\,dr = \int_0^a \omega r B\,dr = \omega B\left[\frac{r^2}{2}\right]_0^a$$

$$\therefore \quad e = \frac{1}{2} B a^2 \omega \text{ volts.} \quad ...(6.47)$$

Example 6.10: *In a Faraday disc generator, a copper disc of 50 cm diameter is rotated at 3500 rpm on a horizontal axis passing through the centre and perpendicular to the plane of the disc lying in the magnetic meridian. What is the value of the induced e.m.f., if the horizontal component of the earth's magnetic field is 0.03 m Wb/m² ?*

Solution: Given diameter of the disc = 50 cm

radius of the disc, a = 25 cm

$= 25 \times 10^{-2}$ m = 0.25 m

Flux density B = 0.03 m Wb/m² and speed of rotation of the disc, N = 3500 r.p.m.

$$\therefore \text{ Angular velocity } \omega = \frac{2\pi N}{60} \text{ radians/second}$$

$$= \frac{2\pi \times 3500}{60} \text{ rad/s} = 116.66\,\pi \text{ rad/s.}$$

The e.m.f. induced in a Faraday disc generator is given as

$$e = \frac{1}{2}\omega a^2 B \text{ Volts}$$

Substituting the given values, we get

$$e = \frac{1}{2} \times 116.66 \times (0.25)^2 \times (0.03 \times 10^{-3}) \text{ volts}$$

$$= 0.3434 \text{ mV.}$$

Example 6.11: *A copper disc of 0.6 m diameter is rotated at a constant speed of 1500 r.p.m. on a horizontal axis perpendicular to the disc through its center. Two brushes make contact with the disc-one at the edge and the other at the center. If the horizontal component of the earth's magnetic field is 0.02 milli Tesla, calculate the e.m.f. induced between the brushes.*

Solution: This is a problem on Faraday disc generator.

Given diameter of the disc = 0.6 m and therefore radius a = 0.3 m.

Magnetic flux density $B = 0.02 \times 10^{-3}$ Wb/m²

The e.m.f. induced between the brushes is given as:

$$e = \frac{1}{2}\omega a^2 B \text{ volts,}$$

Where, ω = angular velocity

We have, $$\omega = \frac{2\pi N}{60} \text{ radians/second}$$

$$= \frac{2\pi \times 1500}{60} = 157 \text{ radians/sec.}$$

$$e = \frac{1}{2} \times 157 \times 0.3^2 \times (0.02 \times 10^{-3}) = 0.1413 \text{ mV.}$$

Example 6.12: *An infinite conductor carries a current of 3 A in the z-direction. Find the magnitude of the force on 2 m length of the conductor, if the field in which the conductor is placed, is given as* $B = (0.2u_x - 0.3u_y)$ *Tesla.*

Solution: Given current, $I = 3$ A, length, $L = 2$ m and the magnetic flux density

$$B = (0.2u_x - 0.3u_y) \text{ Wb/m}^2. \text{ [1 Tesla = 1 Wb/m}^2\text{]}$$

The force exerted on a current-carrying conductor placed in a magnetic field is given by the expression $F = BIL \sin \theta$ newtons.

The conductor is lying along the z-axis and the B field has components in the x-and y-directions. Hence $\theta = 90°$ and $\sin \theta = \sin 90° = 1$

Magnitude of $B = \sqrt{0.2^2 + 0.3^2} = 0.3605$ Wb/m².

$\therefore$ Force/metre length of the conductor = 0.3605 × 3 × 2 = 2.1633 newton.

Example 6.13: *Two long parallel wires separated 7 m apart carry curry currents of 55 A and 105 A respectively in the same direction. Determine the magnitude and direction of the force between them per unit length.*

Solution: Given $I_1 = 55$ A, $I_2 = 105$ A and $d = 7$ m. Since the currents flow in the same direction, the force between the parallel wires is an attractive force.

We have force $F = \dfrac{\mu_0 I_1 I_2 L}{2\pi d}$ newtons, in free space.

Substituting the given values, we have

$$F = \frac{(4\pi \times 10^{-7})55 \times 105 \times L}{2\pi \times 7} \text{ N} = 1650 \text{ L } \mu \text{ N}$$

$$\therefore \quad \text{Force/meter length} = \frac{1650L}{L} = 1650 \,\mu\text{N}.$$

Example 6.14: *If a point charge of 4 coulombs moves with a velocity of* $(5u_x + 6u_y - 7u_z)$ *m/s, find the force exerted (a) if the electric field intensity is* $(15u_x + 8u_y - 5u_z)$ *v/m (b) if the flux density is* $(5u_x + 7u_y - 9u_z)$ *Wb/m².*

Solution: (*a*) Given $Q = 4$ coulombs, velocity $v = (5u_x + 6u_y - 7u_z)$ m/s and

$$E = (15u_x + 8u_y - 5u_z) \text{ v/m}$$

So, the force exerted can be written as $F = QE$

$$\therefore \quad F = 4(15u_x + 8u_y - 5u_z) \text{ N}$$
$$= (60u_x + 32u_y - 20u_z) \text{ N}$$

Magnitude of the force $= \sqrt{60^2 + 32^2 + (-20)^2} = 70.88$ N

Here the direction of F is same as that of E.

(*b*) Given $B = (5u_x + 7u_y + 9u_z)$ wb/m², $Q = 4$ coulombs, velocity $v = (5u_x + 6u_y - 7u_z)$ m/s

So, the expression for force can be written as $F = Q(v \times B)$

$$v \times B = \begin{vmatrix} U_x & U_y & U_z \\ 5 & 6 & -7 \\ 5 & 7 & 9 \end{vmatrix}$$

$$= 103u_x - 80u_y + 5u_z$$

So, $F = Q(v \times B) = 4\,(103u_x - 80u_y + 5u_z)$

$= 412u_x - 320u_y + 20u_z$ N

Magnitude of $F = \sqrt{412^2 + (-320)^2 + 20^2} = 522.05$ N.

Example 6.15: *A conductor of 290 cm long carries a current of 35 amps at right angles to a uniform field produced by the pole core of an electrical machine. Now if core has a circular cross-section of 150 mm diameter and total flux in core is 25 m Wb, then find (a) force developed on the conductor (b) power required to move conductor at a speed of 30 m/sec in a plane at right angles to the field.*

Solution: Given, the length of conductor = 290 cm = 2.9 m and current through the conductor = 35 A.

Diameter of pole core = 150 mm = 0.15 m

So, cross-sectional area of the core $= \frac{\pi}{4} \times (0.15)^2 = \frac{3.14}{4} \times 0.0225 = 0.01766 \text{ m}^2$

∴ Total magnetic flux = 25 m Wb

So, flux density $B = \frac{\text{Total flux}}{\text{polecorearea}} = \frac{0.025}{0.01766} = 1.4156 \text{ Wb/m}^2$

(*a*) So, force on conductor is $BIL = 1.4156 \times 35 \times 2.9 = 143.68$ N

(*b*) Speed of the conductor = 30 m/sec

Power required = force × distance traveled per sec

= 143.68 × 30 = 4310.4 watts.

Example 6.16: *An electron ($q_e = -3.42218 \times 10^{-19}$ C, $m = 8.55339 \times 10^{-31}$ kg) is moving at const, velocity $\bar{v} = 4.5 \times 10^7 U_x$ m/s along the negative x-axis. At the origin, it encounters the uniform magnetic field. $\bar{B} = -2.5 \times 10^{-3} U_y$ T and remains in it up to x = 3 cm. If we assume that electron remains on x-axis while it is in magnetic field, find its y and z-coordinate values where x = 0.4 cm.*

Solution: We know that $\bar{F} = Q\bar{v} \times \bar{B}$

$= -3.42218 \times 10^{-19}\,[(4.5 \times 10^7 U_x) \times (-2.5 \times 10^{-3} U_y)$

$= 3.8499 \times 10^{-14}$ U_z N

Now $\bar{F} = m\,\bar{a}$

∴ $\bar{a} = \frac{\bar{F}}{M} = \frac{3.8499 \times 10^{-14} U_z}{8.55339 \times 10^{-31}} = 4.501 \times 10^{16} U_z$ m/s.

6.14 MAGNETIC DIPOLE AND DIPOLE MOMENT

6.14.1 Magnetic Dipole

A magnetic dipole is created when a negative magnetic charge '$-Q_M$' and a positive magnetic charge '$+Q_M$' are placed at a small separation 'l'.

6.14.2 Dipole Moment

A small bar magnet having pole strength Q_m and length l may be visualized as Magnetic dipole with Magnet dipole moment, $m = Q_m\, l$ and its direction is from '$- Q_m$ to $+ Q_m$'. The line joining $- Q_m$ to $+ Q_m$ is called axis of dipole. If such a magnetic dipole is linked to a current loop with area A and carrying current I ; then that loop has a magnetic moment 'm' = IA

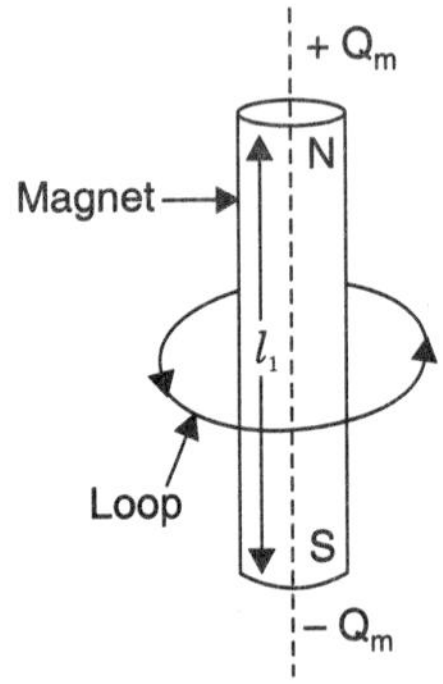

Fig. 6.9 (*a*)

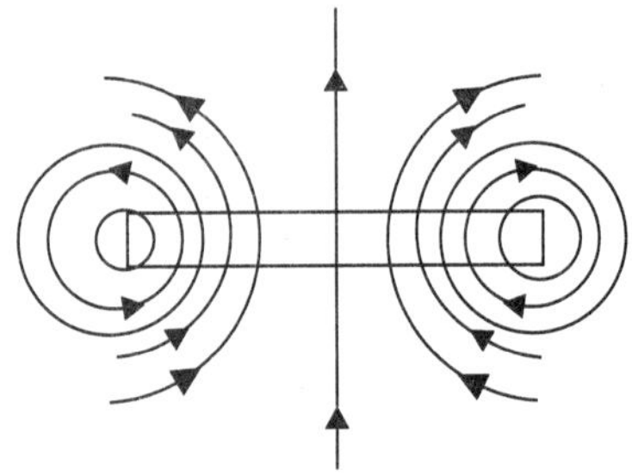

Fig. 6.9 (*b*)

Fig. 6.9 (*a*) shows a dipole moment $Q_m\, l$ and equivalent current loop moment $m = IA$. So, it can be written as

$$Q_m l = IA \qquad ...(6.48)$$

Field lines produced by a current loop behave like a magnetic dipole shown in Fig. 6.9 (*b*).

Let a bar magnet be considered which is composed of large number of magnetic dipoles. If A is taken as the area of cross-section of bar magnet and 'l' its axial length, the volume of bar magnet whose pole strength is Q_m and dipole moment $Q_m l$ is $A\ l$, so that net dipole moment per unit volume is

$$\frac{Q_m l}{Al} = \frac{Q_m}{A}$$

and it is called as magnetization which can be represented as vector M directed as $\bar{l}$ from S to N pole.

$$\bar{M} = \frac{Q_m l}{Al} \cdot \bar{u}_l = \frac{Q_m}{A} \cdot \bar{u}_l \qquad ...(6.49)$$

Where $\bar{u}_l$ is a unit vector directed from S to N pole. The vector $\bar{M}$ can be compared with polarization vector $\bar{P}$.

$$\text{Dimension of } M = \frac{\text{dipole moment}}{\text{volume}} = \frac{\text{ampere–meter}^2}{m^3}$$

$$= \text{ampere per meter (or) as } H$$

When a magnetic field is applied to a material the magnetic induction B (flux density) is given by the sum of effect of vaccum and that on the material.

In a case of free space $B = \mu_0 H$ and it is flux density produced by magnetizing force 'H'.

Due to magnetization the flux density is increased due to M. Consequently

$$\overline{B} = \mu_0 \overline{H} + \mu_0 \overline{M} = \mu_0 \mu_r \overline{H} \qquad [\because B = \mu H]$$

$$\Rightarrow \qquad \mu_0 \overline{M} = \mu_0 (\mu_r - 1) \overline{H}$$

Then

$$\overline{M} = (\mu_0 - 1) \overline{H}$$

(or)

$$\overline{M} = \chi_m \overline{H} \qquad ...(6.50)$$

Where $\chi_m = (\mu_r - 1)$ is magnetic susceptibility, a dimensionless quantity. M and H have same unity as A/m.

The magnetic susceptibility is also expressed as

$$\chi_m = \frac{\overline{M}}{\overline{H}} \text{ which is analogous to electric susceptibility } \chi_e = \frac{\overline{P}}{\overline{E}}$$

6.15 TORQUE ON A DIFFERENTIAL CURRENT LOOP PLACED IN MAGNETIC FIELD (B)

The loop lies in the x-y plane and the sides of loop are parallel to x and y-axis and are of length dx and dy. The value of magnetic field at center of loop is taken as B_0. Since the loop is of differential size, the value of B at all points on the loop may be taken as B_0. The total force on loop is therefore zero and we are free to choose the origin for torque at center of loop. The vector force on side 1 is

$$dF_1 = I\, dx\, U_x \times B_0 \qquad ...(6.51)$$

(or)

$$dF_1 = I\, dx\, (B_{0y} U_z - B_{0z} U_y)$$

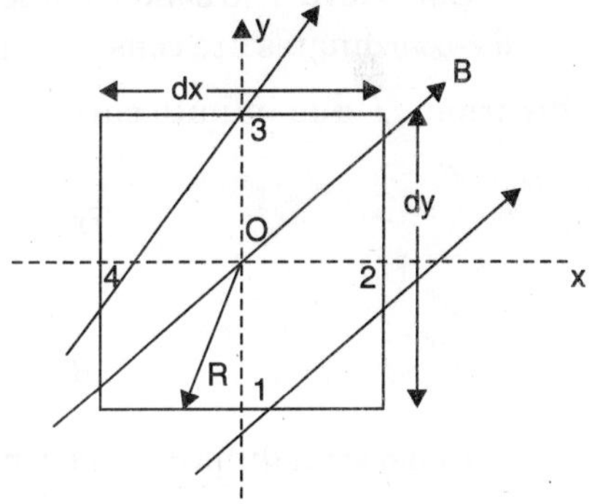

Fig. 6.10

From this side of loop the lever arm R extends from origin to the mid-point of side $R_1 = -\frac{1}{2}\, dy\, U_y$ and the contribution to the total torque is

$$dT_1 = R_1 \times dF_1$$

$$= -\frac{1}{2}\, dy\, U_y \times I\, dx\, (B_{0y} U_z - B_{0z} U_y)$$

$$= -\frac{1}{2}\, dx\, dy\, I\, (B_{0y} U_x) \qquad ...(6.52)$$

The torque contribution on side 3 is found to be the same

$$dT_3 = R_3 \times dF_3$$

$$= \frac{1}{2}\, dy\, U_y \times (-I\, dx\, U_x \times B_0)$$

$$= -\frac{1}{2}\, dx\, dy\, I\, B_{0y} U_x = dT_1 \qquad ...(6.53)$$

And $\quad dT_1 + dT_3 = -\, dx\, dy\, I\, B_{0y} U_x$

Similarly evaluating the torque on side 2 and 4, we get

$$dT_2 + dT_4 = dx\, dy\, I\, B_{0x}\, U_y$$

And the total torque is given by

$$dT = I\, dx\, dy\, (B_{0x}\, U_y - B_{0y}\, U_x)$$

The quantity within parenthesis may be represented by a cross product

$$dT = I\, dx\, dy\, (U_z \times B_0)$$

$$dT = I\, ds \times B \qquad ...(6.54)$$

Where ds is the vector area of differential current loop and now the product of loop current and vector area of the loop can be defined as differential magnetic dipole moment dm. Thus

$$dm = I\, ds \quad \text{and} \quad dT = dm \times B.$$

Example 6.17: *In an infinite filament carrying $14U_z$ Amps lies on the z-axis and a straight solid conductor segment extends from P_A (2, – 2, 0) to P_B (2, 2, 0) carrying $3\overline{U}_y$ Amps. Find the torque on the conductor segment about an origin at P_0 (2, 0, 0) as a function of l?*

Solution: The solid conductor extends from P_A (2, – 2, 0) to P_B (2, 2, 0). In these points x and z-coordinates are constant. Hence general point on solid conductor is given by $(2, y, 0)$. At this point $\overline{H}$ due infinite current of $14U_z$ Amps on the z-axis is given by

$$\overline{H} = \frac{14}{2\pi\sqrt{4+y^2}}\left\{\overline{U}_z \times \left[\frac{2\overline{U}_x + yU_y}{\sqrt{4+y^2}}\right]\right\} = \frac{7(-yU_x + 2\overline{U}_y)}{\pi(4+y^2)} \text{ A/m}$$

$$\therefore \quad B = \mu_0\, \overline{H} = \frac{6\mu_0(-yU_x + 2U_y)}{\pi(4+y^2)} \text{ Wb/m}^2$$

Consider a differential current element $IdL = 3\, dy\, U_y$ A-m located at point $(2, y, 0)$. Then

$$d\overline{F} = I\, \overline{dL} \times \overline{B}$$

$$= 3\, dy\, \overline{U}_y \times \left[6\,\mu_0 \frac{(-yU_x + 2U_y)}{\pi(4+y^2)}\right]$$

$$= \frac{18\mu_0\, y\overline{U}_z\, dy}{\pi(4+y^2)} \text{ N}$$

$$d\overline{T} = \overline{r} \times d\overline{F},$$

Where, $\overline{r}$ = a vector directed from (2, 0, 0) towards $(2, y, 0) = y\overline{U}_y$

$$d\overline{T} = y\overline{U}_y \times \frac{18\mu_0\, yU_z}{\pi(4+y^2)}\, dy$$

$$= \frac{18\mu_0\, y^2\overline{U}_x}{\pi(4+y^2)}\, dy \text{ N-m}$$

$$\overline{T} = \int d\overline{T} = \int_{-2}^{2} \frac{18\mu_0\, y^2\overline{U}_x}{\pi(4+y^2)} = \frac{18\mu_0\, \overline{U}_x}{\pi}\left[\frac{y}{2} - \tan^{-1}\frac{y}{2}\right]_{-2}^{2}$$

$$= 14.4\,[1 - \tan^{-1} 1]\, \overline{U}_x\ \mu \text{ N-m.}$$

6.16 BOUNDARY RELATIONS IN MAGNETIC FIELDS

The relationships that have been derived so far are applicable everywhere except for points at the surface. Therefore either they must be modified to take into account of the surface magnetizing current or limiting procedure must be adopted to include the discontinuity in dipole moment $\overline{M}$. However instead of dealing with the surface currents they can be accounted for by specifying certain boundary conditions for the field vectors $\overline{B}$ and $\overline{H}$ at the boundary between two different media. There are three different boundary conditions as described below and the boundary relationships are derived by considering separately the field components normal to the boundary and tangential to the boundary.

6.16.1 First Boundary Condition

The Fig. 6.11 shows a boundary between two isotropic homo-genous magnetic materials with permeability μ_1 and μ_2.The boundary condition of normal components of B is determined by allowing the surface to cut a small cylindrical Gaussian surface.

Applying Gauss law of magnetic field

$$\nabla \cdot \overline{B} = 0 \quad \text{(or)} \quad \oint \overline{B} \cdot ds = 0$$

If B_{n_1} and B_{n_2} are the normal components of flux densities, then we have $B_{n_1} \cdot u_{n_1}\ \Delta s - B_{n_2} \cdot \Delta s u_{n_2} = 0$

$$B_{n_1} \cdot \Delta s - B_{n_2} \cdot \Delta s = 0$$

$$B_{n_1} = B_{n_2}$$

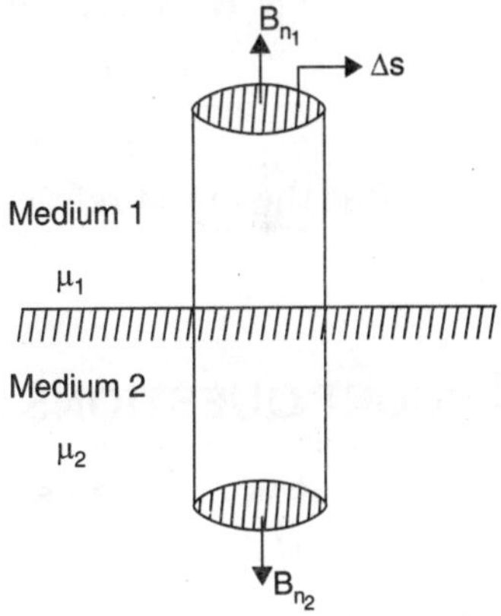

Fig. 6.11

Flux density and the surface are orthogonal *i.e.*, the normal components of magnetic flux densities are equal at the boundary between the two magnetic media.

6.16.2 Second Boundary Condition

Consider a rectangular area in the boundary. At the boundary, there exists a sheet current of density H A/m. Applying Ampere's law at the boundary to the small closed path in the plane.

$$\oint \overline{H} \cdot d\bar{l} = I_{enclosed}$$

$$H_{t_1}\ \Delta l - H_{t_2} \Delta l = J\ \Delta l$$

$$\therefore \quad H_{t_1}\ \Delta l - H_{t_2} \Delta l = 0$$

[If the current density at the boundary is 0]

$$\therefore \quad H_{t_1} = H_{t_2}$$

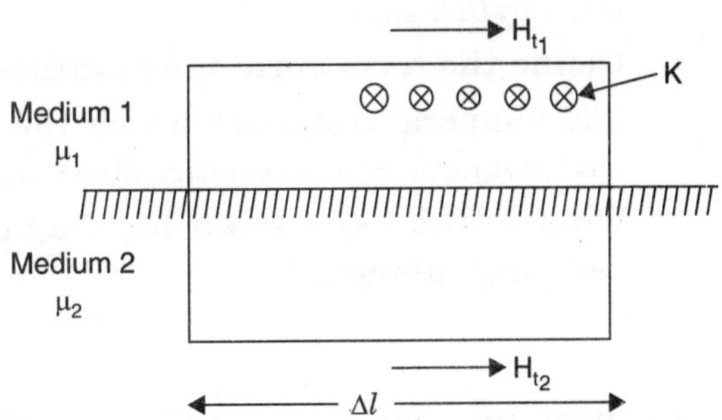

Fig. 6.12

i.e., the tangential components of field intensities are equal.

6.16.3 Refraction

We know that,

$$B_{n_1} = B_{n_2}$$

[using the first boundary condition]

$$\Rightarrow \quad B_1 \cos \theta_1 = B_2 \cos \theta_2$$

$$\mu_1 H_1 \cos \theta_1 = \mu_2 H_2 \cos \theta_2 \qquad \text{...(6.55)}$$

From second boundary condition,

$$H_{t_1} = H_{t_2}$$

$$H_1 \sin \theta_1 = H_2 \sin \theta_2 \qquad \text{...(6.56)}$$

Dividing equation (6.56) by equation (6.55), we get

$$\frac{H_1 \sin \theta_1}{\mu_1 H_1 \cos \theta_1} = \frac{H_2 \sin \theta_2}{\mu_2 H_2 \cos \theta_2}$$

$$\frac{\tan \theta_1}{\mu_1} = \frac{\tan \theta_2}{\mu_2}$$

It is the law of refraction for magnetic flux lines at a boundary with no surface current.

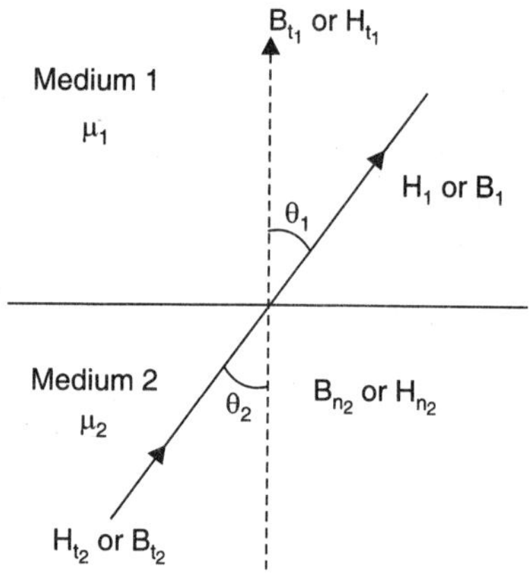

Fig 6.13

SHORT QUESTIONS AND ANSWERS

1. **What is Maxwell's third equation?**

   Curl $H = J$ (or) $\nabla \times H = J$

2. **What is meant by Lorentz force equation?**

   The force on a moving particle due to combined electric and magnetic filed is obtained easily by superposition which is given by $F = Q\,(E + V \times B)$. This equation is called as Lorentz force equation.

3. **What is the expression for force on a straight current carrying conductor in a magnetic field?**

   The expression for force on a straight current carrying conductor in a magnetic field is given byBIL sin θ.

4. **Define the term curie temperature.**

   The temperature above which the elements like iron and nickel lose all their ferromagnetic characteristics above a temperature called the Curie temperature.

5. **What is the expression for magnetic field intensity due to a long current carrying filament?**

$$H = \frac{I}{2\pi d}$$

6. **Write the expression for cyclotron frequency.**

$$f = \frac{QB}{2\pi M}$$

7. Write the expression for force between two parallel current carrying conductors of finite length.

$$F = \frac{\mu_0 I_2 I_1 L}{2\pi d}$$

8. Write the expression for e.m.f. generated in a Faraday's disc generator.

$$e = \frac{1}{2} B\, a^2\, \omega \text{ volts}$$

9. Write the expression which explains Stoke's theorm.

$$\oint H \cdot dL = \int_S (\nabla \times H) \cdot ds$$

10. What is the expression for force between two current carrying wires?

$$F_1 = \frac{\mu_0 I_1 I_2}{2\pi D} = F_2$$

MULTIPLE CHOICES

1. The ratio of intensity of magnetization to the magnetizing force is known as ....
   (*a*) susceptibility (*b*) flux density
   (*c*) relative permeability (*d*) absolute permeability
2. The substances having permeability less than the permeability of free space are known as ....
   (*a*) bipolar (*b*) paramagnetic
   (*c*) diamagnetic (*d*) ferromagnetic
3. The forces between two long conductors in parallel is inversely proportional to.......
   (*a*) distance between conductors (*b*) current in one conductor
   (*c*) radius of conductor (*d*) product of current of two conductors
4. It is difficult to magnetize steel because.......
   (*a*) high permeability (*b*) low permeability
   (*c*) high specific gravity (*d*) corrodes easily
5. Magnetic intensity H and flux density B are related to each other by.......
   (*a*) $B = \frac{\mu_r H}{\mu_0}$ (*b*) $B = \mu H$
   (*c*) $B = \frac{\mu_0 H}{\mu_r}$ (*d*) $B = \frac{H}{\mu_0 \mu_r}$
6. The force experienced by a current carrying conductor lying parallel to magnetic field is ....
   (*a*) HIL (*b*) $BIL \sin \theta$
   (*c*) zero (*d*) BIL
7. The direction of induced e.m.f. can be determined by..........
   (*a*) Flemings right hand rule (*b*) Lenz law
   (*c*) Coulomb law (*d*) Kirchhoff's voltage law

8. The current carried by each of two long parallel conductors is doubled. If their separation is also doubled, then the force between them would be ....

(*a*) remains the same (*b*) increases two folds

(*c*) increases four fold (*d*) becomes half

9. Pick out the wrong statement:

Magnetic leakage is undesirable in electric machines because ....

(*a*) it leads to increased weight (*b*) increases cost of manufacture

(*c*) produces fringing (*d*) lowers their power efficiency

10. Permeability in a magnetic circuit corresponds to .... in a circuit.

(*a*) conductivity (*b*) resistivity

(*c*) conductance (*d*) resistance

11. The expression for Lorentz force equation is given by..........

(*a*) $F = Q\,(E - v \times B)$ (*b*) $F = Q\,(E + v \times B)$

(*c*) $F = QE$ (*d*) $F = v \times B$

12. Tangential component of H is continuous but tangential component of B is not continuous across a surface.

(*a*) false (*b*) true

(*c*) can't be determined (*d*) none of these

13. The force acting on a current element is ....

(*a*) $F = \oint Idl \times B$ (*b*) $F = \int_S Kds \times B$

(*c*) $F = \int Jdv \times B$ (*d*) all of the above

14. The force acting on a stationary charge in magnetic field is ....

(*a*) zero (*b*) $Q(V \times B)$

(*c*) Qe (*d*) $Q(V \times B)$

15. When a charged particle having charge Q travels with velocity V in a magnetic field, it experiences a force F_m given by

(*a*) $F_m = QVB \sin \theta$ (*b*) $F = Q(V \times B)$

(*c*) none of these (*d*) both (*a*) and (*b*)

16. If a particle is at rest in magnetic field, it will experience.......

(*a*) force (*b*) torque

(*c*) no force (*d*) both (*a*) and (*b*)

17. The force per unit length between two straight long and parallel current carrying conductors is given by.........

(*a*) $\dfrac{F}{l} = \dfrac{\mu_0 I_1 I_2}{2\pi d}$ (*b*) $\dfrac{F}{l} = \dfrac{\mu_0 I_1^{\,2} I_2}{2\pi d}$

(*c*) $\dfrac{F}{l} = \dfrac{\mu_0 I_2}{2d}$ (*d*) $\dfrac{F}{l} = \dfrac{\mu_0 I_2}{2\pi d}$

18. Normally, charges in motion produce a ....

(*a*) electric field (*b*) magnetic field

(*c*) electromagnetic field (*d*) magneto static field

19. The temperature above which the elements like iron and nickel lose their ferromagnetic characteristics is called as ......

(*a*) curie temperature (*b*) room temperature

(*c*) ambient temperature (*d*) none of these

20. The expression for force between two parallel current carrying conductors of finite length is given by.......

(*a*) $F = \dfrac{\mu_0 I_2 I_1}{2\pi d}$ (*b*) $F = \dfrac{\mu_0 I_2 I_1 L}{2\pi d}$

(*c*) $F = \dfrac{\mu_0 I_2 I_1 L}{2\pi d^2}$ (*d*) $F = \dfrac{\mu_0 I_2 L}{2\pi d}$

21. The expression for e.m.f. generated in a Faraday's disc generator is........

(*a*) $\dfrac{1}{2} B a^2$ volts (*b*) $\dfrac{1}{2} a^2 \omega$ volts

(*c*) $\dfrac{1}{2} B a^2 \omega$ volts (*d*) $\dfrac{1}{2} B \omega$ volts

22. The expression for Stoke's theorm is given by ....

(*a*) $\oint H \cdot dL = \int_S (\nabla \times S) \cdot ds$ (*b*) $\oint B \cdot dL = \int_S (\nabla \times H) \cdot ds$

(*c*) $\oint H \cdot dL = \int_S (\nabla \times H) \cdot ds$ (*d*) $\oint B \cdot dL = \int_S (\nabla \times B) \cdot ds$

23. Maxwell's third equation is given by ....

(*a*) Curl $H = J$ (*b*) $\nabla \times H = J$

(*c*) none of these (*d*) both (*a*) and (*b*)

24. The expression for force between two current carrying wires is given by ....

(*a*) $F = \dfrac{\mu_0 I_1 I_2}{2\pi D}$ (*b*) $F = \dfrac{\mu_0 I_1 I_2 l}{2\pi D}$

(*c*) $F = \dfrac{\mu_0 I_2 l}{2\pi D}$ (*d*) none of these

KEY

1. (*a*)	**2.** (*c*)	**3.** (*a*)	**4.** (*b*)	**5.** (*b*)	**6.** (*c*)
7. (*b*)	**8.** (*b*)	**9.** (*d*)	**10.** (*a*)	**11.** (*b*)	**12.** (*b*)
13. (*d*)	**14.** (*a*)	**15.** (*d*)	**16.** (*c*)	**17.** (*a*)	**18.** (*b*)
19. (*a*)	**20.** (*b*)	**21.** (*c*)	**22.** (*c*)	**23.** (*d*)	**24.** (*a*)

REVIEW QUESTIONS

1. If a stationary charge is placed in magnetic field, what is the force experienced by it?
2. Two parallel conductors carry currents in the same direction. Is the force experienced by them attractive or repulsive?
3. What is the significance of $\nabla \cdot \vec{B} = 0$?
4. Can we use the force equation to define current? If yes, how?
5. Is any energy required to maintain a steady magnetic flux after it is established? Explain.
6. In the design of magnetic circuits we always strive to set up the required flux density with as few mmf as possible. What does this mean in terms of (a) the permeability of the magnetic material and (b) the reluctance of the magnetic material?
7. Is it possible for the mmf to be numerically equal to the current producing it?
8. What happens to the reluctance of a ferromagnetic material if the flux density is increased?
9. Does the permeability of a ferromagnetic material increase as the flux density in the material decreases?
10. Does the permeability of a nonmagnetic material change with the flux density? How about the reluctance?
11. Can we apply Ohm's law for the magnetic field to ferromagnetic materials?
12. What do we mean when we say that a magnetic material is saturated?
13. What is hysteresis? What is meant by hysteresis loss? Is the hysteresis loss more for hard steel than for soft steel?
14. A nickel–cobalt–manganese alloy known as Perminvar is found to have constant permeability over a wide range of flux densities. Why is it important to find such a material?
15. Bismuth is not a ferromagnetic material; however, it changes its resistance when place in a magnetic field. Do you think we can use it to measure the magnetic flux density in a region? Explain.
16. There are some alloys for which the Curie point is quite low. Each of these alloys has been found to show appreciable change in its permeability for a small change in temperature. Do you think we can use such alloys to measure temperature? Explain.
17. What is residual flux density?
18. What is the force on a test charge q moving through a magnetic field $\overline{B}$ with a velocity $\overline{u}$? When is the force maximum? When is the force minimum?
19. What is the net force on a charge q moving with a velocity $\overline{u}$ in a region where $\overline{E}$ and $\overline{B}$ fields are at right angles to each other?
20. A current carrying conductor has zero net charge. Why, then, does a magnetic field exert a force on it?

21. If a charged particle moves in an electric field, there is a change in its kinetic energy. If the same charged particle moves in a magnetic field, the kinetic energy remains uncharged. Why?

22. What is a magnetic dipole? How does a magnetic dipole differ from an electric dipole?

EXERCISE PROBLEMS

1. Find the magnitude of magnetic flux density in a material for which (*a*) The magnetization is 2.8 A/m and magnetic susceptibility is 0.0025 (*b*) The magnetic field intensity is 1300 A/m and relative permeability is 1.006 (*c*) These are 8.2×10^{28} atoms per cubic meter each having a dipole moment of 3×10^{-30} A-m^2 is the same direction and magnetic susceptibility is 2×10^{-4}.

2. If $\overline{B} = 0.05 \times \overline{U}_y$ T in a mentioned for which $X_m = 2.5$; find (*a*) μ_R (*b*) μ (*c*) $\overline{H}$ (*d*) $\overline{M}$ (*e*) $\overline{J}$ (*f*) $\overline{J}_6$.

3. Determine energy density stored is free space by field :

(*a*) $\overline{H}_A = 10^3 U_x$ A/m

(*b*) $\overline{H}_B = 10^3 U_x + 10^3 U_y$ A/m

(*c*) $H_C = 10^3 U_x - 10^3 U_y$ A/m

(*d*) $\overline{H}_B + \overline{H}_C$

(*e*) $10^3 U_x$ A/M + $10^{-3} \overline{U}_y$ T

4. A current filament on the *z*-axis produces the field. $\overline{H} = -2U_x$ | *Y* A/M at *P* (0, *y*, *z*) is free space (*a*) Find the force excited on a current strip lying between *y* = 1 and 3 m is the *x* = 0 plane carrying a current density $\overline{k} = -2\pi\, \overline{U}_z$ A/m (*b*) What force is exerted on the current filament by the current strip?

5. A 0.2 C point charge, moving through the field $\overline{B} = 20\overline{U}_z$ mT is located at the origin at $t = 0$ with a velocity $V = 10^5 U_y$ m/s. Let $m = 2 \times 10^{-16}$ kg. Develop appropriate differential equation and solve them to evaluate at $t = 2 \times 10^{-6}$ s (*a*) The position *p* (*x*, *y*, *z*) of charge (*b*) Its velocity V (*c*) Its kinetic energy.

6. Find the magnetic field intensity within a magnetic material where (*a*) *M* = 150 A/m and $\mu = 1.5 \times 10^{-5}$ H/m. (*b*) $\overline{B} = 300\ \mu$ T and $X_M = 15$ (*c*) There are 8.2×10^{28} atoms /m^3, each atoms has a dipole moment of 5×10^{-27} A-m^2 and $\mu_R = 30$.

7. The magnetic flux density in a magnetic material with $X_M = 6$ is given in a certain region as $\overline{B} = 0.005\, y^2 \overline{U}_x$ T. At *y* = 0.4 m find magnitude of (*a*) $\overline{J}$ (*b*) $\overline{J}_b$ (*c*) $\overline{J}_T$.

8. A Faraday disc generator of diameter 10 cm is kept at the center of solenoid of length 1 m, diameter 12 cm, number of turns 500 and current 5 A. If the disc is rotated at a speed of 1000 rpm, find the voltage induced between the center and the edge of the disc.

9. State and explain Ampere's circuital law. Apply the law and determine the magnetic field intensity both inside and outside a conductor of radius '*a*'. The conductor carries a current of 1 ampere. Sketch the fields.

10. Show that the force on a point charge Q coulomb moving in a uniform magnetic field of B Tesla with a velocity of v m/sec is given by $Q\ (\bar{v} \times \bar{B})$.

11. An aluminium disc of radius 9 cm is located at the center of solenoid of length = 100 cm, radius = 10 cm, number of turns = 100 and current = 5 amp. The axis of the disc is along that of solenoid and rotates at 1000 rpm. Calculate the voltage induced between the center and edge of the disc.

12. (*a*) Derive expression for the torque acting on a circular current ring placed in a uniform magnetic field. (*b*) Calculate the force on a straight conductor of length 0.30 m carrying a current of 5.0 A in the z-direction where the magnetic field is $B = 3.50 \times 10^{-3}\ (u_x - u_y)$ Tesla. (u_x and u_y are unit vectors). (c) Find the maximum torque on an 85 turn, rectangular coil 0.2 m by 0.3 m, carrying a current of 20 A and field B = 6.5 Wb/m^2.

CHAPTER

7

Magnetic Potential and Inductance

7.1 INTRODUCTION

In the previous chapter, the various magnetic forces were studied and also the concept of Faraday's disc generator was introduced. In this unit the concept of scalar and vector magnetic potential is studied. Scalar magnetic potential is used to obtain relation between H and magnetic potential V_M and vector magnetic potential is used in regions where the current density is zero (or) almost-zero. Next the concept of self and mutual inductance is introduced. Here the concept of Neumann's formula is introduced which is useful in finding out the mutual inductance between two closed loops. Next the concept of self inductance is introduced and effort is made to define them by different means in the case of solenoid, toroid and for different types of transmission lines.

7.2 SCALAR MAGNETIC POTENTIAL

The solution to the electrostatic field problem is simplified by use of scalar electrostatic potential. Now electric field intensity E at any point in electric field is related to electric potential V at a point by

$$E = -\nabla \cdot V \qquad ...(7.1)$$

This relationship is of immense help in obtaining field intensity in several cases. In magneto statistics there is a similar relationship obtaining between H and magnetic potential V_M.

$$H = -\nabla V_M \qquad ...(7.2)$$

Since, V_m is a scalar and also to distinguish it from a vector, V_m is referred to as scalar magneto potential. Now H is negative potential gradient which holds good provided the current density J is zero everywhere in field for which V_m is defined.

Consider $\quad H = -\nabla V_M$

Applying Curl on both sides

$$\text{Curl } H = \text{Curl } (-\nabla V_M)$$

$$\nabla \times H = \nabla \times (-\nabla V_M) \quad ...(7.3)$$

But $\nabla \times H = J$

Where J = current density

But R.H.S. of equation (7.3) is equal to '0' *i.e.*, $\nabla \cdot (-\nabla V_M) = 0$

As a result $\nabla \times H = 0$

i.e., $J = 0$

Hence, $H = -\nabla V_M$ provided $J = 0$ ...(7.4)

Limitations:

A scalar magnetic potential can't be uniquely defined. It means magneto static field is generally non-conservative and becomes conservative only when $J = 0$. In the presence of steady currents which are sources of B, a scalar magnetic potential can't be defined.

7.3 VECTOR MAGNETIC POTENTIAL

This vector field is one which is extremely useful in studying radiation from antennas and radiation leakage from transmission lines, wave guides etc. The vector magnetic potential may be used in region where the current density is zero (or) almost-zero and we are also able to extend it to time varying cases and it is denoted by A.

We studied earlier that $\nabla \times B = 0$

'A' is so defined that B is curl of A

i.e., $B = \text{Curl } A$

or $B = \nabla \times A$ ...(7.5)

We have $B = \mu_0 H$ in free space

$$H = \frac{B}{\mu_0} = \frac{1}{\mu_0} \nabla \times A \quad ...(7.6)$$

$$\text{Curl } H = \nabla \times H = \nabla \times \left(\frac{1}{\mu_0} \nabla \times A\right)$$

$$= \frac{1}{\mu_0} \nabla \times \nabla \times A \quad ...(7.7)$$

But $\nabla \times \text{H} = J$

$$\therefore \quad J = \frac{1}{\mu_0} (\nabla \times \nabla \times A)$$

But, $\nabla \times \nabla \times A$ is curl of vector which is not zero.

Now $\nabla \times \nabla \times A = \nabla (\nabla \cdot \text{A}) - \nabla^2 A$

So $$J = \frac{1}{\mu_0} [\nabla (\nabla \cdot \text{A}) - \nabla^2 \text{A}]$$

But $\nabla (\nabla \cdot A) = 0$

$$\therefore \quad \nabla^2 \text{A} = -\mu_0 J \quad ...(7.8)$$

It can be compared with Poisson's equation of electrostatic field which can be written as

$$\nabla^2 V = \frac{-\rho_v}{\varepsilon_0} \quad ...(7.9)$$

Comparing equations (7.8) and (7.9) we can infer that

(*i*) A is analogous to V.

(*ii*) J is analogous to ρ_V and μ_0 is analogous to $\frac{1}{\varepsilon_0}$ and at a distance R from differential current element Idl.

It can be written as

$$A = \oint \frac{\mu_0 \, Idl}{4\pi R} \text{ Wb/m in free space} \quad ...(7.10)$$

For a current sheet

$$A = \int_S \frac{\mu_0 \, k \, ds}{4\pi R} \text{ Wb/m} \quad ...(7.11)$$

For volume with current density J we have

$$A = \int_{vol} \frac{\mu_0 \, Jdv}{4\pi R} \text{ Wb/m} \quad ...(7.12)$$

Example7.1: *Find the vector magnetic potential for an infinitely long solenoid with T turns per unit length, radius 'a' and current I.*

Solution: Let A be the vector magnetic potential which is being asked for :

By Stoke's theorm it can be written as

$$\oint A \cdot dl = \iint_S \nabla \times A \cdot n \, ds$$

$$= \iint_S B \cdot n \, ds = \phi \text{ (flux)}$$

Within the radius of the solenoid *i.e.*, $r < a$, we have

$$\oint A \cdot dl = A(2\pi r) = \iint_S B \cdot n \; ds = \phi$$

$$\oint A \cdot dl = A(2\pi r) = \mu_0 \; TI \; (\pi r^2)$$

$$A = \frac{\mu_0 TIr}{2} u_\phi \text{ (for } r < a)$$

Outside the solenoid, the flux is

$$\iint_S B \cdot n \; ds = \mu_0 \; TI(\pi a^2) \text{ (since the field extends to 'a' only)}$$

Thus, $$\oint A \cdot dl = A(2\pi r) = \mu_0 \; TI \; (\pi a^2)$$

Consequently, $$A = \frac{\mu_0 \; TI(\pi a^2)}{2\pi r}$$

In vector form, $$A = \frac{\mu_0 \; TIa^2}{2r} \mu_\phi \text{ (for } r > a)$$

Example 7.2: *If the vector magnetic potential is given as* $A = \frac{20}{x^2 + y^2 + z^2} u_x$. *Obtain the magnetic flux density in vector form.*

Solution: Given $A = \frac{20}{x^2 + y^2 + z^2} u_x$

We have $B = \nabla \times A$

it can be written as

$$B = \begin{vmatrix} U_x & U_y & U_z \\ \frac{\partial}{\partial x} & \frac{\partial}{\partial y} & \frac{\partial}{\partial z} \\ \frac{20}{x^2 + y^2 + z^2} & 0 & 0 \end{vmatrix}$$

$$= u_x (0) - u_y \left\{0 - \frac{\partial}{\partial z}\left(\frac{20}{x^2 + y^2 + z^2}\right)\right\} + u_z \left\{0 - \frac{\partial}{\partial y}\left(\frac{20}{x^2 + y^2 + z^2}\right)\right\}$$

$$= 0 + u_y \left(\frac{20}{x^2 + y^2 + z^2}\right) 2z - u_z \left(\frac{20}{x^2 + y^2 + z^2}\right) 2y$$

$$B = \left(\frac{40}{x^2 + y^2 + z^2}\right)(zu_y - yu_z).$$

Example 7.3: *Find the vector potential* A_z *due to two parallel straight currents I flowing in the + z and – z directions respectively. Find the equations for the equipotential and evaluate B. Show that B satisfies B · grad* $A_z = 0$ *and thus the lines of B are in the direction of* A_z = *constant.*

Solution: The vector potential due to two parallel infinite straight currents I flowing in the ± z directions can be written as

$$A = A_z u_z = \frac{\mu_0 I}{2\pi} \ln \frac{r_2}{r_1}$$

Where, $r_1^2 = (x - a)^2 + y^2 \; ; r_2^2 = (x + a)^2 + y^2$

$$\therefore \quad A = A_z u_z = \frac{\mu_0 I}{2\pi} \ln \left\{\frac{(x + a)^2 + y^2}{(x - a)^2 + y^2}\right\}^{1/2}$$

$$B = \nabla \times A = \frac{\partial A_x}{\partial y} u_x - \frac{\partial A_z}{\partial x} u_y = B_x u_x + B_y u_y$$

$$\therefore \quad B_x = \frac{\partial A_z}{\partial y} = \frac{\mu_0 I}{4\pi} \frac{(x - a)^2 + y^2}{(x + a)^2 + y^2} \times \frac{2y\{(x - a)^2 + y^2\} - 2y\,\{(x + a)^2 + y^2\}}{\{(x - a)^2 + y^2\}^2}$$

$$= \frac{\mu_0 I}{4\pi} \frac{(-8xya)}{(r_1 r_2)^2}$$

$$B_y = -\frac{\partial A_z}{\partial x} = -\frac{\mu_0 I}{4\pi}\frac{(x-a)^2+y^2}{(x+a)^2+y^2} \times \frac{2(x+a)\{(x-a)^2+y^2\} - 2(x-a)/\{(x+a)^2+y^2\}}{\{(x-a)^2+y^2\}^2}$$

$$= \frac{\mu_0 I}{4\pi}\frac{4a(x^2-a^2)+y^2}{(r_1 r_2)^2}$$

$$\text{grad } A_z = \nabla A_z = \frac{\partial A_z}{\partial y} u_x + \frac{\partial A_z}{\partial y} u_y$$

$$\therefore \quad B \cdot \text{grad } A_z = \left\{\frac{\partial A_z}{\partial y} u_x + \left(-\frac{\partial A_z}{\partial x}\right) u_y\right\}\left\{\frac{\partial A_z}{\partial x} u_x + \frac{\partial A_z}{\partial y} u_y\right\} = 0$$

Hence, B and grad A_z are orthogonal, *i.e.*, mutually perpendicular. In other words B is at right angles to the direction of maximum rate of change of A_z.

7.4 SELF AND MUTUAL INDUCTANCE

While defining inductance we need to introduce the concept of flux linkage. Whenever current flowing in a coil changes, an e.m.f. is induced in the coil and direction of induced e.m.f. is always such that it opposes the very cause producing it. If current is increasing, the induced e.m.f. acts to oppose this increase. If however current is decreasing, the e.m.f. acts to oppose this decrease. The property of the coil by which it opposes changes of current increase or decrease is called as inductance and we define inductance as ratio of total flux linkages to current that they link which is given by

$$L = \frac{N\phi}{I}$$

But if two coils are placed side by side such that flux produced by a current in first coil links the second coil, the coils are said to be magnetically coupled. If the current in first coil is changing, there is an e.m.f. induced in the second coil.

Let I_1 be current in first coil of N_1 turn and let N_2 be number of turns of second coil.

If current I_1 in first coil sets up magnetic flux ϕ_1 and all this flux links the second coil we have flux linkage of second coil = $N_2\phi_1$ Wb-turns, current in the first coil producing flux $\phi_1 = I_1$ amperes. The ratio $\frac{N_2\phi_1}{I_1}$ is termed as mutual inductance and M_{12}.

Now, $$M_{12} = \frac{N_2\phi_1}{I_1} \text{ H} \qquad ...(7.13)$$

And $$M_{21} = \frac{N_1\phi_2}{I_2} \text{ H} \qquad ...(7.14)$$

Example 7.4: *Current in a coil is increased from zero to 15 amps at a uniform rate in 6 seconds. It is found that this coil develops self-induced e.m.f. of 150 volts whereas an emf of 25 volt is produced in a neighbouring coil. Compute self-inductance of the first coil and the mutual inductance between the two coils.*

Solution: The self-induced e.m.f., $e = L\dfrac{di}{dt}$

The rate of change of current in the first coil, $\dfrac{di}{dt} = \dfrac{15-0}{6} = 2.5$ A/s

$\therefore$ The self-inductance of the first coil $L = \dfrac{e}{(di/dt)} = \dfrac{150}{2.5} = 60$ Henrys

The e.m.f. induced in the second coil, $e_2 = M\dfrac{di_1}{dt}$

$\therefore$ The mutual inductance between the two coils is

$$M = \frac{e_2}{\left(\dfrac{di_1}{dt}\right)} = \frac{25}{2.5} = 10 \text{ Henrys.}$$

Example 7.5: *Two coils having the same self-inductance are connected in series. When a current I flows through the coils, the magnetic energy stored in their fields is W joules. If the connection of one coil are interchanged and the current is reduced to* $\left(\dfrac{1}{2}\right)$ *I, the energy stored is again W joules. Calculate the ratio of the mutual inductance and the self-inductance.*

Solution: The effective inductances of the circuit, for the two connections are:

$$L_A = L + L - 2M$$
$$= 2L - 2M$$

and

$$L_B = L + L + 2M$$
$$= 2L + 2M$$

$\therefore$ Stored energy in the circuit 1, $W_A = \dfrac{1}{2}L_A I_1^2$

and the stored energy in the circuit 2, $W_B = \dfrac{1}{2}L_B I_2^2$

Now, we are given that $I_2 = \dfrac{1}{2}I_1 = \dfrac{1}{2}I$ and $W_A = W_B$

$$\therefore \quad \frac{1}{2}L_A I^2 = \frac{1}{2}L_B\left(\frac{1}{2}I\right)^2$$

$$\therefore \quad \frac{1}{2}L_A = \frac{1}{8}L_B \text{ or } L_A = \frac{1}{4}L_B$$

It can be written as $2(L - M) = \left(\dfrac{1}{4}\right)2(L + M)$

$\therefore$ $4(L - M) = L + M$ or $3L = 5M$

$$\therefore \quad \frac{M}{L} = \frac{3}{5}.$$

Example7.6: (*a*) *Two coils are wound on a common magnetic circuit of 55 sq cm in cross section and mean length of 120 cm.One coil has 90 turns and the other 600. Calculate the mutual inductance of the coils, if the relative permeability of the iron path is 1400. (b) If a current in the 90 turns coil is changing from 4 A to zero in 0.01 second, find the induced e.m.f. in the second coil.*

Solution:

(*a*) It is given that, $N_1 = 90, N_2 = 600$

Now $$\mu = \mu_r\mu_0 = 1400 \times 4\pi \times 10^{-7}$$
$$l = 120 \times 10^{-2} \text{ and } A = 55 \times 10^{-4}$$

Mutual inductance of the coils,

$$L_{21} = N_1N_2\frac{\mu A}{l}$$
$$= \frac{90 \times 600 \times 1400 \times 4\pi \times 10^{-7} \times 55 \times 10^{-4}}{120 \times 10^{-2}}$$
$$= 435.204 \text{ mH.}$$

The induced e.m.f. in the second coil is given by

$$V_2 = L_{21}\frac{dI}{dt}$$
$$= 435.204 \times 10^{-3}\left(\frac{4}{0.01}\right) = 174.08 \text{ volts.}$$

7.5 NEUMANN'S FORMULAE

Consider 2 closed loops C_1 and C_2 of thin wire carrying currents I_1 and I_2 respectively as shown in Fig. 7.1.

Surfaces corresponding to loop C_1 and loop C_2 are S_1 and S_2 respectively. Let the loops C_1 and C_2 be separated by a distance of '*r*' meters.

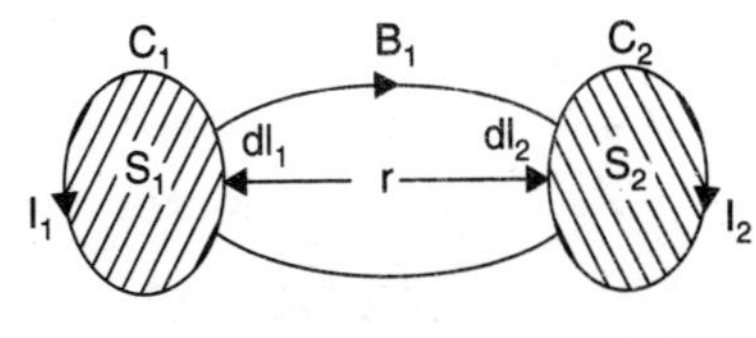

Fig. 7.1

The magnetic field B, produced by I_1 is given by

$$B_1 = \nabla \times A_1 \qquad ...(7.15)$$

We know that vector magnetic potential is given by

$$A_1 = \frac{\mu}{4\pi}\oint_{C_1}\frac{I_1 dl_1}{r} \qquad ...(7.16)$$

Let ϕ_{21} be the flux linked by coil 2 due to change of current in coil 1, therefore

$$\phi_{21} = B_1 \cdot ds_2$$

Therefore, the total flux linkage through loop C_2 is

$$\psi_{21} = \int_{S_2} B_1 \cdot ds_2$$
$$= \int_{S_2} (\nabla \times A_1) \cdot ds_2 \qquad ...(7.17)$$

By Stoke's theorem

$$\int_{S_2} (\nabla \times A_1) \cdot ds_2 = \oint_{C_2} A_1 \cdot dl_2 \qquad ...(7.18)$$

$$\therefore \qquad \psi_{21} = \oint_{C_2} A_1 \cdot dl_2 \qquad ...(7.19)$$

Where, dl_2 is element of length along loop C_2

Substituting equation (7.19) in equation (7.20), we get

$$\psi_{21} = \oint_{C_2} \frac{\mu}{4\pi} \oint_{C_1} \frac{I_l dl_1}{r} \cdot dl_2$$

$$= \frac{\mu I_1}{4\pi} \oint_{C_2} \oint_{C_1} \frac{dl_1 dl_2}{r} \qquad ...(7.20)$$

From definition of mutual inductance, we obtain

$$M_{21} = \frac{\psi_{21}}{I_1} = \frac{\mu}{4\pi} \oint_{C_2} \oint_{C_1} \frac{dl_1 dl_2}{r} \qquad ...(7.21)$$

Similarly,

$$M_{12} = \frac{\psi_{12}}{I_2} = \frac{\mu}{4\pi} \oint_{C_2} \oint_{C_1} \frac{dl_1 dl_2}{r} \qquad ...(7.22)$$

$dl_1\, dl_2$ is positive if both loops are oriented in same direction, otherwise $dl_1\, dl_2$ is negative.

7.6 SELF INDUCTANCE OF SOLENOID

Consider a solenoid of N turns carrying a current of I amps as shown in Fig. 7.2.

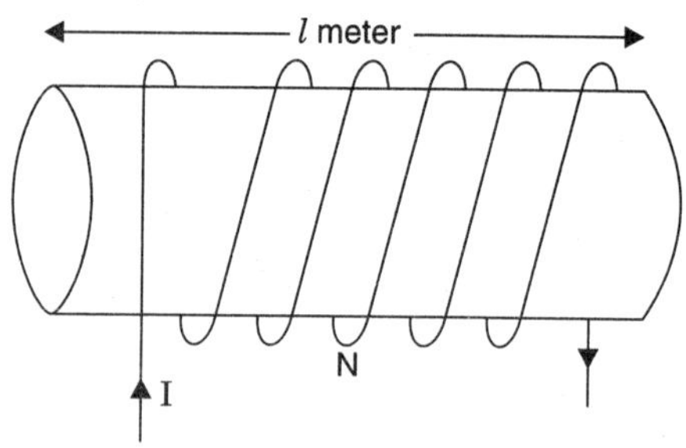

Fig 7.2

Let total flux be ϕ webers, l = length of solenoid and a = cross-sectional area.

According to amperes circuital law

$$\oint H \cdot dl = NI \qquad ...(7.23)$$

$$H \cdot \oint dl = NI$$

$$Hl = NI$$

$$H = \frac{NI}{l}$$

We have flux linkages of coil = $N\phi$ Wb-turns ...(7.24)

But $\phi = a\,B$, where B is the flux density

$$\phi = a\,\mu\,H \qquad ...(7.25)$$

Substituting equation (7.24) in equation (7.25), we get

Flux linkage of coil, $\phi = Na\mu H$

Self inductance (L) $= \dfrac{N\phi}{I} = \dfrac{Na\mu H}{I}$

But, $H = \dfrac{NI}{l}$

$$\therefore \qquad L = \frac{Na\mu}{I}\left(\frac{NI}{l}\right) = \frac{N^2 a\mu}{l} = \frac{N^2 a\mu_0\mu_r}{l}$$

Inductance of the solenoid, $L = \dfrac{N^2 a\mu_0\mu_r}{l}$ Henry. ...(7.26)

7.7 SELF INDUCTANCE OF TOROID

A toroid is a circular cylinder bent into the form of a closed ring. When current I is passed through the coil wound round the toroid, magnetic flux is set up and all the flux produced is confined to body of ring, so that mean circumference of ring may be taken as length of magnetic circuit.

We know that $\oint dL = 2\pi R$ ...(7.27)

Where, R is the mean radium

$$\oint H \cdot dl = NI \quad \text{(Since, } N = \text{number of turns)}$$

$$H\,(2\pi R) = NI$$

$$\therefore \qquad H = \frac{NI}{2\pi R} \text{ A/m} \qquad ...(7.28)$$

Now, $B = \dfrac{\mu NI}{2\pi R}$

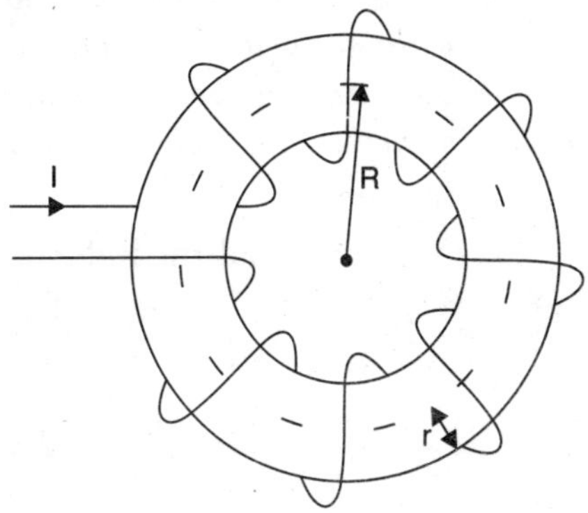

Fig 7.3

Flux linkage of toroid = $N\,\phi$ Wb-turns

Flux, $\phi = a\,B$

$$= a\left(\frac{\mu NI}{2\pi R}\right) = \frac{\mu NIa}{2\pi R}$$

Flux linkage, N ϕ $= N\dfrac{\mu NIa}{2\pi R} = \dfrac{N^2\mu Ia}{2\pi R}$ ...(7.29)

Self-inductance, $L = \dfrac{N\phi}{I} = \dfrac{N^2\mu Ia}{2\pi RI} = \dfrac{N^2\mu a}{2\pi R}$

Substituting $a = \pi r^2$ in the above equation, we get

$$L = \frac{N^2\,\mu(\pi r^2)}{2\pi R} = \frac{\mu N^2 r^2}{2R}$$

Where, r = radius of cross-section

$$L = \frac{\mu N^2 r^2}{2R} \qquad ...(7.30)$$

Example 7.7: *Calculate the inductance of (a) A 20 m length of co-axial cube filled with material for which ε_R =16, $\sigma = 0$ and $\mu_r = 70$ and having dimensions of $a = 2$ mm, $b = 6$ mm (b) A solenoid 9 cm is length, 4 cm is radius having $\mu_r = 130$ and carrying 1000 turns of wire.*

Solution:

(*a*)
$$L = \frac{\mu_0 \mu_r}{2\pi} \ln\left(\frac{b}{a}\right) \text{ H/m}$$

$$= \frac{4\pi \times 10^{-7} \times 70}{2\pi} \ln\left(\frac{6}{2}\right) = 15.3805 \ \mu \text{ H/m}$$

For 20 m length, inductance $L = 15.3805 \times 20 = 307.61$ μH

(*b*)
$$L = \frac{\mu_0 \mu_r N^2 s}{d} = \frac{4\pi \times 10^{-7} \times 130 \times (1000)^2 [\pi(4 \times 10^{-2})^2]}{9 \times 10^{-2}} = 9.12 \text{ H.}$$

Example 7.8: *A toroidal coil of 400 turns is wound on a steel ring of 0.4 m mean diameter and 3 × 10⁻³ m² cross-sectional area. An excitation of 4500 Am⁻¹ produces a flux density of 1 tesla. Find the inductance of the coil. If a 15 mm long gap is cut in the ring, find the current required to maintain the flux density at 1 tesla. Also find the inductance under these new conditions. Neglect all leakage and fringing.*

Solution:

(*a*) We know that the inductance of the coil is given by

$$L = \frac{N\phi}{I} = \frac{N}{I} B \times \text{area}$$

$$= \frac{N}{I} \times \mu H \times \text{area} = \frac{N}{I} \times \frac{\mu NI}{2\pi r} \times \text{area}$$

$$= \frac{\mu N^2}{2\pi r} \times \text{area} \qquad \left(\text{where, } \mu = \frac{B}{H} = \frac{1}{4500} = 0.22 \times 10^{-3} \text{ H/m}\right)$$

$$\therefore \quad L = \frac{0.22 \times 10^{-3} \times 400 \times 400}{2\pi \times 0.2} \times 3 \times 10^{-3} = 84.07 \text{ mH}$$

(*b*) Now $\quad NI = H_i l_i + H_a l_a,$

Where, suffix '*i*' stands for iron path and suffix '*a*' stands for air path.

$$\therefore \quad NI = \frac{B_i}{\mu_i} l_i + \frac{B_a}{\mu_a} l_a$$

Where, $\quad l_i = 0.4 \times \pi$ m, $l_a = 15 \times 10^{-3}$ m

$$\therefore \quad NI = \frac{1 \times 0.4 \times \pi}{0.22 \times 10^{-3}} + \frac{1 \times 15 \times 10^{-3}}{4\pi \times 10^{-7}}$$

$$I = 44.129 \ A$$

(*c*)
$$L = \frac{N\phi}{I} = \frac{N \times B \times A}{I} = \frac{400 \times 1 \times 3 \times 10^{-3}}{44.129} = 27.193 \text{ mH.}$$

Example 7.9: *Calculate the inductance of a 450 turn coil wound on a toroidal core having an outer diameter of 25 cm, mean diameter of 15 cm, a square cross-section and permeability of 100. What error will be introduced by assuming that the magnetic flux density was equal to the flux density at the mean diameter multiplied by the area?*

Solution: *H* at the radius '*r*' is given by

$H \cdot 2\pi r = NI = 450\ I$, the direction of H being circumferential.

$$\therefore \quad H_\phi = \frac{450I}{2\pi r} = \frac{NI}{2\pi r}$$

and
$$B_\phi = \frac{\mu NI}{2\pi r}, \mu = \mu_0\mu_r$$

$$\therefore \quad \text{Flux linked per turn} = \int_{r=2}^{r=3} B_\phi\, dr = \frac{\mu NIa}{2\pi} \ln r\Big]_2^3$$

$$= \frac{\mu NIa}{2\pi} \ln \frac{3}{2}$$

$$\therefore \quad \text{Flux linkages in 450 turns(= } N \text{ turns)}, \phi = \frac{\mu N^2 Ia}{2\pi} \ln \frac{3}{2}$$

$$\therefore \quad \text{Inductance,} \qquad L = \frac{\phi}{I} = \frac{\mu N^2 a}{2\pi} \ln \frac{3}{2}$$

Example7.10: *Find the mutual inductance between two toroidal windings which are closely wound on iron core of relative permeability 900. The mean radius of the core is 5 cm and radius of its cross-section is 5 cm. Each winding has also 800 turns.*

Solution: It is given that, $N_1 = N_2 = 800$; $\mu = \mu_r\mu_0 = 900 \times 4\pi \times 10^{-7}$

$$r = 5 \times 10^{-2} ; l = 2\pi r = 2\pi \times 5 \times 10^{-2}$$

and,
$$r = 5 \times 10^{-3} ; A = \pi r^2 = \pi(5 \times 10^{-3})^2$$

The mutual inductance is

$$L_{21} = N_1 N_2 \frac{\mu A}{l}$$

Putting the given values, we have

$$L_{21} = \frac{800 \times 800 \times 900 \times 4\pi \times 10^{-7} \times \pi(5 \times 10^{-3})^2}{2\pi \times 5 \times 10^{-2}}$$

$$= \frac{64 \times 18 \times \pi \times 25 \times 10^{-7}}{5 \times 10^{-2}} = 3.14 \times 576 = 10^{-4} = 0.1809\text{H}.$$

7.8 INDUCTANCE OF TRANSMISSION LINES

7.8.1 Inductance of Co-axial Transmission Line with Conducting Cylinders

Let us consider a co-axial transmission line with conducting cylinders of radius '*a*' (internal) and '*b*' (external) shown in Fig. 7.4.

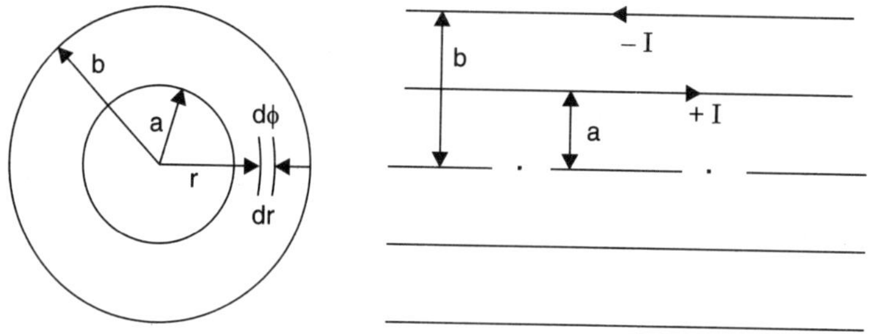

Fig. 7.4

Let 'b' be the inner radius of the outer conductor and let $+I$ and $-I$ be the currents in the inner and the outer conductors. Flux density at any point 'r' from a transmission line is given by

$$B = \frac{\mu_0 I}{2\pi r}$$

Consider a meter length of the line and radial thickness 'dr' at distance 'r' from the inner conductor axis. Flux passing through an area $(1 \times dr)$ is given by

$$d\phi = B \times (1 \times dr) = \frac{\mu_0 I dr}{2\pi r}$$

Hence, total flux linkage up to a large finite distance S is given by

$$\lambda_1 = \int_a^s d\phi = \int_a^s \frac{\mu_0 \, I dr}{2\pi r} = \frac{\mu_0 I}{2\pi} \ln \frac{s}{a} \qquad \text{...(7.31)}$$

The flux linkage set up by $-I$ flowing in the outer conductor is

$$\lambda_2 = \int_b^s d\phi = \int_b^s \frac{\mu_0 \, I}{2\pi r} = \frac{-\mu_0 I}{2\pi} \ln \frac{s}{b} \qquad \text{...(7.32)}$$

Adding equation (7.31) and equation (7.32)

$$\lambda = \lambda_1 + \lambda_2 = \frac{\mu_0 I}{2\pi} \ln \frac{b}{a} \qquad \text{...(7.33)}$$

Therefore inductance per meter length of the co-axial transmission line is

$$L = \frac{\lambda}{I} = \frac{\mu_0}{2\pi} \ln \frac{b}{a} \text{ H/m} \qquad \text{...(7.34)}$$

7.8.2 Inductance of Co-axial Line with Solid Inner Conductor

This is similar to the **section.7.8.1,** but a solid conductor of radius 'a' is incorporated within. Assuming the current I is uniformly distributed over the whole cross-section, the field intensity at any radius $(r \leq a)$ is given by

$$H' = \frac{I'}{2\pi R} = \frac{1}{2\pi r} \times \left(I \times \frac{r^2}{a^2} \right) = \frac{Ir}{2\pi a^2}$$

Consequently, B within the conductor at any radial distance 'r' from the axis is given by

$$B = \frac{\mu_0 I r}{2\pi a^2} \quad \text{(for non-magnetic conductors)}$$

$$= \frac{\mu_0 \mu_r I r}{2\pi a^2} \text{ (for Ferromagnetic conductors) where } (0 \le r \le a)$$

At distance 'r' $$B = \frac{\mu_0 I r}{2\pi a^2}$$

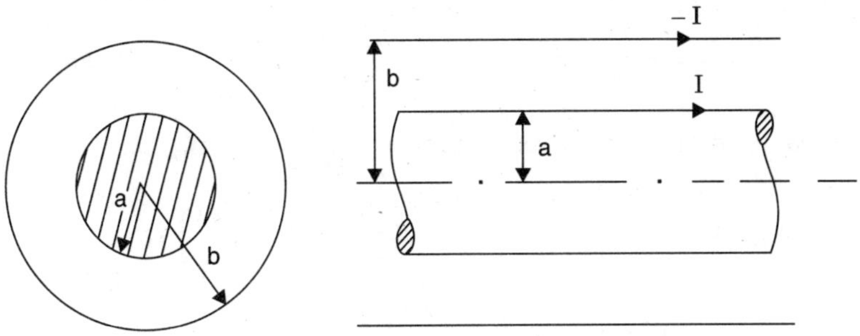

Fig. 7.5

For meter length, $$d\phi = B \times (1 \times dr) = Bdr = \frac{\mu_0 I r}{2\pi a^2} dr$$

This links with the current, $I' = I \times \dfrac{r^2}{a^2}$

$$d\lambda' = \frac{\mu_0 \mu_r I r}{2\pi a^2} \times \frac{r^2}{a^2}\, dr = \frac{\mu_0 \mu_r I r^3}{2\pi a^4}\, dr$$

$$\lambda' = \int d\lambda' = \int_0^a \frac{\mu_0 \mu_r I r^3\, dr}{2\pi a^4} = \frac{\mu_0 \mu_r I}{8\pi} \quad \text{...(7.35)}$$

The value of λ between $r = a$ to $r = b$ is

$$\lambda = \frac{\mu_0 I}{2\pi} \ln \frac{b}{a} \quad \text{...(7.36)}$$

So, the total flux linkage can be obtained by adding equation (7.35) and equation (7.36)

$$\lambda' + \lambda = \frac{\mu_0 \mu_r I}{8\pi} + \frac{\mu_0 I}{2\pi} \ln \frac{b}{a}$$

$$= \frac{\mu_0 I}{4\pi} \left(\frac{1}{2} \mu_r + 2 \ln \frac{b}{a} \right)$$

Therefore, the inductance of the co-axial line is given by the expression

$$L = \frac{\lambda' + \lambda}{I} = \frac{\mu_0}{4\pi} \left(\frac{\mu_r}{2} + \ln \frac{b}{a} \right) \quad \text{...(7.37)}$$

7.9 TWO WIRE TRANSMISSION LINE

Consider two conductors A and B of radii 'a' and 'b' respectively and with spacing 'D' as shown in Fig. 7.6.

Assume $D >> a$ or b. Consider a meter length of the transmission line. The external flux linkage with the conductor 'A' is

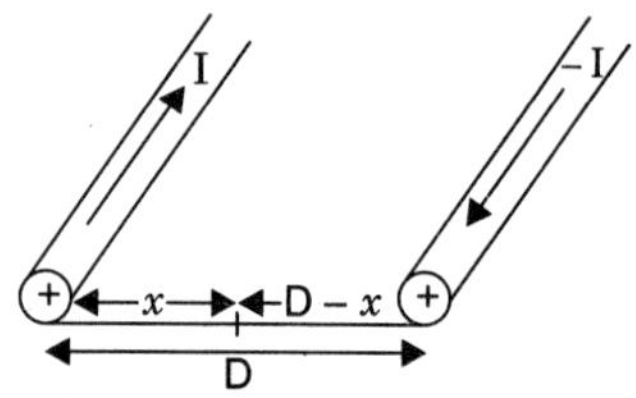

Fig. 7.6

$$\lambda_A = \frac{\mu_0 I}{2\pi r} \ln \frac{D}{a}$$

$$\therefore \quad L_A = \frac{\lambda_A}{I} = \frac{\mu_0}{2\pi} \ln \frac{D}{a}$$

The inductance due to internal flux linkage $= \frac{\mu_0 \mu_r}{8\pi} = \frac{\mu_0}{8\pi}$, if $\mu_r = 1$

$\therefore$ Total inductance of conductor A, $L_A = \left(\frac{\mu_0 \mu_r}{8\pi} + \frac{\mu_0}{2\pi} \ln \frac{D}{a} \right)$ H/m

$$= \frac{\mu_0}{4\pi} \left(\frac{\mu_r}{2} + 2 \ln \frac{D}{a} \right) \text{ H/m}$$

Similarly, $$L_B = \frac{\mu_0}{4\pi} \left(\frac{\mu_r}{2} + 2 \ln \frac{D}{a} \right) \text{ H/m}$$

So, the total loop inductance is given by the expression

$$L_A + L_B = \frac{\mu_0}{4\pi} \left[\mu_r + 2 \ln \frac{D}{a} + 2 \ln \frac{D}{b} \right] = \frac{\mu_0}{4\pi} \left[1 + 4 \ln \frac{D}{\sqrt{ab}} \right], \text{ if } \mu_r = 1$$

If $a = b = r$, then $L = L_A + L_B = 0.4 \ln \frac{D}{r'}$ mH/km $\left(\text{where } r' = e^{-\frac{1}{4}} r \right)$

7.10 MUTUAL INDUCTANCE BETWEEN A STRAIGHT LONG WIRE AND A SQUARE LOOP WIRE IN SAME PLANE

Consider a straight long conductor which carries a current I in the positive y-direction as shown in Fig. 7.7. The square loop is also shown in the same plane. The flux density at any point P in the x-y plane at a distance 'x' from the wire is given by expression

$$B = \frac{\mu_0 I}{2\pi x} \qquad \text{...(7.38)}$$

B is in a direction perpendicular to x-y plane and increased as shown in figure. Mutual inductance is given by flux linkage with loop per unit current in the straight wire.

If ψ is the flux linkage with the coil and I be the current in the wire, then

$$M = \frac{\psi}{I} = \frac{\lambda}{I}$$

Now, flux linkage $\lambda = \psi = \int_S \text{B} \cdot ds \qquad \text{...(7.39)}$

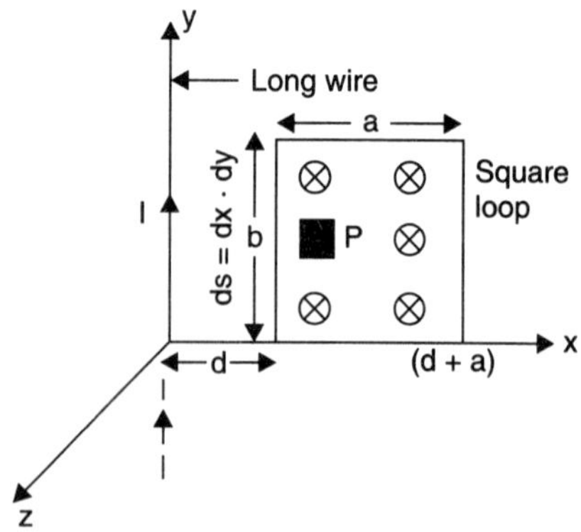

Fig. 7.7

Where, ds is surface element around any point $P(x, y)$ in the loop, plane surface given by $ds = dx\,dy$.

As the flux density $\overline{B}$ is a vector normal to plane of coil, we can write equation (7.39) as

$$\psi = \int_S \int \mu_0 \frac{I}{2\pi x}\, dx\, dy$$

$$= \frac{\mu_0 I}{2\pi} \int_{y=0}^{b} \int_{x=d}^{d+a} \frac{1}{x}\, dx\, dy$$

$$= \frac{\mu_0 I b}{2\pi} \int_{x=d}^{d+a} \frac{1}{x}\, dx = \frac{\mu_0 I b}{2\pi} \ln\left(\frac{d+a}{d}\right)$$

$$\therefore \quad M = \frac{\psi}{I} = \frac{\mu_0 b}{2\pi} \ln\left(\frac{d+a}{d}\right) \text{ henrys} \qquad \text{...(7.40)}$$

7.11 ENERGY STORED IN MAGNETIC FIELD

Just as energy is stored in electrostatic field, energy is also stored in magnetic field of inductor. Now work must be done in order to force a current through an inductor, since inductor opposes changes in current magnitude owing to its inherent property of inductance.

Consider a coil of self-inductance L Henry, let it be required to establish a current I through coil. As current increases gradually from zero, the flux linkages and e.m.f. induced is given by the following expression.

$$e = \frac{N d\phi}{dt} \qquad \text{...(7.41)}$$

$$= N \frac{d}{dt}\left(\frac{\phi}{i} \cdot i\right)$$

$$= \frac{N\phi}{i} \frac{di}{dt} \qquad \left(\text{Since } \frac{\phi}{i} \text{ is fixed magnitude}\right)$$

$$e = L \frac{di}{dt} \qquad \text{...(7.42)}$$

Let differential work done in time dt be dw

So, $\quad dw = e\, i\, dt$ joules

$$= L \frac{di}{dt}\, i\, dt$$

$$= Li\, di \qquad \text{...(7.43)}$$

Total work done in establishing current I is

$$W_M = \int dw = \int_0^I Li\, di$$

$$= \frac{1}{2} LI^2 \qquad \text{...[7.43(a)]}$$

7.12 ENERGY DENSITY

The above expression *i.e.*, equation [7.43(*a*)] is the energy stored in magnetic field of inductor. Assume that this energy is evenly distributed throughout the magnetic field.

Consider an elemental volume dv inside magnetic field. It may be volume of small cube of side Δl such that $\Delta v = (\Delta l)^3$, let ΔI be elemental current associated with cube.

Energy associated with this volume is

$$\Delta W_M = \frac{1}{2} \Delta L \, (\Delta I)^2 \text{ joules} \quad ...(7.44)$$

$$H = \frac{\Delta I}{\Delta l} \qquad \left(\text{since, } H = \frac{NI}{l} = \frac{I}{l} = \frac{\Delta I}{\Delta l} \text{ when } N = 1\right)$$

Also inductance of small cube $\Delta L = \mu \Delta l$

Putting these values in equation (7.44), we get

$$\Delta W_M = \frac{1}{2} \mu H^2 (\Delta l)^3$$

$$= \frac{1}{2} \mu H^2 \Delta v \quad ...(7.45)$$

Energy density is given by $\underset{\Delta v \to 0}{lt} \dfrac{\Delta W_M}{\Delta v}$

Magnetic energy density is $W_M = \dfrac{1}{2} \mu H^2$ J/m³. ...(7.46)

Example7.11: *Determine energy density in free space by field: (a)* $H_A = 15^3\overline{U}_x$ *A/m (b)* $H_B = [15^3U_x + 15^3U_y]$ *A/m (c)* $H_C = [15^3U_x - 15^3U_y]$ *A/m (d)* $H_B + H_C$ *(e)* $B = 15^{-3}U_x + 15^{-3}u_y$.

Solution:

(*a*) Energy density $= \dfrac{1}{2} \mu_0 H^2$

$$= \frac{1}{2} \mu_0 \times 15^6 = 7.1569 \text{ J/m}^3.$$

(*b*) Energy density $= \dfrac{1}{2} \mu_0 H^2$

$$= \frac{1}{2} \mu_0 [\sqrt{(15^3)^2 + (15^3)^2}\,]^2 = 14.31388 \text{ J/m}^3$$

(*c*) Energy density $= \dfrac{1}{2} \mu_0 [\sqrt{(15^3)^2 + (-15^3)^2}\,]^2$

$$= 14.31388 \text{ J/m}^3.$$

(*d*)
$$H_B + H_C = 15^3U_x + 15^3U_y + 15^3U_x - 15^3U_y$$
$$= 6750U_x \text{ A/m.}$$

Energy density $= \dfrac{1}{2} \mu_0 (6750)^2 = 28.627$ J/m³.

(*e*) Energy density $= \dfrac{B^2}{2\mu_0} = \dfrac{\sqrt{(15^{-3})^2 + (15^{-3})^2}}{2\mu_0} = 0.069862 \text{ J/m}^3$.

Example7.12: *Conducting planes in air at Z = 0 and Z = d carry surface currents of* $\pm k_0$ $\overline{U}_x$ *A/m (a) Find the energy stored in the magnetic field per unit lengths (0 < x < 1) in a width 'w' (0 < y < w) and separation d (0 < z < d) (b) Calculate the inductance per unit length of this transmission line from* $W_H = \frac{1}{2} LI^2$ *where I is the total current in a width 'w' in either conductor (c) Calculate the total flux passing through rectangle 0 < x < 1, 0 < z < d in plane y = 0 and from this, result again find the inductance per unit length.*

Solution:

(*a*) Let $\overline{k} = k_0\overline{U}_x$ at $Z = 0$ and $\overline{k} = -k_0\overline{U}_x$ at $Z = d$

In between the conducting planes

$$\overline{H} = \overline{H}_1 + \overline{H}_2$$

$$= \frac{k_0\overline{U}_x \times \overline{U}_z}{2} + \frac{-k_0\overline{U}_x \times (-U_z)}{2}$$

$$= -0.5\, k_0\overline{U}_y - 0.5\, k_0\overline{U}_y$$

$$= -k_0\overline{U}_y \text{ A/m}$$

$$\therefore \quad W_H = \frac{\mu_0}{2}\int H^2\, dv = \frac{\mu_0}{2}\int_0^1\int_0^w\int_0^d k_0^2\, dx\, dy\, dz$$

$$= 0.5\, \mu_0\, k_0^2\, wd \text{ joule.}$$

(*b*) Current *I* in a width *w* in one conductor only = surface current density × width

$$\therefore \quad I = k_0\, w$$

$$\frac{1}{2}LI^2 = 0.5\, \mu_0\, k_0^2\, wd$$

$$\frac{1}{2}L\,(k_0 w)^2 = 0.5\, \mu_0\, k_0^2\, wd$$

or

$$L = \frac{\mu_0 d}{w} \text{ H/m.}$$

(*c*)

$$\overline{B} = \mu_0\overline{H} = k_0\, \mu_0\, \overline{U}_y$$

$$\overline{dS} = dx\, dz\, \overline{U}_y$$

$$\phi = \int \overline{B}\cdot\overline{dS} = \int K_0\, \mu_0\, dx\, dz\, \overline{U}_y$$

$$= \int_0^1\int_0^d k_0\, dx\, dz = k_0\, \mu_0\, d$$

$$L = \frac{\phi}{I} = \frac{k_0\mu_0 d}{k_0 w} = \frac{\mu_0 d}{w} \text{ H/m.}$$

Example7.13: *A toroid having a square cross-section is defined by the surface z = 0, z = 10 mm, ρ = 20 mm and ρ = 30 mm. The surface at ρ = 20 mm carries a current* $\bar{k} = 20\bar{U}_z$ *k A/m. (a) Find the surface currents on the other 3 surfaces (b) Find* $\bar{H}$ *within the toroid (c) If* μ_R *= 10 within the toroid, find total stored energy.*

Solution:

(*a*) Current on the ρ = 20 mm surface

$$= (20 \times 10^3)(2\pi \times 20 \times 10^{-3}) = 2513.27 \text{ A}$$

Current on all surfaces remain the same

$$\text{Surface current on } Z = 10 \text{ mm surface} = \frac{2513.27}{2\pi\rho} U_\rho$$

$$= \frac{400}{\rho} U_\rho \text{ A/m}$$

Surface current on ρ = 30 mm surface

$$= \frac{2513.27}{2\pi(30 \times 10^{-3})} (-U_z)$$

$$= -13{,}333\ U_z \text{ A/m}$$

$$\text{Surface current on } Z = 0 \text{ surface} = \frac{2513.27}{2\pi\rho} (-U_\rho) = \frac{-400}{\rho} U_\rho \text{ A/m.}$$

(*b*) Consider a circular path of radius ρ such that it lies between ρ = 20 mm and ρ = 30 mm. Applying Amperes circular path *i.e.*, $\oint H \cdot dL = I_{\text{encl}}$. This circular path encloses current $\bar{k}$ on the surface ρ = 20 mm which is directed in U_Z direction. Hence, H_ϕ component of *H* is produced.

$$\text{L.H.S.} = \oint H \cdot dL = \int_0^{2\pi} H_\phi U_\phi \cdot \rho\, d\phi\, U_\phi$$

$$= \int_0^{2\pi} \rho H_\phi\, d\phi = 2\pi\rho H_\phi$$

But, R.H.S. is given by the value 2513.27

$$\therefore \qquad 2\pi\rho H_\phi = 2513.27 \text{ (or) } H_\phi = \frac{400}{\rho}$$

$$\therefore \qquad \bar{H} = \frac{400}{\rho} U_\phi \text{ where } 20 \text{ mm} < \rho < 30 \text{ mm}$$

(*c*)

$$W_H = \frac{\mu_0\mu_r}{2} \int H^2\, dv$$

$$= \frac{\mu_0(10)}{2} \int_{0.02}^{0.03} \int_0^{2\pi} \int_0^{0.01} \left[\frac{400}{\rho}\right]^2 \rho\, d\rho\, d\phi\, dz$$

$$= 8 \times 10^5 \mu_0 \left[\ln \rho\right]_{0.02}^{0.03} \left[\phi\right]_0^{2\pi} \left[z\right]_0^{0.01} = 0.0512 \text{ J.}$$

Example7.14: *Define the gradient of a scalar and the curl of a vector and give their differential notations in the Cartesian notation system. The scalar magnetic potential in a region is given by* $\Omega = C\ ln\sqrt{x^2+y^2}$. *Find the magnetic vector potential which produces the same magnetic field, assuming that the z-component of the vector potential is zero everywhere.*

Solution: Normally it can be written as $\nabla\Omega = U_x\dfrac{\partial\Omega}{\partial x} + U_y\dfrac{\partial\Omega}{\partial y} + U_z\dfrac{\partial\Omega}{\partial z}$

$$\nabla\times A = U_x\left(\frac{\partial A_z}{\partial y} - \frac{\partial A_y}{\partial z}\right) + U_y\left(\frac{\partial A_x}{\partial z} - \frac{\partial A_z}{\partial x}\right) + U_z\left(\frac{\partial A_y}{\partial x} - \frac{\partial A_x}{\partial y}\right)$$

Here in this problem we assume $A_z = 0$

Now, $$H = -\nabla\Omega \text{ and } \Omega = C\ln\sqrt{x^2+y^2} = \frac{C}{2}\ln(x^2+y^2)$$

$$= -U_x\frac{\partial\Omega}{\partial x} - U_y\frac{\partial\Omega}{\partial y} \text{ and } \frac{\partial\Omega}{\partial z} = 0$$

$\therefore$ $$B_x = -\mu\frac{\partial\Omega}{\partial x} = -\frac{\mu C}{2}\frac{2x}{x^2+y^2};\ B_y = -\mu\frac{\partial\Omega}{\partial y} = -\frac{\mu C}{2}\frac{2y}{x^2+y^2}$$

Also, $$B_x = \nabla\times A = U_xB_x + U_yB_y = U_x\left(-\frac{\partial A_y}{\partial z}\right) + U_y\left(-\frac{\partial A_x}{\partial z}\right)$$

$\therefore$ $$B_x = -\frac{\partial A_y}{\partial z} = -\frac{\mu Cx}{x^2+y^2} \text{ and } A_y = \frac{\mu Cxz}{x^2+y^2} + C_1(x, y)$$

$$B_y = -\frac{\partial A_x}{\partial z} = -\frac{\mu Cy}{x^2+y^2} \text{ and hence } A_x = \frac{\mu Cyz}{x^2+y^2} + C_2(x, y)$$

And also, $$B_Z = \frac{\partial A_y}{\partial x} - \frac{\partial A_x}{\partial y} = 0$$

$\therefore$ From the above equation it can be written as $\dfrac{\partial A_y}{\partial x} = \dfrac{\partial A_x}{\partial y}$

Hence, $$A_x = \frac{\mu Cyz}{x^2+y^2} \text{ and } A_y = \frac{\mu Cxz}{x^2+y^2}$$

Example7.15: *A solenoid is wound on a long, square section and containing no magnetic material. It is bent found into a toroid of internal and external radii a and b, respectively. A straight thin cable of infinite length passes along the axis of the toroid at right angles to its plane. Show that the mutual inductance between the cable and the solenoid is* $M = \mu_0 n\dfrac{b^2-a^2}{2}\ln\left(\dfrac{b}{a}\right)$ *henry, where n is the mean number of turns per metre on the solenoid.*

Solution: Due to the infinite long cable, at a point P, $B = \dfrac{\mu_0 I}{2\pi r}$

$\therefore$ Flux linked by the cross-section of the toroid (per turn):

$$\lambda = \int_z^b \frac{\mu_0 I}{2\pi r}(b-a)dr = \frac{\mu_0 I(b-a)}{2\pi}\ln r\Big]_a^b = \frac{\mu_0 I(b-a)}{2\pi}\ln\left(\frac{b}{a}\right)$$

Total number of turns in the toroid = $n \times$ mean length of the toroid

$$= n \times 2\pi\, r_{mean} = 2\, \pi n\, \frac{a+b}{2}$$

$$= \pi n\, (a + b)$$

$\therefore$ Total flux linked $= \lambda_T$

$$= \frac{\mu_0 I(b-a) \cdot \pi n(a+b)}{2\pi} \ln\left(\frac{b}{a}\right)$$

$$= \frac{\mu_0 n I(b^2 - a^2)}{2} \ln\left(\frac{b}{a}\right)$$

$\therefore$

$$M = \frac{\lambda_T}{I} = \frac{\mu_0 n(b^2 - a^2)}{2} \ln\left(\frac{b}{a}\right).$$

SHORT QUESTIONS AND ANSWERS

1. **What is meant by a toroid?**

   A toroid is a circular cylinder bent into the form of a closed ring and so when current I is passed through the coil wound round the toroid, magnetic flux is set up.

2. **What is expression for self-inductance of solenoid?**

$$L = \frac{N^2 a \mu_0 \mu_r}{l} \text{ henry}$$

3. **What is expression for self-inductance of toroid?**

$$L = \frac{\mu N^2 r^2}{2R} \text{ henry}$$

4. **Write the expression for energy density in a magnetic field.**

$$\frac{1}{2} \mu H^2 \text{ J/m}^3$$

5. **Define the term self-inductance.**

   Self inductance of a circuit is the property of the circuit by which any change in the circuit current induces an e.m.f. in the circuit to oppose the change of current.

6. **Define the term mutual inductance.**

   The property by which a changing current in one circuit produces an induced voltage in another circuit is called mutual inductance.

7. **State Faraday's law of electromagnetic induction.**

   It can be stated as 'The e.m.f. induced in a circuit is equal to negative of rate of increase of flux of B through the circuit.' *i.e.*, $V_i = \frac{-d\phi}{dt}$.

8. **What is the differential form of Faraday's law?**

$$\text{Curl } \overline{E} = \frac{d\overline{B}}{dt}$$

9. Where is the concept of vector magnetic potential extensively used?

It is used extensively in regions where current density is zero and it can also be extended to time varying cases.

10. Where are the concepts of scalar magnetic potential can't be defined?

The scalar magnetic potential can't be defined in the presence of steady currents which are sources of B.

11. What is expression for vector magnetic potential in the case of current sheet?

$$A = \int_S \frac{\mu_0 k}{4\pi R}\, ds \text{ Wb/m}$$

12. What is expression for vector magnetic potential in the case of differential current element *Idl*?

$$\text{A} = \oint \frac{\mu_0 I}{4\pi R}\, dl \text{ Wb/m.}$$

MULTIPLE CHOICES

1. Scalar magnetic potential is applicable only for magnetic field produced by
   (*a*) permanent magnets (*b*) diamagnets
   (*c*) both (*a*) and (*b*) (*d*) none of these
2. In the expression $\overline{B} = \overline{\nabla} \times \overline{A}, \overline{A}$ is called as
   (*a*) area of field (*b*) magnetic vector potential
   (*c*) scalar magnetic potential (*d*) none of these
3. In the expression $\nabla^2 A = -\mu$ J, ∇^2 is an operator known as
   (*a*) vector laplacian (*b*) scalar laplacian
   (*c*) both (*a*) and (*b*) (*d*) none of these
4. The property by which a changing current in one coil can induce an e.m.f. in other coil is called
   (*a*) self-inductance (*b*) mutual inductance
   (*c*) self-induced e.m.f. (*d*) mutual induced e.m.f.
5. The total inductance of two series connected coils of self-inductance L_1 and L_2 and mutual inductance is given as
   (*a*) $L_1 + L_2 + 4$ m (*b*) $L_1L_2 \pm M^2$
   (*c*) $L_1 + L_2 \pm 2M$ (*d*) $L_1L_2 \pm M$
6. Two coils of inductance 10 H and 20 H have a mutual inductance 6 H. The maximum and minimum inductance values available from series connection of two coils are
   (*a*) 36H, 24H (*b*) 236H, 174H
   (*c*) 42H, 18H (*d*) 212H, 188H
7. The coefficient of coupling K between the two coils is
   (*a*) $K = M\sqrt{L_1L_2}$ (*b*) $K = \sqrt{\frac{M}{L_1L_2}}$

(c) $K = \frac{M}{\sqrt{L_1 L_2}}$ (d) none of these

8. A coil is known to have twice the self-inductance of second coil. When connected in series one way, their total inductance is 100 mH and when the connections are inter changed, the total inductance is 20 mH. The self-inductance of each coil

(a) 15, 30 (b) 30, 60

(c) 20, 40 (d) none of these

9. In question No. (8), the mutual inductance between the coils is

(a) 10 mH (b) 20 mH

(c) 15 mH (d) 28 mH

10. In question No.(8), the coefficient of coupling between the two coils is

(a) 1.414 (b) 0.0707

(c) 0.5 (d) 0.707

11. The degree of magnetic coupling between magnetically coupled circuits is known as

(a) magnetic coupling coefficient (b) magnetic coupling

(c) coefficient of inductance (d) magnetic coefficient

12. A coil of inductance 2 H and resistance 1 Ω is connected to a 10 V battery having negligible resistance. The amount of energy stored in the magnetic field is

(a) 100 joules (b) 25 joules

(c) 50 joules (d) 8 joules

13. In cases where magnetic coupling coefficient K has a small value, the coils are said to be

(a) rigidly coupled (b) loosely coupled

(c) both (a) and (b) (d) none of these

14. The force of .... is experienced between two parallel conductors carrying current in opposite direction.

(a) attraction (b) repulsion

(c) zero (d) none of these

15. For a toroid, the self-inductance L =

(a) $\frac{\mu_0 N r^2}{2RM}$ (b) $\frac{\mu_0 N^2 r^2}{2RM}$

(c) $\frac{\mu_0 N^2 r^{-2}}{RM}$ (d) $\frac{\mu_0 N^2}{2RM}$

16. For a solenoid, the self-inductance L =

(a) $\frac{\mu_0 N}{l}$ (b) $\frac{\mu_0 N^2}{2l}$

(c) $\frac{\mu_0 N^{-2}}{lM}$ (d) $\frac{\mu_0 N^2}{l}$

17. A toroidal coil of 600 turns has a mean radius of 50 cm and radius for winding is 4 cm. The value of average self inductance with an iron core of $\mu_r = 900$ is

(*a*) 7.6 H (*b*) 12.2 H

(*c*) 7 H (*d*) 9.1 H

18. Scalar magnetic potential is defined for the following:

(*a*) current carrying conductors (*b*) permanent magnets

(*c*) solenoids (*d*) none of these

19. Scalar magnetic potential is a .... quantity.

(*a*) vector (*b*) scalar

(*c*) both (*a*) and (*b*) (*d*) none of these

20. Gradient of scalar magnetic potential is .... quantity.

(*a*) scalar (*b*) vector

(*c*) dimensionless (*d*) none of these

21. Negative gradient of scalar magnetic potential gives ....

(*a*) magnetic flux density (*b*) magnetic field intensity

(*c*) magnetic flux (*d*) none of these

22. Vector magnetic potential is defined for the following:

(*a*) current carrying conductors (*b*) permanent magnets

(*c*) empty conductors (*d*) none of these

23. Vector magnetic potential is a .... quantity.

(*a*) vector (*b*) scalar

(*c*) both (*a*) and (*b*) (*d*) none of these

24. Curl of vector magnetic potential is .... quantity.

(*a*) scalar (*b*) vector

(*c*) both (*a*) and (*b*) (*d*) none of these

25. Always inductor opposes change of ....

(*a*) current (*b*) flux

(*c*) both (*a*) and (*b*) (*d*) none of these

26. Scalar magnetic potential can be defined as follows:

(*a*) $\nabla \times \overline{H} = l$ (*b*) $\overline{E} = -\nabla V$

(*c*) $\overline{H} = -\nabla V_m$ (*d*) $\overline{B} = \nabla \times \overline{A}$

27. A similar definition that can be compared with the definition of scalar magnetic potential in electrostatics is ....

(*a*) $\nabla \cdot \overline{D} = \rho$ (*b*) $\overline{E} = -\nabla V$

(*c*) $\overline{H} = -\nabla V_m$ (*d*) $\overline{B} = \mu_o \overline{H}$

28. The definition of scalar magnetic potential is applicable only for these regions where ....

(*a*) $\overline{H} = 0$ (*b*) $\overline{B} = 0$

(*c*) $\overline{J} = 0$ (*d*) none of these

29. Following expression gives the limitation of scalar magnetic potential:

(*a*) $\nabla \cdot \overline{B} = 0$ (*b*) $\frac{\nabla}{\overline{H}} = \overline{J}$

(*c*) $\overline{J} = 0$ (*d*) $\overline{E} = -\nabla V$

30. Vector magnetic potential can be defined as follows:

(*a*) $\nabla \cdot \overline{H} = 1$ (*b*) $\nabla^2 v = 0$

(*c*) $\overline{B} = \nabla \times \overline{A}$ (*d*) $\overline{E} = -\nabla V$

31. The definition of vector magnetic potential is applicable for those where ....

(*a*) $\overline{B} = 0$ (*b*) $\overline{H} = 0$

(*c*) $\overline{J} \neq 0$ (*d*) $\overline{H} \neq 0$

32. Curl of vector magnetic potential gives ....

(*a*) $\overline{H}$ (*b*) $\overline{B}$

(*c*) $\overline{J}$ (*d*) none of these

33. Vector magnetic potential $|\overline{A}|$ is given by following expressions:

(*a*) $\int \frac{\vec{I}\,\vec{dl}}{4\pi R^2}\overline{u}_x$ (*b*) $\int \vec{B}\,\vec{ds}$

(*c*) $\int \frac{\mu_0 \vec{I}\,\vec{ds}}{4\pi R}$ (*d*) none of these

34. Divergence of curl of $\overline{A}$ is ....

(*a*) $\overline{H}$ (*b*) $\overline{J}$

(*c*) zero (*d*) none of these

35. $\nabla \times (-\nabla V_m) =$ .... where V_M is scalar magnetic potential.

(*a*) $\overline{H}$ (*b*) $\overline{J}$

(c) zero (*d*) none of these

KEY

1. (*a*)	**2.** (*b*)	**3.** (*b*)	**4.** (*b*)	**5.** (*c*)	**6.** (*c*)
7. (*c*)	**8.** (*c*)	**9.** (*b*)	**10.** (*d*)	**11.** (*a*)	**12.** (*a*)
13. (*b*)	**14.** (*b*)	**15.** (*b*)	**16.** (*d*)	**17.** (*c*)	**18.** (*b*)
19. (*b*)	**20.** (*b*)	**21.** (*b*)	**22.** (*d*)	**23.** (*a*)	**24.** (*a*)
25. (*c*)	**26.** (*c*)	**27.** (*b*)	**28.** (*c*)	**29.** (*c*)	**30.** (*c*)
31. (*c*)	**32.** (*b*)	**33.** (*c*)	**34.** (*c*)	**35.** (*c*)	

REVIEW QUESTIONS

1. What do we mean by magnetic vector potential?
2. What is the difference between magnetic vector potential and magnetic scalar potential?
3. Why is there no such thing as electric vector potential in the case of electrostatic fields?
4. Can you derive the Biot-Savart's law from the magnetic vector potential?
5. What do we mean by bound volume current density and bound surface current density?
6. An electric and a magnetic field are at right angles to each other. An electron moving from left to right passes through this region without experiencing any force. What are the directions of the $\overline{E}$ and $\overline{B}$ fields?
7. A proton is deflected upward when it passes through a magnetic field. What is the direction of the magnetic field?
8. Is it possible to define magnetic potential energy for a charged particle moving in a magnetic field?
9. Devise a simple statement of how a current carrying loop will orient itself in a magnetic field?

EXERCISE PROBLEMS

1. (*a*) Use energy relationship to show that internal inductance of a non-magnetic cylindrical wire of radius 'a' carrying a uniformly distributed current I is $\frac{\mu_0}{8\pi}$ H/m. (*b*) Find the internal inductance if the portion of conducts for which $\rho < C < a$ is removed.
2. An infinite filament lies along z-axis in free space and a square coil of N turns lies in the plane $y = 0$ with its corner at $(b, 0, 0)$, $(b + a, 0, 0)$, $(b + a, 0, a)$ and $(b, 0, a)$. Find the mutual inductance between filament and coil in term of a, b, N and μ_0.
3. A toroidal core is compared of a material with $\mu_r = 25$. The boundary surface are $z = 0$, $z = 0.025$, $\rho = 0.025$ and P = 0.04 M. The core is wound symmetrically with 12000 turns of wire such that H is in the U_ϕ direction. Find the inductance.
4. The ratio of inner and action conductors of a coax are 0.3 cm and 2 cm. (*a*) Find the volume enclosed between the conductors in a 1 m length of coax. (*b*) Three quarters of this volume is to have $\mu = \mu_0$, while the remaining 25 per cent has $\mu_R = 10$. Show how the magnetic material should be distributed through out the volume to obtain a maximum inductance per unit length? (*c*) Calculate L_{max}.
5. Conducting planes in air at $z = 0$ and $z = d$ carry surface currents of $\pm k_0\, a_x$ A/m, (*a*) Find the energy stored in the magnetic field per unit length $(0 < x < 1)$ in a width w $(0 < y < w)$ and separation d $(0 < z < d)$. (*b*) Calculate the inductance per unit length of this transmission line from $W_H = 0.5\, LI^2$, where I is the total current in a width w in either conductor. (*c*) Calculate the total flux passing through the rectangle $0 < x < 1$, $0 < z < d$, in the plane $y = 0$, and from this result again find the inductance per unit length.

6. The cones $\theta = 20°$ and $160°$ are conducting surfaces and carry total currents of 50 A. The currents return on a spherical conducting surface of 0.2 m radius. (*a*) Find H in the region $0 < r < 0.2$, $20° < \theta < 160°$, $0 < \phi < 2\pi$ (*b*) How much energy is stored in this region?
7. The B-H curve of a certain ferrous material is represented by the expression $B = (H/200)(0.2 + e^{-H/400})$, where He is in A/m and B is in T. If a magnetic circuit contains a 10 cm length of this material, as well as 0.2 mm air gap, assume a uniform cross-section of 3 cm^2 and find (*a*) The total mmf required to produce a flux of 360 μ Wb (*b*) The flux if the applied mmf is 500 A.T.
8. A toroid has a cross-section with an area of 5 cm^2 and an effective radius of 5 cm. The coil contains 500 turns. The core is composed of two semi circular segments, each with a circular cross-section; one is silicon steel and the other is a linear material for which $\mu_R = 250$. If there is a 0.3 mm air gap at each of the two joints, (*a*) What coil current is required to produce a flux density of 1 T in the core? (*b*) What flux density is produced if the current is 0.25 A?
9. A toroid having a square cross-section is defined by the surface $z = 0$, $z = 10$ mm, $\rho = 50$ mm and $\rho = 60$ mm. The surface at $\rho = 50$ mm carries a current $\bar{k} = 30\bar{u}_z k$ A/m. (*a*) Find the surface currents on the other 3 surface (*b*) Find $\overline{H}$ within the toroid (*c*) If $\mu_R = 10$ within the toroid, find total stored energy.
10. A straight solid wire segment carrying a current 2 A extends from A (0, 2, 1) to B (0, 5, 3) in free space. This wire is subjected to the magnetic field of an infinite current filament lying along the z-axis and carrying 45 A in the $\overline{U}_z$ direction. Find the vector torque on the wire segment about an origin at : (*a*) P_A (0, 4, 2) (*b*) P_B (0, 2, 1)?
11. Find $\vec{H}$ at $P(2, 3, 5)$ in Cartesian coordinates if there is an infinitely long current filament passing through the origin and point C. The current of 50 A is directed from origin to C, where, the location of C is : (*a*) C(0, 0, 1); (*b*) C(0, 1, 0).
12. Determine energy stored in free space by field: (*a*) $H_A = 10^3\overline{U}_x$ A/m (*b*) $H_B = (10^3U_x + 10^3U_y)$ A/m (*c*) $H_C = (10^3U_x - 10^3U_y)$ A/m (*d*) $H_B + H_C$.

CHAPTER

8

Time Varying Fields and Maxwell's Equations

8.1 INTRODUCTION

It can be observed that the basic relationships of the electrostatic and steady magnetic field have been obtained in the previous chapters and now, time varying field should be discussed. The concept of vector analysis and vector calculus is once again used here and some of the relationships derived in the previous chapters are taken unchanged and some other relationships are changed only slightly.

Two new concepts are introduced in this unit:

1. The electric field produced by a changing magnetic field.
2. Magnetic field produced by a changing electric field.

8.2 FARADAY'S LAW

Faraday proposed his law in the belief that if a current could produce a magnetic field, then a magnetic field should be able to produce a current. In terms of field we can now say that a time varying magnetic field produces an electromotive force (e.m.f.) which may establish a current in a suitable closed circuit. An electromotive force is merely a voltage that arises from conductors moving in a magnetic field (or) from changing magnetic fields. So, Faraday's law can be stated as

$$\text{e.m.f.} = -\frac{d\phi}{dt} \text{ volts}$$

The magnetic flux is that flux which passes through any and every surface whose perimeter is the closed path and $\frac{d\phi}{dt}$ is the time rate of change of this flux. If the closed path is that taken by a N-turn filamentary conductor it is often sufficiently accurate to consider the turns as coincident and let

$$\text{e.m.f.} = -N\frac{d\phi}{dt}$$

The negative sign is an indication that e.m.f. is in such direction as to produce a current whose flux if added to the original flux would reduce the magnitude of e.m.f. and ϕ is interpreted as flux passing through any one of N coincident paths

8.3 VECTOR FORM OF FARADAY'S LAW (MAXWELL'S 4th EQUATION)

We define e.m.f. as

$$\text{e.m.f.} = \oint E \cdot dL \qquad ...(8.1)$$

Here the e.m.f. is the voltage about a specific closed path. If any part of the path changes, the e.m.f. in general changes. In electrostatics the line integral leads to a potential difference, with time-varying fields, the result is an e.m.f. (or) a voltage.

Replacing E in equation (8.1) by the surface integral of B, we have

$$\text{e.m.f.} = \oint E \cdot dL = -\frac{d}{dt}\int_S B \cdot ds \qquad ...(8.2)$$

A flux density B in the direction of 'ds' and increasing with time thus produces an average value of E which is opposite to the positive direction about the closed path. We first consider a stationary path. The magnetic flux is the only time varying quantity and a partial derivative may be taken under the integral sign.

$$\text{e.m.f.} = \oint E \cdot dL = -\int_S \frac{\partial B}{\partial t} \cdot ds \qquad ...(8.3)$$

So, in order to obtain the point form of this integral equation, applying Stoke's theorem to the closed line integral, we have

$$\int_S (\nabla \times E) \cdot ds = -\int_S \frac{\partial B}{\partial t} \cdot ds$$

Where, the surface integrals may be taken over identical surfaces. The surfaces are perfectly general and may be chosen as differential

$$(\nabla \times E) \cdot ds = -\frac{\partial B}{\partial t} \cdot ds$$

and,

$$\nabla \times E = -\frac{\partial B}{\partial t} \qquad ...(8.4)$$

Equation(8.4) is one of Maxwell's four equations written in **differential** (or) **point** form. Equation (8.3) is the **integral** form of this equation and is equivalent to Faraday's law as applied to a fixed path.

Example 8.1: *In free space,* $\overline{D} = D_m \sin(\omega t + \beta z)u_x$. *Using Maxwell's equations, show that* $B = -\dfrac{\omega\mu_0 D_m}{\beta}\sin(\omega t + \beta z)u_y$.

Solution: Maxwell's equation in differential form, from Faraday's law is given by

$$\nabla \times E = -\frac{\partial B}{\partial t}$$

From the data,

$$E = \frac{D_m}{\varepsilon_0}\sin(\omega t + \beta z)u_x$$

$$\nabla \times E = \begin{vmatrix} u_x & u_y & u_z \\ \frac{\partial}{\partial x} & \frac{\partial}{\partial y} & \frac{\partial}{\partial z} \\ \frac{D_m}{\varepsilon_0} \sin(\omega t + \beta z) & 0 & 0 \end{vmatrix}$$

$$= \beta \frac{D_m}{\varepsilon_0} \cos(\omega t + \beta z) u_y = -\frac{\partial B}{\partial t}$$

On integration it can be written as $B = -\frac{\beta D_m}{\varepsilon_0} \frac{1}{\omega} \sin(\omega t + \beta z) u_y$

But $\beta = \frac{\omega}{c}; \beta^2 = \frac{\omega^2}{c^2} = \mu_0 \varepsilon_0 \omega^2$

So, it can be shown as $B = -\frac{\beta^2 D_m}{\varepsilon_0 \omega \beta} \sin(\omega t + \beta z) u_y$

$$= -\frac{\mu_0 \varepsilon_0 \omega^2 D_m}{\varepsilon_0 \omega \beta} \sin(\omega t + \beta z) u_y$$

$$= -\frac{\omega \mu_0 D_m}{\beta} \sin(\omega t + \beta z) u_y$$

8.4 INDUCED E.M.F.

Normally induced e.m.f. can be divided into two types (1) Statically induced e.m.f. (2) Dynamically induced e.m.f. In the first case, usually the conductor or the coil remains stationary and flux linked with it changes by simply increasing (or) decreasing the current producing this flux. But in second case the field is stationary and conductors cut across it.

8.4.1 Statically Induced e.m.f.

It can be sub-divided into (*a*) Self-induced e.m.f. and (*b*) Mutually induced e.m.f.

If a conductor (or) circuit is placed in a time varying field *i.e.,* a time rate of change of $\overline{B}$, then according to Faraday's law e.m.f. induced in that conductor (or) circuit is fixed. This e.m.f. is called statically induced e.m.f.

According to Faraday's lav of electromagnetic induction, the e.m.f. induced in a circuit (or) conductor due to a time-varying field is given by

$$e = \oint \overline{E} \cdot \overline{dL} = -N \frac{d\phi}{dt} \text{ volts}$$

Since the flux is alternating it can be written as $\phi = \phi_m \cos \omega t$

∴ The e.m.f. induced in a stationary conductor due to time varying field is given by

$$e = -N \frac{d\phi}{dt}$$

$$= -N \frac{d}{dt} (\phi_m \cos \omega t) = N \phi_m \sin \omega t.$$

$$E = N \phi_m \sin \omega t \text{ volts.} \qquad \text{...(8.5)}$$

8.4.2 Dynamically Induced e.m.f.

The e.m.f. which is generated if a conductor moves with a velocity 'v' through a magnetic field $\overline{B}$ is called as dynamically induced e.m.f. We know that the equation for the force on a current carrying conductor situated in a magnetic field is given by

$$\overline{F} = q(\overline{v} \times \overline{B}) \qquad \text{...(8.6)}$$

Conductor moves in a magnetic field, thus developing an e.m.f. in it given by

$$\overline{E} = \frac{\overline{F}}{q} = (\overline{v} \times \overline{B})$$

According the Faraday's laws of electromagnetic induction, e.m.f. is induced in the conductor which is given by

$$e = \int \overline{E} \cdot \overline{dL} = \int (\overline{v} \times \overline{B}) \cdot \overline{dL}$$

$$= BLv \sin \theta \text{ volts} \qquad \text{...(8.7)}$$

This equation is general expression for motional e.m.f.

8.5 DISPLACEMENT CURRENT

A capacitor (considering parallel plate capacitor) which is connected to an a.c. voltage source is shown in Fig. 8.1.

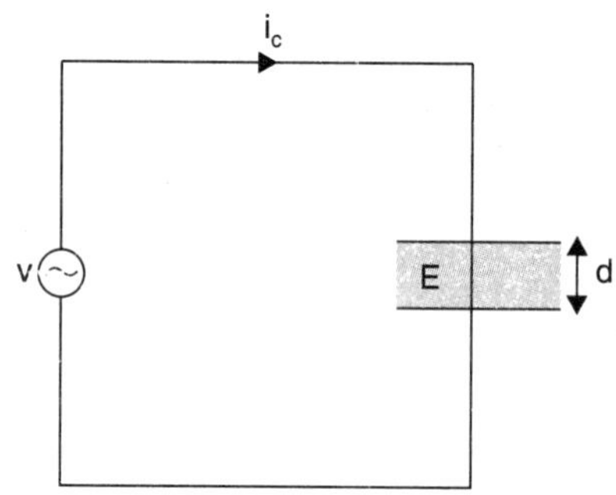

Fig. 8.1

The current through the capacitor is

$$i_c = \frac{dQ}{dt} = C \frac{dv}{dt}$$

And also we know that $C = \dfrac{\varepsilon A}{d}$

Where, A is area of parallel plates, d is the separation between the plates and ε is the permitivity of dielectric material.

$\therefore$ It can be written as $i_c = \dfrac{\varepsilon A}{d} \dfrac{dv}{dt}$

$$= \varepsilon A \frac{dE}{dt} \qquad \left[\because E = \frac{V}{d}\right]$$

Where, E is the electric field intensity

So, $$\frac{i_c}{A} = \varepsilon \frac{dE}{dt}$$

which implies $$J_d = \frac{dD}{dt} \qquad [\because D = \varepsilon E]$$

Vector D may be vary with space, so we can write as

$$J_d = \frac{\partial D}{\partial t} \quad ...(8.8)$$

The displacement current density does not represent a current which directly passes through the capacitor. The term $\frac{\partial D}{\partial t}$ has the dimensions of current density, since it results from a time-varying electric flux density. The displacement current represents a rate at which flow of voltage takes place from electrode to electrode in the external circuit.

Example 8.2: *In a material for which $\sigma = 5.5$ V/m and $\varepsilon_r = 1$, the electric field intensity is $E = 400 \sin 10^9 t$ V/m. Determine the conduction and displacement current densities and the frequency at which they have equal magnitude.*

Solution: Conduction current density, is

$$J_c = \sigma E = 5.5 \times 400 \sin 10^9 t \text{ A/m}^2$$
$$= 2200 \sin 10^9 t \text{ A/m}^2$$

Assuming that the field direction does not vary with time, we have the displacement current density,

$$J_d = \frac{\partial D}{\partial t} = \frac{\partial}{\partial t}(\varepsilon_0 \varepsilon_r 400 \sin 10^9 t) \quad (\because \; D = \varepsilon_0 \varepsilon_r E)$$
$$= 400 \times (8.854 \times 10^{-12})(10^9) \cos 10^9 t$$
$$= 3.541 \cos 10^9 t \text{ A/m}^2$$

$$\frac{\partial D}{\partial t} = \varepsilon \frac{\partial E}{\partial t}$$

If $E = E_m \sin \omega t$, then $J_d = j\,\omega\varepsilon\, E_m \cos \omega t$ and hence $J_d = \omega\varepsilon E$

for $J_c = J_d$ (in magnitude) and $\sigma = \omega\varepsilon$

$$\omega = \frac{\sigma}{\varepsilon} = \frac{5.5}{8.854 \times 10^{-12}} = 6.21 \times 10^{11} \text{ rad/sec}$$

Cyclic frequency $$f = \frac{\omega}{2\pi} = \frac{6.21}{2\pi} \times 10^{11} \text{ Hz} = 98.8 \text{ GHz}.$$

Example 8.3: *Find the frequency at which conduction current density and displacement current density are equal in a medium with $\sigma = 3 \times 10^{-4}$ mho/m and $\varepsilon_r = 75$.*

Solution: Given $\sigma = 3 \times 10^{-4}$ mho/m and $\varepsilon_r = 75$ and also conduction current density $J = \sigma E$, where, E = electric field intensity.

$$\therefore \quad J = 3 \times 10^{-4}\, E \text{ A/m}^2$$

Displacement current density $J_d = \frac{\partial D}{\partial t}$

But $$D = \varepsilon E = \varepsilon_0 \varepsilon_R E$$

Let $E = E_m \sin \omega t$, so $J_d = \frac{\partial D}{\partial t} = \frac{d}{dt}[\varepsilon_0 \varepsilon_r E_m \sin \omega t]$

$$= [\varepsilon_0 \varepsilon_r E_m \sin \omega t] \cdot \omega$$
$$= (8.854 \times 10^{-12}) \times 75 \times \omega\, E_m \cos \omega t$$

Replacing $\omega = 2\pi f$ where f = frequency in Hz

$$J_d = 4170.2 \times 10^{-12} f E_m \cos \omega t$$

Replacing $\cos \omega t = 1$, maximum value of $J_d = 4170.2 \times 10^{-12} f E_m$ A/m^2

Let given value of J be assumed as its maximum value.

If $\quad J_d = J, \quad 4170.2 \times 10^{-12} f E_m = 3 \times 10^{-4} E_m$

from which $\quad f = \dfrac{3 \times 10^{-4}}{4170.2 \times 10^{-12}} = 71.93$ kHz.

Example 8.4: *In a material for which σ = 4 siemens per metre and* $\varepsilon_r = 1$ *the electric field intensity E = 150 sin* 10^{10} *t (V/m). Find the conduction and displacement current densities, and the frequency at which they have equal magnitudes.*

Solution: The conduction current density is

$$J_c = \sigma E,\ \sigma = 4 \text{ s/m},$$
$$E = 150 \sin 10^{10} t \text{ (V/m)}$$

$\therefore$
$$\bar{J}_c = 4 \times 150 \sin 10^{10} t$$
$$= 600 \sin 10^{10} t \text{ A/m}^2.$$

The displacement current density, $\overline{J_d} = \dfrac{\partial \overline{D}}{\partial t}$

Since $\quad E = 150 \sin 10^{10} t,\ \varepsilon_r = 1$

$$D = \varepsilon_0 \varepsilon_r E = \varepsilon_0 \times 150 \sin 10^{10} t$$

$\therefore$
$$J_d = \frac{\partial D}{\partial t} = \frac{\partial}{\partial t}(150\varepsilon_0 \sin 10^{10} t)$$
$$= 150\, \varepsilon_0 \times 10^{10} \cos 10^{10} t = 150\, \varepsilon_0 \omega \cos \omega t$$

Note that the magnitude of conduction current density is independent of frequency. In the problem since $J_c = J_d$, Equating the magnitude of these two components, we have

$$4 \times 150 = 150\, \omega\, \varepsilon_0$$

$\therefore$
$$\omega = \frac{4}{\varepsilon_0} = 4 \times 36\pi \times 10^9 \text{ rad/s}$$
$$2\pi f = 4 \times 36\pi \times 10^9$$

$\therefore$
$$f = 72 \text{ GHz.}$$

Example 8.5: *In a material for which σ = 5.5 mho/m and* $\varepsilon_r = 1$, *the electric field intensity is E = 200 sin* 10^9 *t* U_x *V/m. Determine the conduction and displacement current densities and the frequency at which they have equal magnitude.*

Solution: Conduction current density,

$$J_c = \sigma E = 5.5 \times 200 \sin 10^9 t \text{ A/m}^2$$
$$= 1100 \sin 10^9 t \text{ A/m}^2$$

Assuming that the field direction does not vary with time, we have the displacement current density, $J_d = \dfrac{\partial D}{\partial t} = \dfrac{\partial}{\partial t}(\varepsilon_0 \varepsilon_r\ 200 \sin 10^9\ t)$

$$= 200 \times (8.854 \times 10^{-12})(10^9) \cos 10^9 t$$
$$= 1.7708 \cos 10^9 \text{ t A/m}^2$$

But $\quad \frac{\partial D}{\partial t} = \varepsilon \frac{\partial E}{\partial t}$ and If $E = E_m \sin \omega t$

Then $\quad J_d = j\omega\varepsilon E_m \cos \omega t$

Hence, $\quad J_d = \omega\varepsilon E$

For $\quad J_c = J_d$ (in magnitude)

$$\sigma = \omega\varepsilon \text{ which implies } \omega = \frac{\sigma}{\varepsilon} = \frac{5.5}{8.854 \times 10^{-12}} = 6.211 \times 10^{11} \text{ rad/sec}$$

Cyclic frequency, $\quad f = \frac{\omega}{2\pi} = \frac{6.211}{2\pi} \times 10^{11}\ Hz = 98.9 \text{ GHz}.$

8.6 MODIFICATION OF MAXWELL'S EQUATIONS FOR TIME VARYING FIELDS – MODIFIED AMPERE'S LAW

In the previous context the fourth Maxwell's equation was obtained in differential form as

$$\nabla \times E = -\frac{\partial B}{\partial t}$$

which shows that a time varying magnetic field produces an electric field. Taking into account the point form of Ampere's circuital law as it applies to steady magnetic fields, we get

$$\nabla \times H = J \qquad \text{...(8.9)}$$

and also it shows its inadequacy for time varying conditions.

By taking divergence of each side

$$\nabla \cdot \nabla \times H = \nabla \cdot J \qquad \text{...(8.10)}$$

Since the divergence of Curl is identically zero, so $\nabla \cdot J$ is also zero. However the equation of continuity which is given by $\nabla \cdot J = -\frac{\partial \rho_v}{\partial t}$ shows us that equation (8.9) can be true only if $\frac{\partial \rho_v}{\partial t} = 0$. This is not true and so equation (8.9) must be amended before we can accept it for time-varying fields. Suppose we add on unknown term G to equation (8.9).

$$\nabla \times H = J + G \qquad \text{...(8.11)}$$

Again taking divergence, we have

$$\nabla \cdot \nabla \times H = \nabla \cdot (J + G)$$

Which implies $\quad 0 = \nabla \cdot J + \nabla \cdot G$

Thus $\quad \nabla \cdot G = \frac{\partial \rho_v}{\partial t} \qquad \text{...(8.12)}$

Replacing ρ_v by $\nabla \cdot D$, we get

$$\nabla \cdot G = \frac{\partial}{\partial t}(\nabla \cdot D) = \nabla \cdot \frac{\partial D}{\partial t}$$

From which we obtain the simplest solution for G

$$G = \frac{\partial D}{\partial t}$$ and Amperes circuital law in point form thus becomes

$$\nabla \times H = J + \frac{\partial D}{\partial t} \quad \text{...(8.13)}$$

It is to be noted that equation (8.13) has not been derived but it is a form which we have obtained, which does not disagree with continuity equation. The additional term $\frac{\partial D}{\partial t}$ has the dimensions of current density *i.e.*, amperes per square meter. Since it results from a time varying electric flux density, Maxwell termed it as displacement current density. It is sometimes denoted by J_d.

i.e., $$J_d = \frac{\partial D}{\partial t}$$

So, Maxwell's equations can be summarized as follows for static conditions and time varying cases in both point form and integral forms:

Table 8.1: Maxwell's equations: Static conditions

	Point form	**Integral form**
(*a*) Faraday's law	$\nabla \times E = 0$	$\oint E \cdot dL = 0$
(*b*) Ampere's circuital law	$\nabla \times H = J$	$\oint H \cdot dL = I$
(*c*) Gauss's law (electric)	$\nabla \cdot D = \rho_v$	$\oint D \cdot ds = \int \rho_v \, dv$
(*d*) Gauss's law (magnetic)	$\nabla \cdot B = 0$	$\oint B \cdot ds = 0$

Table 8.2: Maxwell's equation: Time varying case

	Point form	**Integral form**
(*a*) Faraday's law	$\nabla \times E = -\frac{\partial B}{\partial t}$	$\oint E \cdot dL = -\int \frac{dB}{dt} \cdot ds$
(*b*) Ampere's circuital law	$\nabla \times H = \frac{\partial D}{\partial t} + J$	$\oint H \cdot dL = I + \int \frac{dD}{dt} \cdot ds$
(*c*) Gauss's law (electric)	$\nabla \cdot D = \rho_v$	$\oint D \cdot ds = \int \rho_v \, dv$
(*d*) Gauss's law (magnetic)	$\nabla \cdot \mathrm{B} = 0$	$\oint B \cdot ds = 0$

8.7 PHYSICAL INTERPRETATION OF MAXWELL'S EQUATIONS (TIME VARYING CASE)

The physical significance of Maxwell's equations can be readily interpreted from their mathematical statement in the integral form and may be expressed as follows:

(*a*) The equation relating to Gauss's law (electric) $\left(\oint D \cdot ds = \int \rho_v \, dv\right)$ can be interpreted as the total electric flux density $\overline{D}$ through the surface enclosing a volume is equal to the charge Q within the volume.

(*b*) From the equation of Gauss's law (magnetic) $\left(\oint B \cdot ds = 0\right)$, it can be observed that the net magnetic flux emerging out from a closed surface is zero.

(*c*) From Ampere's circuital law $\left(\oint H \cdot dL = I + \int \frac{dD}{dt} \cdot ds\right)$ it can be seen that the magneto motive force around a closed path is equal to the conduction current plus the time derivative of the electric flux density $\overline{D}$ through any surface bounded by the path.

(*d*) From Faraday's law $\left(\oint E \cdot dL = -\int \frac{dB}{dt} \cdot ds\right)$ it can be observed that the electromotive force around a closed path is equal to the time derivative of the magnetic flux density through the surface.

The above physical interpretation of Maxwell's equations is often referred as word statement of Maxwell's equations.

8.8 MAXWELL'S EQUATIONS FOR HARMONICALLY TIME VARYING FIELDS

Let us assume that electromagnetic fields are time harmonic *i.e.*, they vary periodically or sinusoidally with time. So, the phasor form is applied to electromagnetic fields. A phasor is a complex number which can be written as $re^{j\phi}$ *i.e.*, $\phi = \omega t + \theta$, where θ may be function of time or space coordinates.

So, any vector field $\overline{A}$ which is sinusoidally time varying can be represented as $\overline{A}(t) = A_0 \, e^{j(\omega t + \theta)}$. The real part is given by $\overline{A}(t) = \overline{A}_0 \cos(\omega t + \theta)$ and imaginary part is given by $\overline{A}(t) = \overline{A}_0 \sin(\omega t + \theta)$.

While performing mathematical expressions either the real part or imaginary part should be used. By dropping the time factor $e^{j\omega t}$ in $\overline{A}(t)$, the resulting complex term $\overline{A}_0 e^{j\theta}$ is called phasor form of vector field and is denoted by $\overline{A}_s$ *i.e.*, $\overline{A}_s = \overline{A}_0 e^{j\theta}$ where subscript '*s*' denotes phasor form of $\overline{A}(t)$.

Thus $\overline{A}(t) = \overline{A}_0 \cos(\omega t + \theta)$ is instantaneous form and it can be expressed as

$$\overline{A}(t) = Re(\overline{A}_s \, e^{j\omega t})$$

$$\frac{\partial \overline{A}(t)}{\partial t} = \frac{\partial}{\partial t}(Re \, \overline{A}_s \, e^{j\omega t}) = Re(j\omega \, \overline{A}_s \, e^{j\omega t})$$

$$\frac{\partial \bar{A}(t)}{\partial t} = j\omega \bar{A}_s$$

The above equation shows that time derivative of instantaneous quantity is equivalent to multiplying its phasor form by $j\omega$. Similarly, it can be written as $\int \bar{A}\, \partial t = \frac{\bar{A}_s}{j\omega}$. So, Maxwell's equations for fields varying harmonically with time can be represented as follows:

Table 8.3: Maxwell's equations: Harmonically Time varying case

	Point form	Integral form
(*a*) Faraday's law	$\nabla \times E_s = -j\omega B_s$	$\oint E_s \cdot dL = -j\omega \int B_s \cdot ds$
(*b*) Ampere's circuital law	$\nabla \times H_s = j\omega D_s + J_s$	$\oint E_s \cdot dL = \int (J_s + j\omega D_s) ds$
(*c*) Gauss's law (electric)	$\nabla \cdot D_s = \rho_{vs}$	$\oint D_s \cdot ds = \int \rho_{vs}\, dv$
(*d*) Gauss's law (magnetic)	$\nabla \cdot B_s = 0$	$\oint B_s \cdot ds = 0$

8.9 WAVE EQUATION

The four Maxwell's equations along with continuity equation and Lorentz force equation form the basis of electromagnetic wave theory. So, when the wave motion in free space is considered the nature of electromagnetic waves and their propagation in free space is studied. Consider the equation

$$\nabla \times H = J + \frac{\partial D}{\partial t}$$

For a non conducting medium like free space the current density $J = 0$, Also $D = \varepsilon_0 E$. Therefore, for free space, the above equation can be written as

$$\nabla \times H = \frac{\partial}{\partial t}(\varepsilon_0 E) = \varepsilon_0 \frac{\partial E}{\partial t} \quad ...(8.14)$$

The physical interpretation is that if electric field E is changing with time at any point, the magnetic field H has curl at that point, thereby implying that if E is changing with time, then H is also changing with time.

Consider the other equation

$$\nabla \times E = -\frac{\partial B}{\partial t}$$

We have $B = \mu_0 H$ for free space, which implies

$$\nabla \times H = -\frac{\partial}{\partial t}(\mu_0 H) = -\mu_0 \frac{\partial H}{\partial t} \quad ...(8.15)$$

If H is changing with time, E has curl and hence it can be seen that a time changing electric field produces a time changing magnetic field and this time varying magnetic field produces time varying electric field and this process repeats and there is a propagation of energy as electromagnetic waves.

We know that given vector field

$E = E_x U_x$ assuming the component E_x to be given by $E_x = E_{xyz} \cos(\omega t + \psi)$

Where, E_{xyz} is a real function of x, y, z but not time and ψ is a phase angle which also may be a function of x, y, z.

Making use of Euler's identity, we get

$e^{j\omega t} = \cos \omega t + j \sin \omega t$

$E_x = Re\ E_{xyz}\ e^{j(\omega t + \psi)}$ where Re indicates that only the real part should be taken

$= Re\ E_{xyz}\ e^{j\psi}\ e^{j\omega t}$

If we drop Re and suppressing $e^{j\omega t}$ the field E_x becomes a phasor (or) a complex quantity which we identify with a 's' subscript E_{xs}. Thus

$$E_{xs} = E_{xyz}\ e^{j\psi} \quad \text{and} \quad E_s = E_{xs}\ U_x$$

Now
$$\frac{\partial E_x}{\partial t} = \frac{\partial}{\partial t}[E_{xyz} \cos(\omega t + \psi)] = -\omega E_{xyz} \sin(\omega t + \psi)$$

$$= Re\,(j\omega\, E_{xs}\, e^{j\omega t}) \qquad ...(8.16)$$

It is evident that taking the partial derivatives of any field quantity w.r.t. time is equivalent to multiplying the corresponding phasor by $j\omega$.

Now
$$\nabla \times H = \varepsilon \frac{\partial E}{\partial t}$$

The corresponding relationship in terms of phasor is given by

$$\nabla \times H_s = j\omega\, \varepsilon\, E_s$$

and also it can be written as

$$\nabla \times E_s = -\mu_0\, j\omega\, H_s \qquad ...(8.17)$$

The next step is to obtain sinusoidal steady state form of wave equation. The wave equation can be taken as

$$\nabla \times \nabla \times E_s = \nabla \cdot (\nabla \cdot E_s) - \nabla^2 E_s \qquad ...(8.18)$$

Considering the L.H.S. term of equation (8.18) we can write as

$$\nabla \times \nabla \times E_s = \nabla \times (-j\, \mu_0\, \omega\, H_s)$$
$$= -j\omega\, \mu_0\, \nabla \times H_s$$
$$= -j\omega\, \mu_0\, (j\omega\, \varepsilon\, E_s)$$
$$= \omega^2\, \mu_0\, \varepsilon\, E_s \qquad ...(8.19)$$

Now, considering the R.H.S. of equation (8.18), we get

$$\nabla(\nabla \cdot E_s) - \nabla^2 E_s = \nabla \left(\nabla \frac{D_s}{\varepsilon_0} \right) - \nabla^2 E_s$$

$$= \frac{1}{\varepsilon_0} \nabla \cdot (\rho_v) - \nabla^2 E_s$$

but
$$\nabla \cdot \rho_v = 0$$

So, it can be written as

$$\nabla(\nabla \cdot E_s) - \nabla^2 E_s = -\nabla^2 E_s \qquad ...(8.20)$$

Substituting equations (8.19) and (8.20) in equation (8.18)

$$\nabla^2 E_s = -\omega^2 \mu_0 \varepsilon E_s$$

Writing the above equation as

$$\nabla^2 E_s = \gamma^2 E_s \text{ (where } \gamma^2 = -\omega^2 \mu_0 \varepsilon) \quad ...(8.21)$$

and γ = propagation constant

Considering the x component of electric field intensity

$$\nabla^2 E_{xs} = \gamma^2 E_{xs}$$

$$\frac{\partial^2 E_{xs}}{\partial x^2} + \frac{\partial^2 E_{xs}}{\partial y^2} + \frac{\partial^2 E_{xs}}{\partial z^2} = \gamma^2 E_{xs}$$

Assuming that E_{xs} does not vary with both x and y but varies only with z variable we can reduce the equation as

$$\frac{\partial^2 E_{xs}}{\partial z^2} = \gamma^2 E_{xs}$$

$$\frac{\partial^2 E_{xs}}{\partial z^2} = -\omega^2 \mu \varepsilon E_{xs}$$

So, we can write the solution as $E_{xs} = Ae^{-\gamma z}$ ...(8.22)

Multiplying both sides of equation (8.22) by $e^{j\omega t}$ and considering only the real part and thus eliminating the imaginary part

$$Re\,(E_{xs}\, e^{j\omega t}) = Re\,(Ae^{-\gamma z}\, e^{j\omega t})$$

$$E_x = Re\,(Ae^{-\gamma z}\, e^{j\omega t}) \quad ...(8.23)$$

Now, let $\gamma = \alpha + j\beta$

$$= Re\ Ae^{-(\alpha + j\beta)z}\ e^{j\omega t}$$

$$= Re\,[Ae^{-\alpha z}\ e^{j(\omega t - \beta z)}]$$

$$E_x = Ae^{-\alpha z} \cos(\omega t - \beta z) \quad ...(8.24)$$

Now, let E_{x0} be the value of E_x at $t = 0$ and $z = 0$. Then the equation (8.24) can be written as

$$E_{xo} = A \quad \text{which implies}$$

$$E_x = E_{x0}\, e^{-\alpha z} \cos(\omega t - \beta z) \quad ...(8.25)$$

8.10 UNIFORM PLANE WAVE

This is defined as the wave whose value remains constant throughout a plane which traverses in the direction of propagation of wave. If the wave is travelling in U_z direction then the plane z = constant will be perpendicular to the direction of propagation. In the plane z = constant the variables are x and y. Hence for this uniform plane wave, the electric field intensity and magnetic field intensity are independent of x and y-coordinates and are functions of z-coordinates only in free space.

8.10.1 Equation for Wave Motion

From equation (8.21),

$$\nabla^2 E_s = -\omega^2 \mu \varepsilon E_s$$

for E_{xs} we can write it as

$$\frac{\partial^2 E_{xs}}{\partial z^2} = -\omega^2 \mu \varepsilon E_{xs} \quad \text{...(8.26)}$$

and also, $E_x = E_{x_0} e^{-\alpha z} \cos(\omega t - \beta z)$

In complete notation it can be written as

$$E_{xs} = E_{x_0} e^{-\alpha z} e^{-j\beta z} \quad \text{...(8.27)}$$

Let $\gamma = \alpha + j\beta$

Which implies that $E_{xs} = E_{x_0} e^{-\gamma z}$ ...(8.28)

Applying equation (8.28) in equation (8.26), we can write as

$$\gamma^2 E_{x_0} e^{-\gamma z} = -\omega^2 \mu\varepsilon E_{x_0} e^{-\gamma z}$$

which implies that $\gamma^2 = -\omega^2 \mu\varepsilon$

and $\gamma = \pm j\omega\sqrt{\mu\varepsilon}$ ...(8.29)

Comparing equation (8.29) with normal equation of $\alpha + j\beta$

We can write as $\alpha = 0$ and $\beta = \omega\sqrt{\mu\varepsilon}$

Now consider the above equation in various mediums like free space, dielectric and conductors etc. Assuming the various initial parameters of these mediums, it is easier to derive the equation for velocity and intrinsic impedance. So first considering **free space medium** which is characterized by $\alpha = 0$, $\mu = \mu_0$ and $\varepsilon = \varepsilon_0$ we can modify the above equation and can be written as

Now $\alpha = 0$ and $\beta = \omega\sqrt{\mu_0\varepsilon_0}$

Now, the velocity in free space is given by

$$v = \frac{\omega}{\beta} = \frac{\omega}{\omega\sqrt{\mu_0\varepsilon_0}} = 3 \times 10^8 \text{ m/sec}$$

And the intrinsic impedance denoted by η is given by

$$\eta = \sqrt{\frac{\mu_0}{\varepsilon_0}}$$

Case-I: Dielectric

In the case of dielectric the basic parameters are different from those taken in free space and they can be shown as

$$\mu = \mu_0 \mu_r \quad \text{and} \quad \varepsilon = \varepsilon_0 \varepsilon_r$$

So, it can be written as $\alpha = 0$ and $\beta = \omega\sqrt{\mu_0\mu_r\varepsilon_0\varepsilon_r}$ and velocity (v) is given by

$$v = \frac{\omega}{\beta} = \frac{\omega}{\sqrt{\mu_0\mu_r\varepsilon_0\varepsilon_r}}$$

$$\text{Intrinsic impedance } (\eta) = \sqrt{\frac{\mu}{\varepsilon}}$$

$$= \sqrt{\frac{\mu_0\mu_r}{\varepsilon_0\varepsilon_r}}$$

Case-II: Conductors

In the case of conductors the concept of conductivity also needs to be added. So, the effect of including conductivity denoted by σ, is that the factor $j\omega\varepsilon$ has become equal to $(\sigma + j\omega\varepsilon)$.

So,
$$\gamma^2 = -\omega^2 \mu\varepsilon$$
$$= (\sigma + j\omega\varepsilon)^2 j\omega\mu$$
$$\gamma = \pm \sqrt{(\sigma + j\omega\varepsilon)\, j\omega\mu}$$
$$= j\omega\sqrt{\mu\varepsilon}\sqrt{1 - j\frac{\sigma}{\omega\varepsilon}}$$

Since σ value will be very large for conductors *i.e.*, $\frac{\sigma}{\omega\varepsilon} >> 1$, we can write as

$$\gamma = j\omega\sqrt{\mu\varepsilon}\sqrt{-j\frac{\sigma}{\omega\varepsilon}}$$
$$\gamma = j\sqrt{-j\omega\mu\sigma}$$

Now
$$\sqrt{-j} = \sqrt{1\angle{-90^\circ}} = 1\angle{-45^\circ} = \frac{1}{\sqrt{2}} - j\frac{1}{\sqrt{2}}$$

So, it can be written as

$$\gamma = j\left(\frac{1}{\sqrt{2}} - j\frac{1}{\sqrt{2}}\right)\sqrt{\omega\mu\sigma}$$

Now
$$\omega = 2\pi f$$

Substituting the value of ω is the above equation of γ, we get

$$\gamma = (1 + j)\sqrt{\pi f\mu\sigma}$$

Comparing it with $\alpha + j\beta$

$$\alpha = \beta = \sqrt{\pi f\mu\sigma}$$

Intrinsic impedance

The ratio of amplitude of electric field intensity to magnetic field intensity has the dimensions of impedance.

$$\eta = \frac{E}{H} \quad \text{and} \quad E_{xs} = Ae^{-\gamma z}$$

Now
$$\nabla \times E_s = -j\omega\mu H_s \qquad \text{...(8.30)}$$

$$\nabla \times E_s = \begin{vmatrix} U_x & U_y & U_z \\ \frac{\partial}{\partial x} & \frac{\partial}{\partial y} & \frac{\partial}{\partial z} \\ Ae^{-\gamma z} & 0 & 0 \end{vmatrix}$$

$$= -A\gamma e^{-\gamma z} U_y \qquad \text{...(8.31)}$$

From equation (8.30) and equation (8.31), it can be written as

$$-A\gamma e^{-\gamma z} U_y = -j\omega\mu H_s \qquad \text{...(8.32)}$$

$$H_s = \frac{A\gamma e^{-\gamma z}}{j\omega\mu} U_y \qquad ...(8.33)$$

Now $$\eta = \frac{E}{H} = \frac{Ae^{-\gamma z}}{\frac{A\gamma e^{-\gamma z}}{j\omega\mu}} = \frac{j\mu\omega}{\gamma} = \frac{j\mu\omega}{\sqrt{j\mu\omega(\sigma + j\omega\varepsilon)}}$$

Which implies $$\eta = \sqrt{\frac{j\omega\mu}{\sigma + j\omega\varepsilon}} \qquad ...(8.34)$$

Example 8.6: *A uniform plane wave in a medium having $\sigma = 10^{-3}$ s/m, $\varepsilon = 80\ \varepsilon_0$ and $\mu = \mu_0$ is having a frequency of 10 kHz. Calculate the different parameters of wave.*

Solution: Given data is $\sigma = 10^{-3}$ s/km, $\varepsilon = 80\ \varepsilon_0$ and $\mu = \mu_0$

Now $$\frac{\sigma}{\omega\varepsilon} = \frac{10^{-3}}{2\pi \times 10^4 \times 80 \times 8.854 \times 10^{-12}} = 22.47 >> 1$$

The medium is a good conductor, so attenuation constant

$$\alpha = \sqrt{\pi f \mu \sigma} = \sqrt{\pi \times 10^4 \times 4\pi \times 10^{-7} \times 10^{-3}}$$

$$= 2\pi \times 10^{-3} \text{ rad/m}$$

Phase constant $\beta = \alpha = 2\pi \times 10^{-3}$ rad/m

Intrinsic impedance $$\eta = \sqrt{\frac{\omega\mu}{\sigma}} \angle 45° = \sqrt{\frac{2\pi \times 10^4 \times 4\pi \times 10^{-7}}{10^{-3}}} \angle 45°$$

$$= 2\pi\,(1 + j) \text{ ohm}$$

Wave length $$\lambda = \frac{2\pi}{\beta} = 1000\text{m}$$

Velocity of wave $$v = \frac{\omega}{\beta} = \frac{2\pi \times 10^4}{2\pi \times 10^{-3}} = 10^7 \text{ m/sec.}$$

Example 8.7: *The magnetic field intensity of a uniform plane wave in air is 30 A/m in the U_y direction. The wave is propagating in the U_z direction at a frequency of 3×10^9 rad/sec. Find (a) the wave length (b) the frequency (c) the period (d) the amplitude of $\overline{E}$.*

Solution:

(*a*) $$\beta = \omega\sqrt{\mu_0\varepsilon_0} = 3 \times 10^9 \sqrt{4\pi \times 10^{-7} \times 8.854 \times 10^{-12}} = 10 \text{ rad/m}$$

∴ Wave length, $$\lambda = \frac{2\pi}{\beta} = \frac{2\pi}{10} = 0.628\text{m}$$

(*b*) Frequency, $$f = \frac{\omega}{2\pi} = \frac{3 \times 10^9}{2\pi} = 477.464 \text{ MHz}$$

(*c*) Time period, $$T = \frac{1}{f} = 2.094 \text{ ns}$$

(*d*) $$\eta = \frac{E}{H} = \sqrt{\frac{\mu_0}{\varepsilon_0}} = 376.734$$

∴ Magnitude of $E = 376.734 \times 30 = 11302.03$ V/m.

Example 8.8: *Let* $\overline{E}_s = [500 \angle{-40°}\ U_y + (200 - j600)\ U_z]\ e^{-j0.5x}$ *V/m in free space. Find (a)* ω *(b) E at (1, 2, 3) t = 0 (c)* $\overline{E}$ *at (1, 2, 3), t = 10 ns (d)* $\overline{E}$ *at (2, 3, 5), t = 20 ns.*

Solution:

(*a*) $$\beta = 0.5 = \omega\sqrt{\mu_0 \varepsilon_0}$$

$$\omega = \frac{0.5}{\sqrt{4\pi \times 10^{-7} \times 8.854 \times 10^{-12}}} = 119.918 \times 10^6 \text{ rad/s}$$

(*b*) $$E_s = [500 \angle{-40°}\ U_y + (200 - j\,600)\ U_z]\ e^{-j0.5x}$$

$$= 500 \angle{-40°}\ U_y + 632.456 \angle{-71.65°}\ U_z]\ e^{-j\,0.5x}$$

$$= 500\ e^{-j(0.5x+40°)}\ U_y + 632.456\ e^{-j(0.5x+71.65°)} U_x$$

$$E(t) = 500\ Re\ [e^{-j(0.5x+40)}\ e^{j\omega t}]\ U_y + 632.456\ Re\ [e^{-j(0.5x+71.65)}\ e^{j\omega t}]U_z$$

$$E(t) = 500\ Re\ (e^{-j(\omega t - 0.5x - 40°)})]U_y + 632.456\ Re(e^{-j(\omega t - 0.5x - 71.65°)})]\ U_z$$

$$E(t) = 500 \cos (\omega t - 0.5x - 40°)\ U_y + 632.456 \cos (\omega t - 0.5x - 71.65°)\ U_z$$

E at (1, 2, 3) at (t = 0) = 500 cos ((– 0.5 × 1) – 40°) U_y + 632.456 cos (– 0.5 × 1 – 71.65°) U_z

$$= [380.20\ U_y - 193.86\ U_z] \text{ V/m.}$$

(*c*) $\overline{E}$ at (1, 2, 3) at (t = 10 ns)

$$= 500 \cos (119.918 \times 10^6 \times 10 \times 10^{-9} - 0.5 \times 1 - 40°)\ U_y$$
$$+\ 632.456 \cos (119.918 \times 10^6 \times 10 \times 10^{-9} - 0.5 \times 1 - 71.56°)\ U_z$$
$$= [388.5\ U_y + 210.48\ U_z] \text{ V/m.}$$

(*d*) $\overline{E}$ at (2, 3, 5) at (t = 20 ns)

$$= 500 \cos (119.918 \times 10^6 \times 20 \times 10^{-9} - 0.5 \times 2 - 40°)\ U_y$$
$$+\ 632.456 \cos (119.918 \times 10^6 \times 20 \times 10^{-9} - 0.5 \times 2 - 71.565)\ U_z$$
$$= [393.99\ U_y + 220.79\ U_z] \text{ V/m.}$$

Example 8.9: *An* $\overline{E}$ *field in free space is given as* $\overline{E} = 900 \cos (10^9\, t - \beta y)\ U_z$ *V/m. Find (a)* β *(b)* λ *(c)* $\overline{H}$ *at P (0.1, 0.5, 0.4) at t = 10 ns.*

Solution:

(*a*) $\omega = 10^9$; $\beta = \omega\sqrt{\mu_0\varepsilon_0} = 10^9\sqrt{4\pi \times 10^{-7} \times 8.85 \times 10^{-12}} = 3.334$ rad/m

(*b*) $$\lambda = \frac{2\pi}{\beta} = 1.883 \text{ m}$$

(*c*) For free space $\eta = \dfrac{E}{H} = 376.7343\ \Omega$; $E = 900$ V/m

$\therefore$ $$H = \frac{E}{376.7343} = \frac{900}{376.73} = 2.388 \text{ A/m}$$

Now $\overline{P} = \overline{E} \times \overline{H}$ and hence the direction of $\overline{H}$ is given by $\overline{U}_x$

$\therefore$ $$\overline{H} = 2.388 \cos (10^9\, t - \beta y)\ U_x \text{ A/m}$$

$\overline{H}$ at P (0.1, 0.5, 0.4) at t = 10 ns

$$= 2.388 \cos (10^9 \times 10 \times 10^{-9} - 3.334 \times 0.5)\ U_x$$
$$= 2.388 \cos (9.8332)\ U_x$$
$$= 2.3538\ U_x \text{ A/m.}$$

Example 8.10: *A uniform plane wave in free space is given by* $E_s = (250 \angle 30°)\ e^{-j300z}\ U_x$ *V/m. Find (a)* β*, (b)* ω*, (c)* f*, (d)* λ*, (e)* η*, (f)* H*, (g)* $|E|$ *at z = 10mm, t = 8psec.*

Solution:

(*a*) The coefficient of Z represents the value of β which is equal to 300 rad/m.

(*b*) $\beta = \omega\sqrt{\mu_0\varepsilon_0} = 300$ rad/m.

Hence, $$\omega = \frac{300}{\sqrt{\mu_0\varepsilon_0}} = 89.93 \text{ G rad/sec}$$

(*c*) $\omega = 2\pi f$, Hence, $f = \dfrac{\omega}{2\pi} = 14.31$ GHz

(*d*) $\lambda = \dfrac{2\pi}{\beta} = \dfrac{2\pi}{300} = 20.94$ mm

(*e*) $\eta = 376.73$ (a const for free space)

$$\frac{E}{H} = \eta = 376.7345\ \Omega.$$

$$H = \frac{E}{376.734} = \frac{250}{376.734} = 0.6635 \text{ A/m}$$

$\therefore$ $$H_s = 0.6635 \angle 30°\ e^{-j300z}\ U_y = 0.6635\ e^{j30°}\ e^{-j300z}\ U_y$$
$$= 0.6635\ e^{-j300z + j30°}\ U_y \text{ A/m}$$

(*g*) $$E_s = 250 \angle 30°\ e^{-j300z}\ U_x = 250\ e^{j30°}\ e^{-j300z}\ U_x$$
$$= 250\ e^{j(-300z+30°)}\ U_x$$

$$\overline{E} = 250\ Re\ [e^{j\omega t}\ e^{j(-300z+30°)}]\ U_x$$
$$= 250\ Re\ [e^{j(\omega t - 300z+30°)}]\ U_x$$
$$= 250 \cos (\omega t - 300z + 30°)\ U_x \text{ V/m}$$

So, $\overline{E}$ at Z = 10 mm ; $t = 8p$ sec = 8×10^{-12} s

$$= 250 \cos [89.93 \times 10^9 \times 8 \times 10^{-12} - 300 \times 0.01 + 30°]\ U_x$$
$$= 250 \cos [-2.28 + 30°]\ U_x$$
$$= 200 \cos [-87.72° + 30°]\ U_x = 133.514\ U_x \text{ V/m}$$

$$|\overline{E}| = 133.514 \text{ V/m.}$$

Example 8.11: *The electric field intensity of 400 MHz uniform plane wave in free space is given as* $E_s = (30 + j60)(U_x + 2U_y)\ e^{-j\beta z}$ *V/m (a) Find* ω*,* λ*, V and* β *(b) Find* $\overline{E}$ *at t = 2 ns, z = 20 cm (c) What is* $|H|_{max}$*?*

Solution:

(*a*) $$\omega = 2\pi f = 2\pi \times 400 \times 10^6 = 2.513 \text{ G rad/s}$$

$$\beta = \omega\sqrt{\mu_0\varepsilon_0} = 8.3823 \text{ rad/m and } \lambda = \frac{2\pi}{\beta} = 0.7495 \text{ m}$$

$$v = \frac{1}{\sqrt{\mu_0\varepsilon_0}} = 3 \times 10^8 \text{ m/sec}$$

(*b*) $$E_s = (30 + j60)\,(U_x + 2U_y)\, e^{-j\beta z} \text{ V/m}$$

$$= 67.082\, e^{j68.43^\circ}\,(U_x + 2U_y)\, e^{-j\beta z}$$

$$= 67.082\,(U_x + 2U_y)\, e^{j(68.43^\circ - \beta z)}$$

$$\bar{E} = 67.082\,(U_x + 2U_y)\,\text{Re}\,[e^{j(68.43^\circ - \beta z)}\, e^{j(\omega t)}]$$

$$= 67.082\,(U_x + 2U_y)\cos(\omega t - \beta z + 68.43^\circ)$$

$\bar{E}$ at $t = 2$ ns ; $z = 20$ cm

$$= 67.082\,(U_x + 2U_y)\cos(2.513 \times 10^9 \times 2 \times 10^{-9} - 8.3823 \times 0.2 + 68.43^\circ)$$

$$= 67.082\,[U_x + 2U_y] \times 0.3126$$

(*c*) $$|E|_{max} = 20.975\sqrt{(1)^2 + (2)^2} = 46.9 \text{ V/m}$$

$$|H|_{max} = \frac{46.9}{\eta} = \frac{46.9}{376.73} = 0.125 \text{ A/m.}$$

Example 8.12: *Determine the propagation constant at 400 kHz for a medium in which $\mu_R = 2$, $\varepsilon_R = 17$ and $\sigma = 0$. At what velocity will an electromagnetic wave travel in this medium?*

Solution: Given $f = 400$ kHz, $\mu_r = 2$, $\varepsilon_r = 17$ and $\sigma = 0$.

The medium is good dielectric material, since $\sigma = 0$ and ε_R is very high, we have

(*a*) $$\alpha = \frac{\sigma}{2}\sqrt{\frac{\mu}{\varepsilon}} \text{ for a good dielectric medium} = 0 \qquad (\because \; \sigma = 0)$$

$$\beta = \omega\sqrt{\mu\varepsilon}\left[1 + \frac{\sigma^2}{8\,\omega^2\varepsilon^2}\right] \text{ for a good dielectric medium.}$$

$$= \omega\sqrt{\mu\varepsilon} \qquad (\because \; \sigma = 0)$$

$$= 2\pi f\sqrt{\mu\varepsilon}$$

$$= 2\pi f\sqrt{(\mu_0 \times \mu_r)(\varepsilon_0 \times \varepsilon_r)}$$

$$= 2\pi \times 400 \times 10^3 \times \sqrt{(4\pi \times 10^{-7} \times 2)(8.854 \times 10^{-12} \times 17)} = 0.0484$$

We have propagation constant $\gamma = \alpha + j\beta$

$$\gamma = 0 + j0.0484$$

(or) $$\gamma = 0.0484\,\angle 90^\circ$$

Now velocity of propagation $$v = \frac{1}{\sqrt{\mu\varepsilon}} = \frac{1}{\sqrt{\mu_0\varepsilon_0}} \cdot \frac{1}{\sqrt{\mu_r\,\varepsilon_r}} = 3 \times 10^8\,\frac{1}{\sqrt{2 \times 17}}$$

$$= 0.5145 \times 10^8 \text{ m/sec.}$$

Since $\frac{1}{\sqrt{\mu_0 \varepsilon_0}} = c$, velocity of light in free space = 3×10^8 m/sec.

Example 8.13: *A wave propagating in a loss less dielectric has the components $E = 450 \cos (10^8 - \beta_z) U_x$ V/m and $H = 2.2 \cos (10^8 - \beta_z) U_y$ A/m. If the wave is travelling at $v = 0.6\, c$, find: (a) μ_r (b) ε_r (c) β (d) λ.*

Solution: Given $E = 450 \cos (10^8 - \beta_z)\, U_x$ and $H = 2.2 \cos (10^8 - \beta_z)\, U_y$. From this it is obvious that the wave is propagating in the z-direction.

Given $\omega = 10^8$, $\nu = 0.6\, c$, where c = velocity of light.

(*a*) $$\nu = \frac{\omega}{\beta} \text{ or } \beta = \frac{\omega}{\nu} = \frac{10^8}{0.6\, c} = \frac{10^8}{0.6 \times 3 \times 10^8} \quad (\text{Since } c = 3 \times 10^8 \text{ m/s.})$$

$$= 0.555 \text{ radian/metre}$$

(*b*) $$\beta = \frac{2\pi}{\lambda} \text{ or } \lambda = \frac{2\pi}{\beta} = \frac{2\pi}{0.555} = 11.31 \text{ m}$$

(*c*) $$\frac{E}{H} = \sqrt{\frac{\mu}{\varepsilon}} = \sqrt{\frac{\mu_0 \mu_r}{\varepsilon_0 \varepsilon_r}} = \sqrt{\frac{\mu_0}{\varepsilon_0}} \sqrt{\frac{\mu_r}{\varepsilon_r}}$$

$$= 377 \sqrt{\frac{\mu_r}{\varepsilon_r}} \qquad \left(\because \sqrt{\frac{\mu_0}{\varepsilon_0}} = 377 \right)$$

$$\therefore \quad \frac{450}{2.2} = 377 \sqrt{\frac{\mu_r}{\varepsilon_r}} \text{ or } \sqrt{\frac{\mu_r}{\varepsilon_r}} = \frac{450}{2.2 \times 377} = 0.542$$

$$\therefore \quad \frac{\mu_r}{\varepsilon_r} = 0.542^2 = 0.293 \text{ or } \mu_r = 0.293\, \varepsilon_r \qquad ...(i)$$

We have $$\nu = \frac{c}{\sqrt{\mu_r \varepsilon_r}}$$

$$\therefore \quad 0.6\, c = \frac{c}{\sqrt{\mu_r \varepsilon_r}}$$

$$\therefore \quad \sqrt{\mu_r \varepsilon_r} = 1.66 \text{ or } \mu_r\, \varepsilon_r = (1.66)^2 = 2.755 \qquad ...(ii)$$

Putting $\mu_r = 0.293\, \varepsilon_r$ from equation (*i*) in equation (*ii*)

$$\varepsilon_r^2 = 9.403 \text{ from which } \varepsilon_r = 3.066$$

and $$\mu_r = 0.293 \times 3.066 = 0.898.$$

Example 8.14: *Given that $E = 50\, \pi e^{j(\omega t - \beta z)} (U_x)$ and $H = H_m\, e^{j(\omega t - \beta z)} (U_y)$ in free space, where $\omega = 10^9$. Evaluate H_m and $\beta (\beta > 0)$.*

Solution: The example deals with a plane wave. As seen before,

$$\frac{\omega}{\beta} = c = 3 \times 10^8 \text{ m/sec and } \frac{E}{H} = 120\, \pi \text{ ohm (constant)}$$

Thus for the given wave, $\beta = \frac{\omega}{c} = \frac{10^9}{3 \times 10^8} = \frac{10}{3}$ rad/m

$$H_m = \pm \frac{50\pi}{120\pi} = \pm \frac{5}{12} \text{ A/m}$$

To determine the sign of H_m, applying $\nabla \times E = -\frac{\partial B}{\partial t}$

$$\nabla \times E = \begin{vmatrix} u_x & u_y & u_z \\ \frac{\partial}{\partial x} & \frac{\partial}{\partial y} & \frac{\partial}{\partial z} \\ 50\pi e^{j(\omega t - \beta z)} & 0 & 0 \end{vmatrix}$$

$$= -50\,\pi\beta\, e^{j(10^9 t - \beta z)} u_y \qquad ...(i)$$

$$B = H = H_m\, e^{j(\omega t - \beta z)} u_y$$

$$\frac{\partial B}{\partial t} = j\,10^9\,\mu_0 H_m\, e^{j(10^9 t - \beta z)} u_y \qquad ...(ii)$$

Equating equations (*i*) and (*ii*), it can be written as

$$-50\,\pi\,\beta e^{j(10^9 t - \beta z)} = -j10^9 \mu_0 H_m e^{j(10^9 t - \beta z)}$$

It can be clearly seen above that H_m is positive.

Example 8.15: *The complex electric field vector of a uniform plane wave propagated in free space is given by* $E = (-2i - 3\sqrt{3}\,j + \sqrt{3}\,k)\ e^{-j0.04\pi(\sqrt{2}x - 3y - 3z)}$ *volt/metre. (a) What is the direction of propagation ? (b)Find the wavelength in the direction of propagation. (c) Find the frequency of the wave. (d) Find the apparent wave-lengths and the apparent phase velocities along X, Y and Z-axes.*

Solution:

(*a*) The expression is of the general form $E = A\ e^{-j\beta\gamma}$ which may be written as

$$E = Ae^{-j(\beta_x x + \beta_y y + \beta_z z)}$$

Let $(\beta')^2 = \beta_x^{\ 2} + \beta_y^{\ 2} + \beta_z^{\ 2}$. In this problem, $\beta_x = \sqrt{2}, \beta_y = -3, \beta_z = -3$

Therefore, $(\beta')^2 = (2 + 9 + 9) = 20$; $\beta' = 4.47$

$$\beta = 0.04\pi\beta'$$

$$= 0.04\,\pi \times 4.47$$

(*b*) Wavelength, $\lambda = \frac{2\pi}{\beta} = \frac{2\pi}{0.04\pi \times 4.47}$

$$\lambda = 11.18 \text{ m.}$$

(*c*) Frequency of the wave, $f = \frac{c}{\lambda} = \frac{3 \times 10^8}{11.18}$ Hz

$$= 26.83 \text{ MHz}$$

(*d*) Apparent wavelength, $\lambda_x = \frac{2\pi}{0.04\pi\sqrt{2}} = 35.355$ m

$$\lambda_y = \frac{2\pi}{0.04\pi \times 3} = \frac{50}{3} = 16.667 \text{ m}$$

$$\lambda_z = \frac{2\pi}{0.04\pi \times 3} = \frac{50}{3} = 16.667 \text{ m}$$

Apparent phase velocities,

$$v_{px} = f\lambda_x = 26.83 \times 10^6 \times 35.355 = 9.48 \times 10^8 \text{ m/sec.}$$

$$v_{py} = f\lambda_y = 26.83 \times 10^6 \times 16.667 = 4.47 \times 10^8 \text{ m/sec.}$$

$$v_{pz} = f\lambda_z = 26.83 \times 10^6 \times 16.667 = 4.47 \times 10^8 \text{ m/sec.}$$

8.11 SKIN EFFECT

When a conductor is carrying d.c. current, this current is uniformly distributed over whole cross-section of conductor. But if alternating current flows through conductor, it has the tendency to concentrate near the surface of conductor called as skin effect. So, the tendency of alternating current to concentrate over the surface of conductor is called **skin effect.**

Due to skin effect the effective area of cross-section of conductor through which current flows is reduced. Consequently the resistance of conductor increases, slightly. A solid conductor may be thought of consisting a large number of strands carrying a small part of current. The inductance of each strand will vary according to its position. Thus the strands near the center are surrounded with greater magnetic flux and hence have large inductance near the surface. The high reactance causes a.c. to flow near surface of conductor.

Normally the skin effect depends on the following parameters:

(*a*) Nature of material

(*b*) Diameter of wire

(*c*) Frequency

(*d*) Shape of wire.

8.12 DEPTH OF PENETRATION (SKIN DEPTH)

In a medium of high conductivity, the wave is attenuated as it progresses due to losses which occur in the medium. In a good conductor the rate of attenuation is very great and the wave may penetrate only a very short distance before being reduced to a negligible small percentage of its original strength. A term that has significance is called depth of penetration (or) skin depth. The depth of penetration, δ, is defined as depth in which the wave has been attenuated to $\frac{1}{e}$ or approximately 37% of the original value.

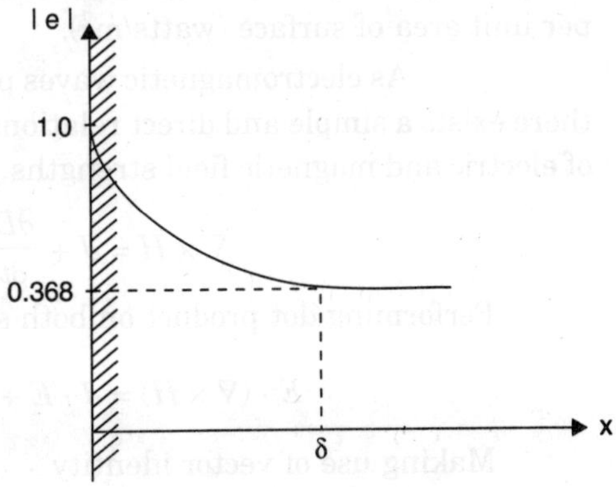

Fig 8.2

The amplitude of the wave decreases by a factor $e^{-\alpha x}$ where α is attenuation constant. It is apparent that a distance x which makes $\alpha x = 1$, the amplitude is only $\frac{1}{e}$ times its value at $x = 0$. This distance is equal to δ, the depth of penetration. We have

$$\alpha x = 1 \quad \text{(or)} \quad \alpha\delta = 1$$

$$\delta = \frac{1}{\alpha} \qquad ...(8.35)$$

In the case of conductors α is given by

$$\alpha = \sqrt{\pi F \mu \sigma} \qquad ...(8.36)$$

So,

$$\delta = \frac{1}{\alpha} = \sqrt{\frac{1}{\pi f \mu \sigma}} \qquad ...(8.37)$$

Example 8.16: *Determine the skin depth for copper with conductivity of 6.6M mho/ meter at 15 MHz frequency.*

Solution: Given conductivity σ = 6.6 M mho/m and f = 15 MHz. Skin depth in a conductor material is given as $\delta = \frac{1}{\sqrt{\pi f \mu \sigma}}$

We have $\mu = \mu_0\mu_r = \mu_0$ since $\mu_r = 1$ for copper which is a non-magnetic medium.

$\therefore \quad \mu = \mu_0 = 4\pi \times 10^{-7}$

Substituting the given values, we get

$$\delta = \frac{1}{\sqrt{\pi \times (15 \times 10^6) \times (4\pi \times 10^{-7}) \times (6.6 \times 10^6)}}$$

$$= 5.06 \times 10^{-5} \text{ m.}$$

8.13 POYNTING THEOREM AND ITS INTERPRETATIONS

An important aspect of wave propagation is the flow of power through space. Travelling waves carry energy with it. At imaginary surface in space at any instant there is flow of power through per unit area of surface (watts/m^2).

As electromagnetic waves propagate through the space from the source to receiver, there exists a simple and direct relationship between the rate of energy transfer and amplitude of electric and magnetic field strengths. The relation can be obtained form Maxwell's equation.

$$\nabla \times H = J + \frac{\partial D}{\partial t} \qquad ...(8.38)$$

Performing dot product on both sides with E, we get

$$E \cdot (\nabla \times H) = J \cdot E + E \cdot \frac{\partial D}{\partial t} \qquad ...(8.39)$$

Making use of vector identity

$$\nabla \cdot (E \times H) = -E \cdot \nabla \times H + H \cdot \nabla \times E \qquad ...(8.40)$$

Substituting equation (8.40) in equation (8.39), we get

$$H \cdot \nabla \times E - \nabla \cdot (E \times H) = J \cdot E + E \cdot \frac{\partial D}{\partial t} \qquad ...(8.41)$$

But, $\nabla \times E = -\dfrac{\partial B}{\partial t}$ (from Maxwell's 4th equation) ...(8.42)

Using the equation (8.42) in equation (8.41).

$$-H \cdot \frac{\partial B}{\partial t} - \nabla \cdot (E \times H) = J \cdot E + E \cdot \frac{\partial D}{\partial t} \qquad ...(8.43)$$

$$-\nabla \cdot (E \times H) = J \cdot E + \varepsilon E \cdot \frac{\partial E}{\partial t} + \mu H \cdot \frac{\partial H}{\partial t}$$

But it can be written as

$$-\nabla \cdot (E \times H) = J \cdot E + \frac{\partial}{\partial t}\left(\frac{\varepsilon E^2}{2} + \frac{\mu H^2}{2}\right) \qquad ...(8.44)$$

Since, $$\varepsilon E \cdot \frac{\partial E}{\partial t} = \frac{\varepsilon}{2}\frac{\partial E^2}{\partial t} = \frac{\partial}{\partial t}\left(\frac{\varepsilon E^2}{2}\right)$$

and $$\mu H \cdot \frac{\partial H}{\partial t} = \frac{\partial}{\partial t}\left(\frac{\mu H^2}{2}\right)$$

Finally if we integrate equation (8.44) through a volume

$$-\int_{\text{vol}} \nabla \cdot (E \times H)\, dv = \int_{\text{vol}} J \cdot E\, dv + \frac{\partial}{\partial t}\int_{\text{vol}}\left(\frac{\varepsilon E^2}{2} + \frac{\mu H^2}{2}\right) dv \qquad ...(8.45)$$

Applying divergence theorem to equation (8.45), we get

$$-\oint_S (E \times H) . ds = \int_{\text{vol}} J \cdot E\, dv + \frac{\partial}{\partial t}\int_{\text{vol}}\left(\frac{\varepsilon E^2}{2} + \frac{\mu H^2}{2}\right) dv \qquad ...(8.46)$$

If we assume that there are no sources within the volume, then R.H.S. of equation (8.46) is total ohmic power dissipated within volume. If sources are present then result of integrating over volume of source will be positive if power is delivered to source and negative if it is by the source.

The equation (8.46) shows the total power flowing out of volume is $\oint E \times H \cdot ds$ and negative sign is that power is flowing out of volume and Poynting vector $P = E \times H$.

Interpretations

Since in equation (8.46) the quantity on the left hand side shows the power flowing out of the volume, so equation (8.46) can be modified as

$$\oint P \cdot ds = \int_{\text{vol}} J \cdot E\, dv + \frac{\partial}{\partial t}\int_{\text{vol}}\left(\frac{\varepsilon E^2}{2} + \frac{\mu H^2}{2}\right) dv \qquad ...(8.47)$$

The first integral on the right hand side involving E and J is responsible for the loss of energy in the form of heat which is equivalent to I^2R loss in the volume. Here E and J are in the same direction. The second integral represents the time rate of increase of energy storage in the volume v carried by electric and magnetic fields. The integral having Poynting's vector term is the source on the left hand side which is responsible for the energy inflow and hence transports energy into the volume.

8.14 POYNTING VECTOR AND FLOW OF POWER

When electromagnetic waves travel from one point to another, there will be energy flow across surface involved. Poynting theorm states that the cross product of E and H at any point is a measure of rate of energy flow per unit area at that point, *i.e.*,

$$P = E \times H \text{ watts/m}^2$$

Poynting vector, P is defined as

$$P = E \times H \text{ watts/m}^2$$

If E and H are instantaneous, then P is also instantaneous.

Example 8.17: *In a perfect dielectric medium, the electric field progressing in the z-direction is given by the equation* $E_x = E_{xs} \cos(\omega t - \beta z)$*, and the associated magnetic field by* $H_y = \dfrac{E_x}{\eta}$ *where* E_{xs} *is the peak value of* E_x *at* $t = 0$*, and* $z = 0$*, and* η *is the intrinsic impedance of the dielectric. Prove that the average power flowing through an area S normal to the z-axis is given by* $P_{z,av} = \dfrac{1}{2}\dfrac{E_{xs}^2}{\eta} S$.

Solution: By Poynting theorem, we have $P = E \times H$.

The electric field E is along x-axis and the magnetic field H is along the y-axis. P must therefore be along the z-axis. *i.e.*, in the direction of propagation.

$$\therefore \quad P_z = E_x H_y$$

$$= E_x \left(\frac{E_x}{\eta}\right)$$

or

$$P_z = \frac{E_x^{\ 2}}{\eta} = \frac{E^2_{\ xs}}{\eta} \cos^2(\omega t - \beta z)$$

The average power P_{av} is obtained by integrating P_z over one full cycle and dividing by the period.

$$P_{z,av} = \frac{1}{T}\int_0^T \frac{E^2_{\ xs}}{\eta} \cos^2(\omega t - \beta z)\, dt, \text{ where } T = \text{period}$$

$$= \frac{E^2_{\ xs}}{\eta}\left[\int_0^T \frac{1 + \cos 2(\omega t - \beta z)}{2}\, dt\right]$$

$$= \frac{E_{xs}^2}{2\eta T}\left[t + \frac{\sin 2(\omega t - \beta x)}{2\omega}\right]_0^T$$

$$= \frac{1}{2}\frac{E_{xs}^2}{\eta}, \text{ on simplification.}$$

$\therefore$ $$P_{z,av} = \frac{1}{2}\frac{E_{xs}^2}{\eta} \text{ is the average power flow/unit area.}$$

Average power flowing through an area S normal to the z-axis is given by

$$P_{z,av} \times S = \frac{1}{2}\frac{E_{xs}^2}{\eta} S \text{ watts.}$$

Example 8.18: *A coaxial cable is carrying a current of 15 A at 150 V. Find the value of power carried by the cable using Poynting theorm.*

Solution: Given I = 15 A and V = 150 volts. The Poynting vector, which represents power density, is given by the expression $P = E \times H$, where E is the electric field intensity and H is the magnetic field intensity. In a coaxial cable, power flow is along the length of the cable, and it is evident that the vectors H and B are mutually perpendicular. Hence the angle between them is 90°.

We have $P = E \times H$ (vector product or cross product)

or $P = EH \sin\theta = EH$ since $\theta = 90°$.

The expressions for E and H for a coaxial cable are $E = \dfrac{V}{r \ln (b/a)}$ at a radial distance r from the centre. Let a and b are radii of the inner and outer conductors, and $b > r > a$.

Also, $$H = \frac{1}{2\pi r}$$

Where, I = current. The magnitude of the Poynting vector is therefore obtained as

$$P = \left(\frac{V}{r\ln\left(\frac{b}{a}\right)}\right)\left(\frac{1}{2\pi r}\right) \text{ W/m}^2$$

The total power flow is obtained as $W = \int_S EH\, ds$ where $ds = 2\pi\, rdr$, the area of the strip of thickness dr at a distance r from the centre. Since, the flow of power is through the dielectric, r ranges from a to b. Hence, the integration must be carried between a and b.

$\therefore$ $$W = \int_a^b \left(\frac{V}{r\ln(b/a)}\right)\left(\frac{I}{2\pi r}\right) 2\,\pi r\, dr$$

$$= \frac{VI}{\ln(b/a)}\int_a^b \frac{dr}{r}$$

$$W = \frac{VI}{\ln (b/a)} [\log r]_a^b$$

$$= \frac{VI}{\ln b/a} \times \ln b / a = VI \text{ Watts.}$$

The total power flow, as obtained on the basis of Poynting theorem, is given as $W = VI$ watts which according to the data provided can be written as $W = 15 \times 150 = 2250$ watts.

Example 8.19: *Show that, in a capacitor, the conduction current and displacement current are equal.*

Solution: Consider a parallel plate capacitor of capacitance C. Let it be connected to a source of a.c. voltage. The sinusoidally-varying voltage may be expressed mathematically as $V = V_0 \cos \omega t$.

The capacitor gets charged and discharges alternately, and a current flows through the circuit. This is the conduction current. Let I denote the current.

We have $$I = C \frac{dv}{dt}$$

$$= C \frac{d}{dt} (V_0 \cos \omega t)$$

$$= CV_0 \frac{d}{dt} (\cos \omega t), \text{ (since } V_0 = \text{constant, being the peak value of voltage)}$$

$$= - C \omega V_0 \sin \omega t$$

For a parallel plate capacitor, capacitance C is given as $C = \frac{\varepsilon S}{d}$ farads, where S = area of each plate and d = distance between the plates.

Substituting this value of C, we get

$$I = -\left(\frac{\varepsilon S}{d}\right) \omega V_0 \sin \omega t = -\left(\frac{\varepsilon S \omega V_0}{d}\right) \sin \omega t$$

As the capacitor charges and discharges, there is transfer of charge from one plate to the other, but since the intervening medium is a dielectric, there is no conduction current through the capacitor. A displacement current I_d may be visualized through the capacitor.

We have charge density $D = \varepsilon E$.

But $E = -$ (potential gradient) $= -\frac{v}{d}$ volt/metre.

$$\therefore \qquad D = \varepsilon \left(\frac{V_0 \cos \omega t}{d}\right) = \frac{\varepsilon V_0}{d} \cos \omega t$$

Displacement current $= \frac{\partial D}{\partial t} = \frac{\partial}{\partial t}\left(\frac{\varepsilon V_0}{d} \cos \omega t\right)$ per unit area

$$= -\frac{\varepsilon V_0 \omega}{d} \sin \omega t \text{ per unit area.}$$

Total displacement current $I_d = S\left(-\frac{\varepsilon \omega V_0}{d}\right) \sin \omega t$

Consequently, $c = \sqrt{\dfrac{1}{\varepsilon_0\mu_0}} = \sqrt{\dfrac{1}{(10^{-9}/36\pi)(4\pi\times10^{-7})}} = 3\times10^8$ (m/s)

Moreover, $\dfrac{E}{H} = \dfrac{\omega\mu_0}{\beta} = \sqrt{\dfrac{\mu_0}{\varepsilon_0}} = 120\,\pi.$

Example 8.23: *For a uniform plane wave $E_y = 15.4e^{-j(\beta x - 2\pi\times10^8 t)}$ μV/m. Find (a) direction of propagation (b) direction of H (c) phase constant β (d) phase velocity v (e) expression for H.*

Solution:

(*a*) Given $E_y = 15.4e^{-j(\beta x - 2\pi\times10^8 t)}$ μ V/m. So a study of wave equation reveals that the wave is in negative *x*-direction.

(*b*) The wave is traveling in *x*-direction and *E* is in the *y*-direction. Hence, *H* must be in the *z*-direction.

(*c*) We have phase constant $\beta = \dfrac{\omega}{v}$, since $v = \dfrac{\omega}{\beta}$

$$\therefore \quad \beta = \frac{\omega}{v} = \frac{2\pi f}{v}$$

$$= \frac{2\pi\times10^8}{3\times10^8}, \text{ assuming medium is free space}$$

$$= 2.09 \text{ rad/m.}$$

(*d*) Phase velocity = velocity of propagation

= 3×10^8 m/s in free space.

(*e*) We have $\dfrac{E}{H} = \eta$ and also $E = E_y U_y$ and $H = H_z U_z$

$$\therefore \quad \frac{E}{H} = \frac{E_y}{H_z} = \eta \quad \text{(or)}\ H_z = \frac{E_y}{\eta}$$

η for free space = 377 Ω

$$\therefore \quad H_z = \frac{1}{377}\left[15.4e^{-j(\beta x - 2\pi\times10^8 t)}\right]\mu V/m$$

(or) $$H = 0.0408\, e^{-j(\beta x - 2\pi\times10^8 t)}\ \mu V/m.$$

Example 8.24: *A travelling electromagnetic wave has a maximum value for E for 15 V/m. The medium is a perfect dielectric with $\mu_r = 1$ and $\varepsilon_r = 5$. Find (a) Velocity of the wave (b) Peak Poynting vector (c) Average Poynting vector (d) Impedance of the medium (e) Peak value of the magnetic field.*

Solution: Given $E_m = 15$ V/m, $\mu_r = 1$ and $\varepsilon_r = 5$.

(*a*) We have velocity of propagation, $v = \dfrac{1}{\sqrt{\mu\varepsilon}} = \dfrac{1}{\sqrt{(\mu_0\mu_r)(\varepsilon_0\varepsilon_r)}}$

$$= \frac{1}{\sqrt{(\mu_0\varepsilon_0)}}\cdot\frac{1}{(\mu_r\varepsilon_r)}$$

$$= \frac{3 \times 10^8}{\sqrt{1 \times 5}} = 1.341 \times 10^8 \text{ m/s.}$$

Since $\frac{1}{\sqrt{(\mu_0 \varepsilon_0)}} = C$, the velocity of light in free space is 3×10^8 m/s.

(*b*) We have peak Poynting vector $= \frac{E_m{}^2}{\eta}$

We have $$\eta = \sqrt{\frac{\mu}{\varepsilon}} = \sqrt{\frac{\mu_0 \mu_r}{\varepsilon_0 \varepsilon_r}} = \sqrt{\frac{\mu_0}{\varepsilon_0}} \cdot \sqrt{\frac{\mu_r}{\varepsilon_r}}$$

$$= 377\sqrt{\frac{1}{5}} \qquad \left[\text{since } \eta_o = \sqrt{\frac{\mu_0}{\varepsilon_o}} = 377\ \Omega\right]$$

$\therefore$ Peak Poynting vector $= \frac{15^2}{168.6} = 1.3345 \text{ W/m}^2.$

(*c*) Average Poynting vector $= \frac{E_{rms}^2}{\eta} = \frac{1}{2}\frac{E_m{}^2}{\eta}$

$$\therefore \qquad P_{avg} = \frac{15^2}{2 \times 168.6} = 0.667 \text{ W/m}^2$$

(*d*) Intrinsic impedance $\eta = 168.6\ \Omega$, (Calculated earlier)

(*e*) We have $$\eta = \frac{E}{H} \text{ or } H = \frac{E}{\eta}$$

$$\therefore \qquad H_{max} = \frac{E_{max}}{\eta} = \frac{15}{168.6} = 0.0889 \text{ A/m.}$$

Example 8.25: *Verify whether the fields* $E = E_m \sin x \sin t\, u_y$ *and* $H = \frac{E_m}{\mu_0} \cos x \cos t\, u_z$ *given above satisfy Maxwell's equations?*

Solution: Given $E = E_m \sin x \sin t\, u_y$ (or) $E_y = E_m \sin x \sin t$

Similarly, $$H = \frac{E_m}{\mu_0} \cos x \cos t\, u_z \text{ (or) } H_z = \frac{E_m}{\mu_0} \cos x \cos t$$

If these equations are to represent a transverse electromagnetic wave propagating in the x-direction, the following equations need to be satisfied:

$$(a)\ \frac{\partial^2 E_y}{\partial t^2} = \frac{1}{\mu\varepsilon}\frac{\partial^2 E_y}{\partial x^2} \qquad (b)\ \frac{\partial^2 H_z}{\partial t^2} = \frac{1}{\mu\varepsilon}\frac{\partial^2 H_z}{\partial x^2}$$

(These equations have been derived on the basis of Maxwell's equations)

Consider $$E_y = E_m \sin x \sin t$$

$$= \frac{E_m}{2}[\cos(x-t) - \cos(x+t)]$$

$$\therefore \quad \frac{\partial E_y}{\partial x} = \frac{E_m}{2}\left[-\sin(x-y)+\sin(x+t)\right]$$

$$\frac{\partial^2 E_y}{\partial x^2} = \frac{E_m}{2}\left[-\cos(x-t)+\cos(x+t)\right] = -E_y$$

$$\therefore \quad \frac{\partial^2 E_y}{\partial x^2} = -E_y \qquad ...(a)$$

Again $\quad E_y = \frac{E_m}{2}[\cos(x-t)-\cos(x+t)]$

$$\therefore \quad \frac{\partial E_y}{\partial t} = \frac{E_m}{2}[\sin(x-t)+\sin(x+t)]$$

$$\frac{\partial^2 E_y}{\partial t^2} = \frac{E_m}{2}\left[-\cos(x-t)+\cos(x+t)\right] = -E_y$$

$$\therefore \quad \frac{\partial^2 E_y}{\partial t^2} = -E_y \qquad ...(b)$$

From equations (*a*) and (*b*), it is evident that $\frac{\partial^2 E_y}{\partial x^2} = \frac{\partial^2 E_y}{\partial t^2}$

Hence, the basic relationship *i.e.*, $\frac{\partial^2 E_y}{\partial x^2} = \mu\varepsilon \frac{\partial^2 E_y}{\partial t^2}$ is **Not** satisfied.

Consider $H_z = \frac{E_m}{\mu_0}\cos x \cos t = \frac{E_m}{2\mu_0}[\cos(x+t)+\cos(x-t)]$

It can similarly be shown that $\frac{\partial^2 H_z}{\partial x^2} = \frac{\partial^2 H_z}{\partial t^2}$

Hence, the other basic relationship *i.e.* $\frac{\partial^2 H_z}{\partial x^2} = \mu\varepsilon \frac{\partial^2 H_z}{\partial t^2}$ is **Not** satisfied. Thus the given electric and magnetic fields do not satisfy Maxwell's equations.

Example 8.26: *Given a non-magnetic material having a relative permittivity of 4.1 and a loss tangent of 0.04, calculate attenuation constant and wave length at frequency of (a) 2 MHz (b) 20MHz.*

Solution: Given $\varepsilon_R = 4.1$, $\tan\theta = 0.04$ and $\mu_R = 1$ (Non-magnetic material). We have $\tan\theta = \frac{\sigma}{\omega\varepsilon}$ where $\sigma =$ conductivity and attenuation constant $\sigma = \frac{\sigma}{2}\sqrt{\frac{\mu}{\varepsilon}}$

Since $\tan\theta = 0.04$, $\frac{\sigma}{\varepsilon} = \omega\tan\theta$ or $\frac{\sigma}{2\varepsilon} = \frac{\omega\tan\theta}{2}$

Substituting this value in the expression for α, we get

$$\alpha = \sqrt{\mu\varepsilon}\left(\frac{\omega\tan\theta}{2}\right)$$

$$= \sqrt{\mu\varepsilon}\left(\frac{2\pi f \tan\theta}{2}\right), \text{ putting } \omega = 2\pi f$$

$$= \sqrt{\mu\varepsilon}\,(\pi f \tan\theta)$$

Putting $\mu = \mu_0\,\mu_r = \mu_0 = 4\pi \times 10^{-7}$ since $\mu_r = 1$ and $\varepsilon = \varepsilon_0\varepsilon_r = 8.854 \times 10^{-12} \times 4.1$

(*a*) For $f = 2\text{MHz}$,

$$\alpha = \sqrt{(4\pi \times 10^{-7}) \times (8.854 \times 10^{-12}) \times 4.1}\;\pi \times 2 \times 10^6 \times 0.04 = 1.696 \times 10^{-3}$$

(*b*) F or $f = 20$ MHz, $\alpha = 16.96 \times 10^{-3}$

To find wave length:

We have velocity of propagation in a medium other than free space

$$v = v_0\left[1 - \frac{1}{8}\left(\frac{\sigma}{\omega\varepsilon}\right)^2\right] \text{ where, } v_0 = \text{velocity of propagation in free space}$$

$$= 3 \times 10^8 \text{ m/s}$$

$$\therefore \qquad v = 3 \times 10^8\left[1 - \frac{1}{8} \times (0.04)^2\right]$$

$$= 3 \times 10^8 \times 0.9998$$

$$= 2.9994 \times 10^8 \text{ m/s.}$$

(*a*) For $f = 2$ MHz, We have wave length $\lambda = \dfrac{v}{f} = \dfrac{2.9994 \times 10^8}{2 \times 10^6} = 149.7\text{m}$

(*b*) For $f = 20$ MHz, we have $\lambda = \dfrac{v}{f} = \dfrac{2.9994 \times 10^8}{20 \times 10^6} = 14.97\text{m}$.

Example 8.27: *The electric field intensity associated with a plane wave traveling in a perfect dielectric medium is given by $E_x(z, t) = 10 \cos(2\pi \times 10^7 t - 0.1\,\pi z)$ V/m. (a)What is the velocity of propagation? (b)Write down an expression for the magnetic field intensity associated with the wave if $\mu = \mu_0$.*

Solution: The general expression is of the form $E_x(z, t) = C\cos(\omega t - \beta z)$ where, $\omega = 2\pi f$ and $\beta = \dfrac{2\pi}{\lambda}$.

Given that $2\pi f = 2\pi \times 10^7$, $f = 10^7$ Hz. which implies $\dfrac{2\pi}{\lambda} = 0.1\,\pi$; $\lambda = 20$ m

(*a*) Velocity of propagation, $v = f\lambda = 10^7 \times 20$ m/sec $= 2 \times 10^8$ m/sec.

Velocity in free space $= 3 \times 10^8$ m/sec.

Hence, $\sqrt{\varepsilon_r\mu_r} = \dfrac{3 \times 10^8}{2 \times 10^8} = 1.5$ as $\mu_r = 1$, $\varepsilon_r = (1.5)^2 = 2.25$

Intrinsic impedance $= \sqrt{\dfrac{\mu}{\varepsilon}} = \dfrac{120\pi}{1.5} = 80\,\pi$.

(*b*) As the propagation is in the positive *z*-direction,

$$H_y = \frac{10}{80\pi} \cos(2\pi \times 10^7 t - 0.1\,\pi z) \text{ A/m.}$$

Example 8.28: *A parallel plate capacitor with plate area of 5 cm² and plate separation of 3 mm has a voltage 50 sin 10³t V applied to its plates. Calculate the displacement current assuming* $\varepsilon = 2\varepsilon_0$.

Solution:

We know that $$D = \varepsilon E = \varepsilon \frac{V}{d} \text{ and } J_d = \frac{\partial D}{\partial t} = \frac{\varepsilon}{d}\frac{dV}{dt}$$

Hence, $I_d = J_d \cdot S = \dfrac{\varepsilon S}{d}\dfrac{dV}{dt} = C\dfrac{dV}{dt}$ which is the same as the conduction current, given by

$$I_c = \frac{dQ}{dt} = S\frac{d\rho_s}{dt} = S\frac{dD}{dt} = \varepsilon S\frac{dE}{dt} = \frac{\varepsilon S}{d}\frac{dV}{dt} = C\frac{dV}{dt}$$

$$I_d = 2 \cdot \frac{10^{-9}}{36\pi} \cdot \frac{5 \times 10^{-4}}{3 \times 10^{-3}} \cdot 10^3 \times 50 \cos 10^3 t$$

$$= 147.4 \cos 10^3 t \text{ nA.}$$

Example 8.29: *An A.C. voltage source* $V = V_0 \sin \omega t$ *is connected across a parallel plate capacitor C. Verify that the displacement current in the capacitor is the same as the conduction current in the wires.*

Solution:

The conduction current in the connecting wire is given by

$$i = C\frac{dv}{dt} = CV_0\, \omega \cos \omega t$$

For a parallel plate capacitor of area A and plate separation d, $C = \varepsilon \dfrac{A}{d}$

Electric field E in the dielectric is $E = \dfrac{v}{d}$

$$D = \varepsilon\, E\, \varepsilon \frac{V_0}{d} \sin \omega t$$

The displacement current $i_d = \displaystyle\int_S \frac{\partial D}{\partial t} \cdot ds = \left(\varepsilon \frac{A}{d}\right) V_0 \omega \cos \omega t$

$$i_d = CV_0\, \omega \cos \omega t = i$$

$\therefore$ $$i = i_d.$$

Example 8.30: *A uniform plane wave in a medium having* $\sigma = 10^{-3}$ *s/m,* $\varepsilon = 80\,\varepsilon_0$ *and* $\mu = \mu_0$ *is having a frequency of 10 kHz. Calculate the different parameters of the wave.*

Solution:

According to the given data, $\sigma = 10^{-3}$ s/km; $\varepsilon = 80\,\varepsilon_0$; $\mu = \mu_0$

$$\frac{\sigma}{\omega\varepsilon} = \frac{10^{-3}}{2\pi 10^4 \times 80 \times 8.854 \times 10^{-12}} = 22.47 >> 1$$

The medium is a good conductor, so attenuation constant

$$\alpha = \sqrt{\pi f \mu \sigma} = \sqrt{\pi \times 10^4 \times 4\pi \times 10^{-7} \times 10^{-3}}$$

$$= 2\pi \times 10^{-3} \text{ Np/m}$$

Phase constant $\beta = \alpha = 2\pi \times 10^{-3}$ rad/m

Intrinsic impedance $\eta = \sqrt{\frac{\omega\mu}{\sigma}} \angle 45°$

$$= \sqrt{\frac{2\pi \times 10^4 \times 4\pi \times 10^{-7}}{10^{-3}}} \angle 45° = 2\pi(1 + j) \text{ ohm}$$

Wave length $\lambda = \frac{2\pi}{\beta} = 1000$ m

Velocity of wave $v = \frac{\omega}{\beta} = \frac{2\pi\, 10^4}{2\pi\, 10^{-3}} = 10^7$ m/s.

Example 8.31: *For a non-magnetic material having* $\varepsilon_0 = 2.25\varepsilon_0$, $\sigma = 10^{-4}$ *s/m. Find (a) the loss tangent (b) attenuation constant (c) phase constant (d) intrinsic impedance for a wave having a frequency of 2.5 MHz. Assume the material to be a good dielectric.*

Solution:

(*a*) Loss tangent $= \frac{\sigma}{\omega\varepsilon}$

$$= \frac{10^{-4}}{2\pi \times 2.5 \times 10^6 \times 2.25 \times 8.854 \times 10^{-12}} = 0.320$$

(*b*) Attenuation constant, $\alpha = \frac{\sigma}{2}\sqrt{\frac{\mu}{\varepsilon}}$

$$= \frac{10^{-4}}{2}\sqrt{\frac{4\pi \times 10^{-7}}{2.25 \times 8.854 \times 10^{-12}}}$$

$$= 0.01256 \text{ Np/m}$$

(*c*) Phase constant, $\beta = \omega\sqrt{\mu\varepsilon}\left(1 + \frac{\sigma^2}{8\omega^2\varepsilon^2}\right)$

$$= 2\pi \times 2.5 \times 10^6 \sqrt{4\pi \times 10^{-7} \times 2.25 \times 8.854 \times 10^{-12}}\left(1 + \frac{0.320^2}{8}\right)$$

$$= 0.0796 \text{ rad/m}$$

(*d*) Intrinsic impedance, $\eta = \sqrt{\frac{\mu}{\varepsilon}}\left(1 + j\frac{\sigma}{2}\right)$

$$= \sqrt{\frac{4\pi \times 10^{-7}}{2.25 \times 8.854 \times 10^{-12}}}\left(1 + j\frac{0.320}{2}\right)$$

$$= 254.35 \angle 9.09° \text{ ohms.}$$

Example 8.32: *Find the skin depth δ at a frequency of 1.6 MHz in aluminum, where σ = 38.2 Ms/m and μ_r = 1. Also find the propagation constant and wave velocity.*

Solution:

We know that $\delta = \frac{1}{\sqrt{\pi f \mu \sigma}} = \frac{1}{\sqrt{\pi \times 1.6 \times 10^6 \times 4\pi \times 10^{-7} \times 38.2 \times 10^6}}$

$$= 0.06438 \text{ mm}$$

Also

$$\alpha = \beta = \delta^{-1}$$

$$\gamma = 15.53 \times 10^3 + j15.5 \times 10^3$$

$$= 21.96 \times 10^3 \angle 45° \text{ m}^{-1}$$

$$\upsilon = \frac{\omega}{\beta} = \omega\delta = 2\pi \times 1.6 \times 10^6 \times 0.06438 \times 10^{-3}$$

$$= 647.2 \text{ m/s.}$$

SHORT QUESTIONS AND ANSWERS

1. **Write boundary conditions at the interface between two magnetic media.**

   Ist Boundary Condition: $B_{n_1} - B_{n_2} = 0$

   IInd Boundary Condition: $H_{t_1} - H_{t_2} = 0$

2. **How is the inconsistency in Ampere's law overcome ?**

   By introducing the displacement current, the Ampere's law is modified as follows to overcome the inconsistency

   $$\nabla \times \overline{H} = \overline{J}_c + \overline{J}_d$$

   Where, $\overline{J}_c$ is the conduction current density.

   J_d is the displacement current density, Also $J_d = \frac{\partial \overline{D}}{\partial t}$

3. **What is the difference between Faraday's theory of magnetic field and Ampere's theory of magnetic field?**

   Faraday's law says that the time varying magnetic field causes the electric field and Ampere's law says that time varying electric field causes magnetic field.

4. State Fleming's right hand rule.

Fleming's right hand rule states that when you stretch your forefinger, middle finger and thumb of your right hand such that they are mutually perpendicular to each other, if the forefinger represents the direction of the magnetic field, thumb represents the direction of motion of the conductor then the middle finger represents the direction of induced e.m.f.

5. State Fleming's left hand rule.

Fleming's left hand rule states that when you stretch your forefinger, middle finger and the thumb of your left hand such that they are mutually perpendicular to each other, if fore finger represents the direction of magnetic field, middle finger represents the direction flow of current then the thumb represents the direction of motion of conductor.

6. Write down the Maxwell's equation corresponding to Faraday's law, in point form.

Faraday's law in point form is as follows:

$$\nabla \times \overline{E} = -\frac{\partial \overline{B}}{\partial t}$$

7. State Faraday's law of electromagnetic induction.

Faraday's law states that electromagnetic force induced in a circuit is equal to the rate of change of magnetic flux linking the circuit.

$$\text{e.m.f.} = -\frac{d\phi}{dt}$$

8. Write down the general integral and point form of Faraday's law.

$$\text{e.m.f.} = L\frac{d\phi}{dt} \text{ (General form)}$$

$$\oint \overline{E}\cdot d\overline{1} = -\iint \frac{\partial \overline{B}}{\partial t} \text{ds (Integral form)}$$

$$\nabla \times \overline{E} = -\frac{\partial \overline{B}}{\partial t} \text{ (Point form)}$$

9. Distinguish between transformer e.m.f. and motional e.m.f.

The e.m.f. induced in a stationary conductor due to charge in flux linked with it, is called transformer e.m.f. or static induced e.m.f.

$$\text{e.m.f.} = -\iint \frac{\partial \overline{B}}{\partial t}\cdot d\overline{s}$$

The e.m.f. induced due to the moment of conductor in magnetic field is called motional e.m.f. or dynamic induced e.m.f.

$$\text{e.m.f.} = \oint_S \overline{v} \times \overline{B} \cdot \overline{dl}$$

10. State Lenz's law.

Lenz's law states that the induced e.m.f. in a circuit produces current, which opposes the change in magnetic flux producing it.

$$\text{e.m.f.} = -\frac{d\phi}{dt}$$

11. Explain, why $\nabla \times \overline{\mathbf{E}} = 0$?

In a region in which there is no time changing magnetic flux, the voltage around the loop would be zero. By Maxwell's equation

$$\nabla \times \overline{E} = -\frac{\partial \overline{B}}{\partial t} = 0 \quad \text{(irrotational)}$$

12. Explain, why $\nabla \cdot \overline{\mathbf{E}} = 0$?

In a free space there is no charge enclosed by the medium. The volume charge density is zero. By Maxwell's equation,

$$\nabla \cdot \overline{D} = \rho$$

$$\nabla \cdot \overline{D} = 0 \quad \text{(For free space)}$$

13. Explain, why $\nabla \times \overline{\mathbf{B}} = 0$?

$\nabla \times \overline{B} = 0$ states that there is no magnetic charge. The net magnetic flux through any closed surface is zero.

MULTIPLE CHOICES

1. What is force on the stationary charged particle in electromagnetic field?
   (*a*) qE (*b*) $q(E + B)$
   (*c*) $Q\,(E + V \times B)$ (*d*) none of these
2. What is the force on the moving charged particle in the electromagnetic field?
   (*a*) $q(E + B)$ (*b*) qE
   (*c*) $q(E + V \times B)$ (*d*) none of these
3. Two conducting coils 1 and 2 are placed in a uniform magnetic field that decreases at a constant rate. If the plane of coils is perpendicular to field lines, which of the following statements is true?
   (*a*) an e.m.f. is induced in both coils (*b*) an e.m.f. is induced in coil 2
   (*c*) equal heating occurs in both coils (*d*) heating does not occur in either coil
4. Which of the following expressions is not Maxwell's equation for time varying fields?
   (*a*) $\nabla \cdot J + \frac{\partial \rho_v}{\partial t} = 0$ (*b*) $\nabla \cdot D = \rho_v$
   (*c*) $\nabla \cdot E = -\frac{\partial B}{\partial t}$ (*d*) $\oint \overline{H} \cdot \overline{dL}$

5. The concept of displacement current was a major contribution attributed to

(*a*) Faraday (*b*) Lenz

(c) Maxwell (*d*) Lorentz

6. A loop is rotating about the y-axis in a magnetic field. $B = B_0 \sin \omega t\, U_x$ Wb/m^2. The voltage induced in the loop is due to

(*a*) motional e.m.f. (*b*) transformer e.m.f.

(*c*) both (*a*) and (*b*) (*d*) none of these

7. Which of the following statements is not true of a phasor:

(*a*) it may be a scalar (or) vector (*b*) it is a time-dependent quantity

(*c*) it is a complex quantity (*d*) none of these

8. If $E_2 = 10e^{j4x}U_y$, which of these is not a correct representation of E?

(*a*) $Re\,(E_s\, e^{j\omega t})$ (*b*) $10 \cos (\omega t + j\, 4x)U_y$

(*c*) imag $(E_s\, e^{j\omega t})$ (*d*) $10 \sin (\omega t + 4x)U_y$

9. The flux through each turn of a 200 turn coil is $(t^3 - 2t)$ M Wb where, t is in seconds. The induced emf at $t = 2s$ is

(*a*) 1 volt (*b*) –9.2 volt

(*c*) 9.2 volt (*d*) –4.6 volt

10. The induced e.m.f. in a moving loop in a static magnetic field is

(*a*) $V_{emf} = \oint_L (V \times B) \cdot dL$ (*b*) $V_{emf} = \oint_L B \cdot dL$

(*c*) $V_{emf} = \oint_S B \cdot ds$ (*d*) none of these

11. The induced e.m.f. in moving loop in time varying magnetic field is

(*a*) $V_{emf} = -\int_S \dfrac{dB}{dt} \cdot ds + \oint_L (V \times B) \cdot dL$ (*b*) $V_{emf} = \int_S \dfrac{dB}{dt} \cdot ds + \oint_L (V \times B) \cdot dL$

(*c*) $V_{emf} = \int_S \dfrac{dB}{dt} \cdot ds \oint_L (V \times B) \cdot dL$ (*d*) $V_{emf} = -\int_S \dfrac{dB}{dt} \cdot ds + \oint_L (V \times B) \cdot dL$

12. In a dielectric, the displacement current is .... the conductor current.

(*a*) just equal to (*b*) just less than

(*c*) greater than (*d*) none of these

13. E and H are related by the equation.

(*a*) $\dfrac{\overline{E}}{\overline{H}} = \sqrt{\dfrac{\varepsilon_0}{\mu_0}}$ (*b*) $\dfrac{\overline{E}}{\overline{H}} = \sqrt{\varepsilon_0 \mu_0}$

(*c*) $\dfrac{\overline{E}}{\overline{H}} = \sqrt{\dfrac{\mu_0}{\varepsilon_0}}$ (*d*) none of these

14. The velocity of electromagnetic waves in a dielectric medium ($\varepsilon_r = 4$) is

(*a*) 3×10^8 metre/second (*b*) 1.5×10^8 metre/second

(*c*) 6×10^8 metre/second (*d*) 12×10^8 metre/second

15. The complex Poynting vector is

(*a*) $E \times H$ (*b*) $\frac{1}{2}(E \times H)$

(*c*) $\frac{1}{2} Re (E \times H)$ (*d*) none of these

16. Inconsistency in the Ampere's circuit law for time varying fields is eliminated by introducing

(*a*) conduction current density

(*b*) transformer e.m.f.

(*c*) a combination of motional and transformer e.m.f.

(*d*) none of these

17. Identify the wrong statement.

(*a*) The magneto-motive force around a closed path is equal to J_c plus J_d

(*b*) The net magnetic flux emerging through any closed surface is non-zero

(*c*) The total electric displacement through the enclosing a volume is equal to the total charge within the volume

(*d*) None of these

18. In a dielectric-dielectric boundary

(*a*) the tangential component of $\vec{E}$ is discontinuous

(*b*) the tangential component of $\vec{E}$ is continuous

(*c*) the normal component of $\vec{D}$ is discontinuous

(*d*) both (*b*) and (*c*).

19. In a dielectric-dielectric boundary

(*a*) the normal component's of both $\vec{E}$ and $\vec{D}$ are continuous

(*b*) the tangential component's of both $\vec{E}$ and $\vec{D}$ are discontinuous

(*c*) the normal component of $\vec{D}$ is continous but $\vec{E}$ is not continuous

(*d*) none of these

20. The magneto motive force a force around a closed path is equal to

(*a*) the conduction current

(*b*) the displacement current

(*c*) sum of conduction and displacement current

(*d*) none of these

21. The electromotive force around a closed path is equal to

(*a*) magnetic displacement

(*b*) time derivative of the magnetic displacement

(*c*) time integral of magnetic displacement

(*d*) electric displacement

22. Identify the correct statement.

(*a*) The net magnetic flux emerging through any closed surface is equal to charge enclosed

(*b*) The net magnetic flux emerging through any closed surface is zero

(*c*) The electromotive force around a closed path is the time integral of the magnetic displacement

(*d*) none of these

23. The displacement current flows

(*a*) in a conductor (*b*) in dielectric

(*c*) both (*a*) and (*b*) (*d*) none of these

24. Inconsistency of Ampere's law is eliminated by

(*a*) displacement current density (*b*) conduction current

(*c*) displacement voltage density (*d*) none of these

25. The magnetic voltage around a closed path is equal to

(*a*) the electric voltage in the path (*b*) the electric current

(*c*) both (*a*) and (*b*) (*d*) none of these

26. Following law explains the principle of d.c. Generator

(*a*) Fleming's right hand rule (*b*) Lenz's law

(*c*) Faraday's law (*d*) none of these

27. Following law explains that the induced e.m.f. opposes the cause producing it

(*a*) Fleming's right hand rule (*b*) Lenz's law

(*c*) Faraday's law (*d*) none of these

28. Electromagnetic induction principles are given by

(*a*) Faraday (*b*) Lenz

(*c*) Hertz (*d*) none of these

29. Electron at rest produces

(*a*) static magnetic field (*b*) static electric field

(*c*) time varying field (*d*) none of these

30. Electron in motion with uniform velocity produces

(*a*) static magnetic field (*b*) static electric field

(*c*) time varying field (*d*) none of these

31. Field around a conductor carrying steady current is

(*a*) static magnetic field (*b*) static electric field

(*c*) time varying field (*d*) none of these

32. What is the kind of field produced when a electron is moving with varying velocity?

(*a*) static magnetic field (*b*) static electric field

(*c*) time varying field (*d*) none of these

33. What is the kind of field around a conductor carrying alternating currents?

(*a*) static magnetic field (*b*) static electric field

(*c*) time varying field (*d*) none of these

34. Static charges produces .... field.
(*a*) electrostatic (*b*) electromagnetic
(*c*) revolving (*d*) none of these

35. Which law states that an e.m.f. will induce in conductor when it links a flux?
(*a*) Lenz's law (*b*) Faraday's law
(*c*) Ohm's law (*d*) none of these

36. E.m.f., $e = Blv \sin \theta$ where, v is ....
(*a*) magnetic flux (*b*) magnetic flux density
(*c*) electric flux (*d*) none of these

37. E.m.f., $e = Blv \sin \theta$, where l is ....
(*a*) velocity (*b*) length of conductor
(*c*) magnetic field (*d*) none of these

38. Transformer e.m.f. is also called as ....
(*a*) static e.m.f. (b) dynamic e.m.f.
(*c*) both (*a*) and (*b*) (*d*) none of these

39. Motional e.m.f. is also called as ....
(*a*) static e.m.f. (*b*) dynamic e.m.f.
(*c*) both (*a*) and (*b*) (*d*) none of these

40. Transformer e.m.f. is due to ....
(*a*) time rate of change of flux (*b*) motion of conductor in field
(*c*) both (*a*) and (*b*) (*d*) none of these

41. Motional e.m.f. is due to ....
(*a*) time rate of change of flux (*b*) motion of conductor in field
(*c*) both (*a*) and (*b*) (*d*) none of these

KEY

1. (*a*)	**2.** (*c*)	**3.** (*a*)	**4.** (*a*)	**5.** (*c*)	**6.** (*c*)
7. (*a*)	**8.** (*b*)	**9.** (*a*)	**10.** (*a*)	**11.** (*a*)	**12.** (*b*)
13. (*c*)	**14.** (*b*)	**15.** (*c*)	**16.** (*d*)	**17.** (*c*)	**18.** (*c*)
19. (*b*)	**20.** (*d*)	**21.** (*c*)	**22.** (*c*)	**23.** (*b*)	**24.** (*b*)
25. (*b*)	**26.** (*c*)	**27.** (*b*)	**28.** (*a*)	**29.** (*b*)	**30.** (*a*)
31. (*a*)	**32.** (*c*)	**33.** (*c*)	**34.** (*a*)	**35.** (*b*)	**36.** (*b*)
37. (*b*)	**38.** (*a*)	**39.** (*b*)	**40.** (*a*)	**41.** (*b*)	

REVIEW QUESTIONS

1. Derive an expression for the motional e.m.f. and state all the assumptions.
2. State Faraday's law of induction.
3. Write Maxwell's equations in (*a*) Point form (*b*) Integral form. Explain the significance of each equation.

4. Can the equation of continuity be derived from Maxwell's equations? If yes, derive it. If not, why?
5. Derive boundary conditions for time-varying fields.
6. Is it necessary for the fields to satisfy Maxwell's equations in order to exist?
7. State Poynting's theorm. What is Poynting's vector?
8. What is average power density?
9. Explain the differences between a plane wave and a uniform plane wave.
10. Can a uniform plane wave exist in 'real life'?
11. What is a uniform medium?
12. A plane wave in a dielectric medium has a $\overline{E}$ component of the form $\overline{E} = E_0 \sin(\beta x - \omega t)\overline{u}_x$ V/m. Derive expressions for the associated $\overline{H}$ field and the instantaneous Poynting vector. What is the average power density in the medium?
13. A uniform plane wave propagates in free space in the x-direction. If the $\overline{E}$ field is in the z-direction and has amplitude of 100 V/m, what must be the associated $\overline{H}$ field? Compute the time-average Poynting vector.
14. Justify that, in a plane wave propagating in a dielectric medium, the average energy stored is equally divided between the electric and magnetic fields.
15. Show that the intrinsic impedance has the dimensions of resistance.
16. Explain the following terms: the propagation constant, the attenuation constant, the phase constant, the wavelength, the intrinsic impedance and the skin depth.
17. Define skin depth and explain surface resistance.

EXERCISE PROBLEMS

1. The electric field intensity of an electromagnetic wave in free space is given by $E_x = E_0 \cos \omega \left(t - \dfrac{z}{v}\right)$; $E_y = 0$; $E_z = 0$. Determine the expression for the components of magnetic intensity H.
2. In a coaxial capacitor, find the expression for displacement current flowing across an arbitrary surface at a radius r between the radii a and b. Applied voltage is $V = V_0 \sin \omega$.
3. A sinusoidal plane wave is transmitted through a medium whose breakdown strength is 50 kV/m and whose relative permeability is 6. Determine the mean possible r.m.s. power flow density and the peak value of the associated magnetising force.
4. Find the displacement current density within a parallel plate capacitor where, $\varepsilon = 200\, \varepsilon_0$, $s = 0.11 m^2$, $d = 0.08$ mm and the capacitor voltage is 200 sin 300 πt volts.
5. The electric field intensity of 500 MHz uniform plane wave in frees space is given as $E_s = (30 + j45)(2U_x + 5U_y)\, e^{-j\beta z}$ V/m (a) Find ω, λ, V and P (b) Find $\overline{E}$ at $t = 3$ ns, $z = 20$ cm (c) What is $|H|$ max.

6. A plane wave is travelling in the positive x-direction in a lossless bounded medium having permeability nine times that of free space. (*a*) Find the phase velocity of the wave.(*b*) If E has only 'y' component with an amplitude of a 20 volts find the amplitude and direction of H.

7. A medium has constants given by $\sigma = 0.6$ mho, $\mu_r = 5$, $\varepsilon_r = 9$. If the constants do not change with frequency find the depth of penetration at (*a*) 20 Hz (*b*) 6 MHz (*c*) 2 GHz.

8. For a uniform plane wave $E_y = 20e^{-j(\beta x - 4\pi \times 10^5 t)}$ μV/m. Find (*a*) direction of propagation (*b*) direction of H (*c*) phase constant β (*d*) phase velocity V (*e*) expression for H.

9. A travelling electromagnetic wave has a maximum value for E for 25 V/m. The medium is a perfect dielectric with $\mu_R = 2$ and $\varepsilon_R = 6$. Find (*a*) Velocity of the wave (*b*) Peak Poynting vector (*c*) Average Poynting vector (*d*) Impedance of the medium (*e*) Peak value of the magnetic field.

10. The electric field intensity of 450 MHz uniform plane wave in free space is given as $E_s = (20 + j65)(U_x + 2U_y)\, e^{-j\beta z}$ V/m (*a*) Find ω, λ, V and β (*b*) find $\overline{E}$ at $t = 3$ ns, $z = 20$ cm (*c*) What is $|H|$ max ?

11. A 6500 MHz uniform plane wave is propagating in polystyrene. If the amplitude of electric field intensity is 20 V/m and the material is assumed to be lossless. Find (*a*) phase constant (*b*) wavelength (*c*) the velocity of propagation (*d*) the intrinsic impedance (*e*) the propagation const (*f*) the amplitude of magnetic field intensity.

Index

A

B

C

D

E

F

G

I

L

M

N

O

P

R

S

T

U

V

W